T0345146

Invisible Light

Invisible Light

The Remarkable Story of Radiology

Adrian Thomas

CRC Press
Taylor & Francis Group
Boca Raton London New York

CRC Press is an imprint of the
Taylor & Francis Group, an **informa** business

First edition published 2022
by CRC Press
6000 Broken Sound Parkway NW, Suite 300, Boca Raton, FL 33487-2742

and by CRC Press
4 Park Square, Milton Park, Abingdon, Oxon, OX14 4RN

CRC Press is an imprint of Taylor & Francis Group, LLC

© 2022 Taylor & Francis Group, LLC

Library of Congress Cataloging-in-Publication Data

Names: Thomas, Adrian, author.
Title: Invisible light: the remarkable story of radiology / Adrian Thomas.
Description: First edition. | Boca Raton: CRC Press, 2022. | Includes bibliographical references and index.
Identifiers: LCCN 2021049174 | ISBN 9780367344269 (hardback) | ISBN 9780367338091 (paperback) | ISBN 9780429325748 (ebook)
Subjects: LCSH: Medical radiology--History. | Radiology--History.
Classification: LCC R895.5.T46 2022 | DDC 616.07/57--dc23/eng/20220106
LC record available at https://lccn.loc.gov/2021049174

ISBN: 978-0-367-34426-9 (hbk)
ISBN: 978-0-367-33809-1 (pbk)
ISBN: 978-0-429-32574-8 (ebk)

DOI: 10.1201/9780429325748

Typeset in Times
by Deanta Global Publishing Services, Chennai, India

Contents

Foreword

The current state of medical practice, with its remarkable scientific and technological achievements, may make us forget that contemporary medicine is the result of a long evolution. Until the beginning of the 19th century, the medicine practised in antiquity was taken seriously into account. For more than 2,000 years, Greek medical treatises were consulted as reliable sources of highly authoritative information. This attitude towards the past of medicine changed radically from the second half of the 19th century, when the medical sciences began to produce breakthroughs and discoveries like never before. Consequently, the study of the past of our profession seemed sterile and unlearned. In this way, the history of medicine was considered the account of the errors and fantasies of the physicians of the past, and the study of their works seemed a waste of time. The brightest intellects among the young graduates were devoted to basic or clinical research, and to the application of new instrumental methods to clinical practice. Techniques which have become an integral part of the standard armamentarium of diagnosis or therapy are passed on to each new generation of physicians or trainees without any sense of the efforts involved in their initial development.

But we know how remarkable it is to get back to the beginnings; how much more complete is our understanding as we trace the struggles of those who innovated the field and faced the problems with intelligence and acumen. Perspective is so often lacking in contemporary life, and often the secret of placing contemporary events and medicine in an appropriate framework lies in familiarity with the past. History teaches us where we come from, where we are in the present moment, and in what direction we are moving.

In this remarkable book, Dr. Adrian Thomas (radiologist, historian, collector, archivist, lecturer, and a prolific writer) present the astonishing growth and development of a discipline that changed the face of medicine forever.

The beginning of only a few medical specialities can be dated as precisely as that of radiology. On 8 November 1895, the German physicist Wilhelm Conrad Röntgen was experimenting with the action of electric energy in partially evacuated glass tubes, a popular scientific activity in the last quarter of the 19th century. Meticulously labouring in a darkened laboratory with the electrically charged tube, he noticed a glow from the far corner of the room. In the days following his discovery of these new, invisible rays, Röntgen experimented doggedly to test its properties. He noted quickly that solid objects placed in the beam between the tube and a fluorescent screen serving as an image receptor attenuated or blocked the beam, depending upon their density and structure. Then, he passed his hand through the beam. As he looked at the screen, the flesh of the hand seemingly melted away, projecting only the outlines of the bones. The hand was intact, unharmed. But on the screen, only the bones showed up. With that observation, the science of medical radiology was born.

Medical imaging has evolved from a basement art capable of answering a few clinical questions to the real hub of much of medical diagnosis and treatment. Since the discovery of X-rays, the realm of diagnostic medical imaging has witnessed continual dramatic changes in the capabilities and extent of visualising the human body. In the first five years following the discovery, the fluoroscopic findings were recorded and correlated with various pulmonary abnormalities and their corresponding clinical findings. Bullets, foreign bodies, and fractures could be evaluated before surgery. Radiographs were used for the evaluation of the skeleton. The normal anatomy and the wide range of normal variants were described. The application of X-rays soon spread to other organs and organ systems through the introduction of appropriate contrast media. The X-rays encouraged a series of associated advances that created medical imaging (including ultrasonography, nuclear medicine, mechanical and later computed tomography, magnetic resonance imaging), radiation therapy, and several new medical subspecialities. It encouraged investment in a variety of medical technologies that have become the scientific basis of medical practice.

The first radiation burn, which was of a hand, was reported hardly a year after the discovery of X-rays. The injury that these rays can cause in tissue is presently well known. Many acute effects were, unfortunately, discovered by painful experience before the end of the 19th century. The discipline of radiation therapy transformed these noxious pollutants into clinically useful tools, and the science of health physics evolved from the need for an understanding of the risks of radiation and from the need to control unnecessary exposure.

To be able to see within the human body had a profound impact on both the medical and lay communities, on medical thought, and on fundamental ideas about the human body. The general public was excited, but apprehensive, with concerns related not to safety, but to the obvious ability of X-rays to penetrate clothing! It was the beginning of a revolutionary change in our understanding of all the physical world. Just as X-rays provided a new window on the interior of the human body, so they provided a radically different perspective on some of the most important and intriguing phenomena known to science, from the structure and function of deoxyribonucleic acid (DNA) to the structure and evolution of some of the largest objects in the universe.

The advent of the X-rays posed questions completely new to medicine. The X-ray apparatus was the first widely used machine to interrupt the traditionally sacred relationship between physician and patient. No longer were history, inspection, touch, percussion, and auscultation the major means of confirming the nature of disease, but an energy-generating electrical machine appeared to be a dramatic advance for the physician.

It became clear that the new technology would be the basis for a new medical speciality, with skilled practitioners and auxiliary personnel, rather than as a service adjunct to already extant hospital departments. The birth of a new medical speciality can be a difficult and torturous labour, involving as it does the inevitable ceding of turf from established fields to the newcomer. The resulting tensions would prevent some major hospitals from establishing true radiology departments until well into the 20th century.

The advent of the X-rays into the medical science changed more than our expectations of medicine. For physicians, it marked the beginning of a most profound revolution in practice: the need for continual adaptation to new and changing technologies. The impact of the X-rays in this relatively static medical setting would set the scene for an era of technological innovation, which would prove a central focus of clinical, political, and financial debate. For the radiologist, as for all other physicians, this tradition of technical innovation has dictated a constant readjustment in expectations, in terms of both skills and the direction of practice.

To understand the present, we must look at the past and the future at the same time. In this book, Dr. Thomas delineated several trends that were and are shaping the present status of medical imaging and its future: from analog to digital imaging; from anatomy and gross pathology to physiology, biochemistry, and metabolism; from qualitative to quantitative analysis; from non-specific to tissue- and disease-specific contrast enhancement; from emphasis on diagnosis to image guided therapy; from generalised to personalised medicine; from human to artificial intelligence and robotics.

This book is an authoritative and engaging history of developments within radiology. Undoubtedly, it will contribute to awakening interest in the past of our speciality and to deepen the study of its historical evolution. It will be of interest not only to those directly working with X-rays, but also to general practitioners and specialists who are confronted with radiologic studies in their practice. This text will also be informative for the interested non-medical readers, despite the unavoidable medical and technical terminology.

Prof. Dr. Alfredo E. Buzzi
Full Professor of Radiology, University of Buenos Aires
Past-President, International Society for the History of Radiology
Past-President, Argentine Society of Radiology
Past-President, Argentine Society for Medical Humanism

Preface

The development of radiology is a complex and fascinating topic, and radiology as a speciality has impacted upon all of our lives. Unlike other medical disciplines, radiology started at a specific point in time with the discovery of the new rays by Wilhelm Conrad Röntgen in Würzburg on 8 November 1895. Röntgen died on 10 February 1923, and this book is being written as the centenary of his death is approaching. With a history of only 125 years, the volume of material related to radiology is surprisingly large and is ever increasing. Any history of radiology will, of necessity, be a personal selection and another author will make a different choice. To determine the significance of events is always difficult, and this is particularly the case as we approach the present moment. What may be historically significant and important during its time may now be medically irrelevant and vice versa, an example being the obsolete technique of kymography. In kymography, a chest radiograph is obtained with a moving slit and relative movements of mediastinal structures can be recorded.[1] There was some popularity of the technique in the 1930s, and it is now difficult to be certain exactly how commonly kymography was used. If an older article is read, was the author describing a common and popular practice, or have we come across this paper by serendipity? Kymography was discussed in the *British Journal of Radiology* in an editorial of December 1934 by "PK", who is the radiologist Peter Kerley (1900–1979),[2] and it is apparent that nothing changes (Figure 0.1). Kerley wrote that it was necessary for the radiologists to look to their laurels and make sure that the method of kymography does not become the prerogative of clinical cardiologists with a consequent damage to the prestige of radiology, and perhaps to the advancement of science. There is therefore nothing new in turf wars, and radiologists and cardiologists were to become rivals again with the newer techniques of ultrasound, angiography, CT scanning, and MRI.

At a Wellcome Trust Symposium on heritage in healthcare in 1999, it was noted that a problem of the present is this significant volume of material.[3] Bruce Madge from the British Library observed that in spite of the growth of computers and electronic media, the last ten years of the 20th century witnessed a 65% increase in the use of paper. The British Library was then adding 8 miles of printed material each year. There was an exponential curve of increase of medical journals, the numbers of titles doubling every 18 years. It is not immediately obvious why quite so many new journals are needed. When Wilhelm Conrad Röntgen described his new rays, he only added three relatively brief publications to his *curriculum vitae*, and it is unimaginable that a contemporary professor of physics would be satisfied with this number of publications. The increase in published material is combined with the development of new electronic media and the Internet. There are major problems with the electronic archive, and particularly as to who will be responsible for the archiving of material. Tilli Tansey from the Wellcome Trust gave the historian's perspective, which was again related to the recurring theme of the plethora of material. Tansey used the word bibliochlothanasia, meaning overcrowding of books. She

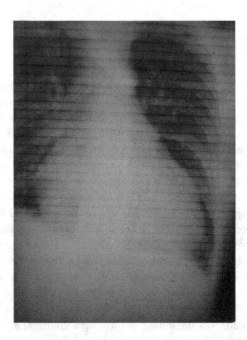

FIGURE 0.1   X-Ray kymogram taken on 2 May 1930 showing movements of the heart in a 62-year-old man following a heart attack ("cardiac infarct"). Author's collection.

also quoted the historian A.J.P. Taylor who had noted that history gets thicker as it approaches recent times with more people, more events, and more books written about them. More evidence is preserved, and possibly too much. Decay and destruction have hardly begun their beneficent work when the historian looks at modern times. And so, the task of writing a book on the history of radiology becomes more difficult the closer we approach the present day. It is more difficult to assess the long-term significance of recent events, and this is complicated by the fact that many of those involved are still alive. The historian has to make choices. Decisions need to be made as to what should be included and what is left out, and as already indicated another writing a similar book to this will make different selections. It is also preferable to cover less in more detail than to attempt to cover too much material.

Perhaps the main purpose of this book is to encourage the readers to read the older material for themselves. In 1944, the British writer C.S. Lewis (1898–1963) wrote an essay "On the reading of old books"[4] in which he encouraged the reading of older material. Lewis wrote that it was a good rule, after reading a new book, never to allow ourselves another new one until we have read an old one in between. If that is too much for us, then we should read one old one to every three new ones. So as an example, in the field of musculoskeletal radiology as well as reading the modern literature, we should most certainly read the works of James Brailsford (1888–1961) who was a radiologist from Birmingham and was a co-describer of the Morquio-Brailsford or Mucopolysaccharidosis Type IV (MPS IV) syndrome. Brailsford wrote a highly influential textbook *The Radiology of Bones and Joints*[5] that went through five editions between 1934 and 1953. Whilst Brailsford was limited in his imaging to conventional radiography, his experience in skeletal radiology was unrivalled, and his words are well worth reading today. It's not that those in the past are necessarily better than we are, but rather that their assumptions and their worldview are different to our own. If we only read our contemporaries, we are reading those who share our own worldview and we are not challenged. When we read older material, we are exposed to a different worldview and to different medical and scientific models. An interesting example is the 2021 paper by Chew and others reviewing low-dose radiation treatment of pneumonia and other inflammations, particularly in the context of the spread of COVID-19 coronavirus (SARS-CoV-2).[6] COVID-19 can result in severe illness with the breakdown of homeostasis. In the first half of the 20th century, there was a considerable interest in low-dose radiotherapy to modify inflammatory and immune-mediated disease such as tuberculosis and asthma and much of this work has been forgotten. Chew reviews treatments to inhibit the inflammatory phase of the illness, which appears to be the main cause of death, and concludes that low-dose radiotherapy may well represent a promising tool in suppressing severe inflammation. In the earliest days of radiology, it was discovered that radiation therapy might influence the immune response for clinical benefit. Max Cushman Rice from Chicago discussed the use of X-rays in the treatment of enlarged neck lymph nodes in tuberculosis at the 1904 American Röntgen Ray Society Annual Meeting.[7] Rice said that no subject was of greater interest to the medical profession than tuberculosis. He perceptively commented that tuberculosis could be arrested, but that "in any instance it must be by increasing the natural resisting power of the tissues". As a result of this, the interest of the medical world "centers itself now about the various forms of light (treatment), such as the Finsen and the Röntgen rays, as a cure for tuberculosis". Radiation was used to treat asthma using a variety of techniques, and especially using splenic radiation with considerable success.[8] Sebastian Gilbert Scott (1879–1941) reviewed the use of radiation in increasing patient resistance and immunity,[9] and particularly promoted the use of low doses for stimulating function. This whole area shows a great deal of potential. At the present day, we usually treat infections using antibiotics and cancers using surgery and chemo-radiotherapy. Ideally, the treatment of any disease should be to stimulate the body's own response by enhancing the tissues' ability to heal itself, which was Scott's main aim. We now enter the world of Sir Ralph Bloomfield Bonington in George Bernard Shaw's 1906 play *The Doctor's Dilemma* when he exclaimed that "There is at bottom only one genuinely scientific treatment for all diseases, and that is to stimulate the phagocytes". In spite of all of our recent advances in medicine, these are perhaps still early days in our understanding of the immune response and disease processes.

We should remember that books about an author or a topic are often more difficult to understand than the author or topic itself. C.S. Lewis noted that it was more difficult to read books on Platonism than was to read the original words of Plato, which is an interesting comment for an academic to make. As has been emphasised, the older literature should be read and a number of factors have made this simpler. A major recent achievement has been the retrodigitisation of many scientific journals, and the full text of the original papers is now readily available. There are also review publications with accompanying essays and biographies for both classical radiology[10] and digital radiology.[11] The development of retrodigitisation of medical journals also helps avoid simple errors. For example, a case report in 1984 purportedly presented the first case of Thorotrast granuloma in the neck seen in the United Kingdom.[12] These granulomata occurred when the contrast medium Thorotrast leaked into the tissues at the site of the intra-arterial injection. In reality, Thorotrast granulomata were well described, and authors had only searched the recent literature and concluded that theirs was the first description.[13] To be fair to the authors, we should remember how difficult literature searches were before the advent of online searches, but we should also ask the question as to why so few publications reference literature that is more than 20 years old?

The greatest justification for knowing history, and for reading the primary material, is to give us a sense of meaning and rootedness. In 1943 whilst with the Free French in London, the French philosopher Simone Weil (1909–1943) wrote her remarkable study *L'Enracinement*.[14] The French words *enracinement* and *déracinement* are difficult to define, and Richard Rees uses the slightly unsatisfactory terms "rootedness" and "uprootedness".[15] Weil sees *le déracinement* – uprootedness or deracination – as an almost universal condition of our time, and if this was the case in the 1940s, then how much more in the 2020s? We have become a people without roots in our environment, and even less within our historical or cultural past. If we become a civilisation without roots, then we become atomised or isolated individuals existing in the present moment, with no awareness of the past and an uncertain future lying ahead of us. It is therefore essential that we are careful with our heritage, since as Weil knows well, the past once destroyed never returns.[16] Weil sees the destruction of the past as perhaps the greatest of all crimes, and the preservation of what little of it remains should become almost an obsession with us.

Determining precedence and priority with scientific and medical discoveries and inventions is difficult, and perhaps ultimately of limited value. What was done, and its consequences, is more important than who actually did it first. It is interesting to observe how often individuals or groups are working independently of each other, and often unaware of each other's existence. This may be seen in the story of tomography, both conventional and computed. It was also the case in the development of MRI, and may result in considerable animosity. Perhaps the most interesting question is why are multiple groups interested in the same problem at the same time? This certainly applied to passing electrical currents across evacuated glass bulbs in the 19th century. It almost seems that the time for a discovery of apparatus had come. In the case of the discovery of X-rays, there were many who could have discovered X-rays, and one wonders why no one did? This is in no way to diminish the honour due to Röntgen.

The introduction of radiology had a profound influence on the practice of medicine and these continue. Perhaps one of the most significant has been the promotion of the hospital and clinic as opposed to private medical practice, and this process has been significantly accelerated by the advances in radiology. The initial radiological apparatus was relatively cheap and easy to maintain and could be purchased by an individual. By the 1920s and 1930s, high-powered and precision apparatus became available with increasingly complicated and expensive techniques. The radiologist needed greater skills and scientific knowledge, and the concentration of resources in hospitals made it possible for previously impracticable investigations to be performed.[17] It was the greater size of the hospital that made such developments possible and financially viable. The 1930s saw the development of specialised apparatus, and these were repeatedly superseded by improved models all adding to the expenses. It therefore became increasingly difficult for one doctor in private practice to afford the expenditure for modern apparatus. If it was difficult for an individual doctor in the 1920s and 1930s, it also became more difficult for even a large hospital to afford the necessary expenditure required for a modern X-ray department. In the period prior to the National Health Service (NHS) in the UK, most hospitals were not state funded, and were funded by voluntary contributions. The London Hospital is a typical example of the Voluntary Hospital Movement. A series of letters from Sydney George Holland (1855–1931), 2nd Viscount Knutsford and Chairman of the London Hospital House Committee (1896–1931), to the radiologist Sebastian Gilbert Scott survive, and they give his response to the requests for the funding of a modern radiology department. On 1 December 1922, Lord Knutsford wrote to Scott, saying

> I very much regret that it is impossible for us to carry out your scheme for a fully equipped RRay Dept (*that is Röntgen Ray Department*) at the London. But with £70000 owing to our Bankers, we are quite unable to embark on any capital expenditure. It does seem wrong to me that the largest Hospital in England, & admittedly a first rate one should not be able to do all that is possible in RRay work.

Knutsford goes on to say that decision was "simply a heartbreaking one to me". Lord Knutsford was called a "Prince of Beggars" from his incessant fundraising, and he essentially continued the ideals of the group that had met in 1740 in the Feathers Tavern in London and had planned the original hospital. The charitably funded Voluntary Hospitals in the UK came under increasing financial stress in the 1930s and it became apparent that modern medical care could not be funded by voluntary donations.[18] In the UK, the immediate financial problems were solved by the introduction of the Emergency Medical Service (EMS) with nationalisation of services in 1939, and the EMS laid the foundation of the National Health Service (NHS). Technology continued to advance and has become relatively more expensive, and satisfactory methods for funding are difficult to achieve.

The introduction of the expensive CT/EMI scanner was in many instances funded by private donations, an example being the National Coal Board (NCB) "King's Mill Scanner Appeal" of the 1980s (Figure 0.2). The fundraising plate illustrates many collieries since King's Mill Hospital served a community of whom many were employed in the coal industry. The plate now serves as a memorial to the Nottinghamshire coal mines which have now closed after a 750-year history.[19] Such appeals, whilst being beneficial, were not always free of problems. The appeal often funded only the machine, and so there were issues with staffing and the ongoing

FIGURE 0.2   China plate commemorating the NCB King's Mill Scanner Appeal, May 1986 (Edwardian Fine bone China). Author's collection.

running costs. Replacement of the scanner could be problematic, and the hospital authorities would be obliged to find the necessary finance. However, such appeals gave local people a sense of ownership which was present in the Voluntary Hospitals and was being progressively eroded by a state-run NHS. The advertising of the scanner appeal was frequently emotive and CT was promoted as a cancer scanner. In 1979, the Royal College of Radiologists noted that the siting of the general-purpose CT scanner was influenced by wishes of private benefactors. The siting was not necessarily the most advantageous for the hospital service nor could it be taken as an indication of a planned policy by the Health Authorities.[20]

This book mainly considers the technical development of the radiological sciences and the people involved; however, the complex technology described needs to be funded, and different countries will have different financial models. There may be multiple sources of revenue and in 2021 and it is interesting that a new hybrid scanner combining a nuclear medicine gamma camera and CT scanner was installed by Sherwood Forest Hospitals at King's Mill Hospital funded by public appeal.[21] This new scanner cost £485,000, and the 1986 scanner appeal had aimed to raise £500,00 and had achieved £370,000 by May 1986 when the plate was being sold. There are issues with whatever funding method is chosen, and an equitable solution needs to be achieved so that all benefit.

RADIOLOGY'S PATRON-SAINT

In Western art, the dragon symbolises Satan or Lucifer, the evil one, who must be trampled underfoot. Many saints are depicted as fighting the dragon. This has been discussed by Grigg who comments that, whilst the most famous dragon-slayer may have been St. George, the undisputed patron of radiology is St. Michael (Figure 0.3).[22] The name Michael means "Who is like El?" or *Quis ut Deus?* ("Who [is] like God?"). There is a cartoon in Grigg's book depicting St. Radio-Michael and the Dragon, which seems appropriate. The nomination of St. Michael as patron-saint was made by professors from the Department of Radiology of the University of Genova in Italy, and was blessed in August 1933 by the local clergy. The Bishop of Cerignola in Italy, Vittorio Consigliere, remarked that there are several kinds of light (or luce in Italian); there is the light that is friendly and conducive to better vision, or alternatively the light may be blinding. Although Lucifer has a name related to luce (or light), he cannot be a patron-saint for obvious reasons. The Archangel Michael was the better choice, and the bishop in his talk said that whereas in theology the light comes from the law, in radiology the law comes from the light. St Michael is represented as a knight in armour and his saint's day is 29 September, called Michaelmas. Pope Pius XII confirmed it in his *Discorsi e Radiomessagi*, and in the *Acta Apostolicae Sedis* of 1941, there was an official announcement: "Sanctus Michaël, Archangelus pro radiologis et radiumtherapeuticis patronus et protector declaratus !" And so, the invisible light that Wilhelm Röntgen discovered in 1895

FIGURE 0.3 The Archangel Michael, patron of radiology. Painting on glass. Author's collection.

may be used for healing and for vision, and will show us ourselves and our world in new and unexpected ways.

NOTES

1. Rubin, M. 1948. *Diseases of the Chest, with Emphasis on X-Ray Diagnosis.* Philadelphia: W.B. Saunders Company.
2. PK (Peter Kerley). 1934. Editorial. *The British Journal of Radiology*, 7, 705–706.
3. Thomas, A. 1999. The Wellcome Trust: A healthy heritage symposium. *The Newsletter of the Radiology History & Heritage Charitable Trust*, No. 11, 1–3.
4. Lewis, C.S. On the reading of old books. In: *C.S. Lewis, Essay Collections and Other Short Pieces*. Ed. L Walmsley. London: Harper Collins.
5. Brailsford, J.F. 1934. *The Radiology of Bones and Joints*. London: J&A Churchill.
6. Chew, M.T., Daar, E., Khandaker, M.U., Jones, B., Nisbet, A., Bradley, D.A. 2021. Low radiation dose to treat pneumonia and other inflammations. *British Journal of Radiology*, 94, 20201265.
7. Rice, M.C. 1904. The Röntgen ray in tuberculous adenitis. Transactions of the American Röntgen Ray Society, 5th Annual Meeting, pp. 50–52.
8. Scott, S.G. 1929. Treatment of asthma by radiation. *British Medical Journal*, 1, 9–11.
9. Scott, S.G. 1940. Wide field Roentgen therapy. *American Journal of Roentgenology and Radiation Therapy*, 43, 1–16.
10. Bruwer, A.J. 1964. *Classic Descriptions in Diagnostic Roentgenology* (2 vols.). Springfield: Charles C Thomas.
11. Thomas, A.M.K., Banerjee, A.K., Busch, U. 2004. *Classic Papers in Modern Diagnostic Radiology*. Berlin: Springer Verlag.
12. Webber, P.A., Milford, C. 1984. Thorotrast granuloma of the neck. *Journal of the Royal Society of Medicine*, 77, 1039–1041.
13. Thomas, A.M.K. 1985. Thorotrast granuloma of the neck. *Journal of the Royal Society of Medicine*, 78, 419.
14. Weil, S. 1999. *Œuvres*. Paris: Quarto Gallimard.
15. Rees, R. 1966. *Simone Weil: A Sketch for a Portrait*. London: Oxford University Press.
16. Weil, S. 1952. *The Need for Roots: Prelude to a Declaration of Duties towards Mankind*. London: Routledge & Kegan Paul.
17. Beath, R.M. 1937. Radiology: its background and its future. *Ulster Medical Journal*, April 1937.
18. Evans, A.D., Howard, L.G.R. (undated c. 1930). *The Romance of the British Voluntary Hospital Movement*. London: Hutchinson & Co.
19. National Coal Board. Records of collieries deposited in the Nottinghamshire Archives Office: https://discovery.nationalarchives.gov.uk/details/r/d0d869ab-0e26-42c8-96f7 -81bb9078686c (accessed 9 August 2021).
20. The Royal College of Radiologists. *Report on the Working Party on CT Scanning*. 12 November 1979.
21. Sherwood Forest Hospitals purchases new gamma scanner following appeal success: https://www.sfh-tr.nhs.uk/news/2020/may-2020/sherwood-forest-hospitals-purchases -new-gamma-scanner-following-appeal-success/ (accessed 9 August 2021).
22. Grigg, E.R.N. 1965. *The Trail of the Invisible Light*. Springfield: Charles C. Thomas.

Acknowledgements

Very many have helped me in the preparation of this book, particularly my wife Johanna, who has been unfailing in her encouragement and in her assistance in research. Having an assistant who researches using a different approach is invaluable. This book is an expanded version of a chapter in a book published by CRC Press,[1] and I am grateful for the support and encouragement of the editorial staff, and of Kirsten Barr in particular. This book is also based on ideas raised during my lectures on radiology for the Diploma on the History of Medicine of the Society of Apothecaries, on material presented at the British Institute of Radiology, and on articles published in Aunt Minnie Europe.

I have been fortunate in my teachers. At school I was taught history by the writer Glen Petrie who at the time was writing his book on Victorian social reformer Josephine Butler (1828–1905), which was published the year before I left school.[2] His enthusiasm for history was contagious. At University College Hospital Medical School, I came under the influence of Edwin Clarke (1919–1996), who was Director of the Wellcome Institute for the History of Medicine. In his teachings to us, Clarke emphasised the need to survey past achievements, to appraise the present situation, and to attempt to forecast the future,[3] which is part of the aim of this book. In 1965, he was a member of the founding committee that established the British Society for the History of Medicine, of which I had the privilege to be President during 2012–2013, which would have pleased him. At that time, Jonathan Miller (1934–2019) was working in the Sub-Department of the History of Medicine at University College London, and lectured myself and medical students. As students we were spellbound by Miller. Miller was working on the life of the phrenologist John Elliotson (1791–1868), which started my interest in the topic,[4] and in 1978 presented his BBC television series *The Body in Question*.[5] Clarke had set up an Intercalated BSc degree in the history of medicine at University College, which I took under the historian William "Bill" Bynum. In an example of synchronicity, Bill Bynum and myself received the FRCP on the same evening ceremony which had not been planned by either of us. As a clinical student, William "Bill" Gooddy (1916–2004), a neurologist at University College Hospital, took his "firm" to the National Hospital in Queen Square to see the new EMI/CT scanner and from then on I was determined to become a radiologist. I received undergraduate teaching in radiology from George Simon (1902–1977) at the National Temperance Hospital and from Peter Bretland at the Whittington Hospital. Both were excellent teachers, and the fourth edition of Simon's classic book on the chest radiograph was being written whilst he was teaching me. The book appeared in 1978, the year after Simon's death and the year I qualified as a doctor.[6] Peter Bretland wrote a book on the essentials of radiology that was also published in 1978, based on the teachings that I and my fellow students had received.[7] I started my radiology training in 1981 under Robert Steiner (1918–2013) at Hammersmith Hospital. Robert had a deep interest in radiological history, as is shown by his presidential address to the British Institute of Radiology in 1972.[8] He encouraged my historical interests, and it has been a privilege to have inherited some of his library. A group became interested in radiology history in the 1980s with an aim to prepare for the 1995 Röntgen Centenary. We formed what became the British Society for the History of Radiology, and through this group I have made many friends, both nationally and internationally. Nationally, I should mention Ian Isherwood (1931–2018), Marion Frank (1920–2010), and Jean Guy (1941–2012), and I spent many hours discussing the history of radiology with all of them. The death of Jean Guy was a particular loss to the history of radiology community, and Jean is still missed. Internationally there have been many deep friendships, including with Alfredo Buzzi in Argentina, who wrote the foreword to this book, and Uwe Busch, who is now the director of the German Röntgen Museum. There has been a steady flow of material between Uwe and myself over the years to our mutual benefit. Uwe has undertaken a major development of the museum with excellent results.

Thanks are due to: Susan Aldworth (for radiology and art), Christos S. Baltas (for material and images of Evangelia "Lia" Farmakidou), Graeme Bydder (information on Northwick Park Hospital, and MRI at Hammersmith Hospital), Stephen Golding (Guy's Hospital and its director Tom Hills), Sumit Patil (information on bone age), Richard Price (darkroom disease), the late David Rickard (for material related to Farnborough Hospital), the late Michael Gilbert Scott (for material relating to his father Sebastian Gilbert Scott), Hugh Turvey (radiology and art), the late Hermann Vogel (the radiation martyrs memorial in Hamburg), Frans W. Zonneveld (information on Hermann Bernard Arnold Bockwinkel).

Thanks are also due to: BD Ltd., The British Institute of Radiology, Philips Medical Systems, and Siemens Healthineers.

NOTES

1. Thomas, A.M.K. 2017. History of radiology, in: *Handbook of X-Ray Imaging: Physics and Technology* (Series in Medical Physics and Biomedical Engineering). Ed. Paolo Russo. Boca Raton: CRC Press.
2. Petrie, G. 1971. *A Singular Iniquity: The Campaigns of Josephine Butler*. Suffolk: Richard Clay (The Chaucer Press) Ltd.
3. Clarke, E. (Ed.). 1971. *Modern Methods in the History of Medicine*. London: The Athlone Press.
4. Miller, J. 1983. A Gower Street scandal. *Journal of the Royal College of Physicians of London*, 17, 181–191.
5. Miller, J. 1978. *The Body in Question*. London: Jonathan Cape.
6. Simon, G. 1978. *Principles of Chest X-Ray Diagnosis*, 4th ed. London: Butterworths.
7. Bretland, P.M. 1978. *Essentials of Radiology, for Medical Students and Others Who Find Looking at Radiographs Difficult*. London: Butterworths.
8. Steiner, R.E. 1973. The impact of radiology on cardiology. *British Journal of Radiology*, 46, 741–753.

Author Bio

Dr Adrian Thomas is a writer and teacher. He has seen the development of modern radiology during his career, entering medical school at University College London in 1972, the year that the CT scanner was announced. He has had an interest in history since his school days, studying medical history with Jonathan Miller, Edwin Clarke and Bill Bynum at University College London for his BSc. Adrian started his radiology training in 1981 at Hammersmith Hospital, which coincided with their pioneering development of the MRI scanner. He is a founder member and past-president of the British Society for the History of Radiology. He has co-authored seven books and written many book chapters. He is past-president of the British Society for the History of Medicine, past-president of the Radiology Section of the Royal Society of Medicine, is Honorary Historian to the British Institute of Radiology, and is a visiting professor at Canterbury Christ Church University.

1 Röntgen's Discovery and Its Background

INTRODUCTION

Röntgen's discovery was momentous and following it nothing was quite the same again. It is now difficult to put ourselves into the mindset of those whose worldview was formed before the discovery. The reverberations were felt in the scientific, medical, and artistic communities and helped to bring in the modern world. There were two streams necessary for the discovery to take place, and these were the electrical and the photographic, and Röntgen was expert in both. In the medical discipline of radiology, these two streams continued separately for many decades, coming together with the advent of digital imaging.

Röntgen made his discovery whilst passing an electrical current across an evacuated glass bulb, a discharge tube. His discovery was the result of systematic and careful research by Röntgen and those who went before him. The discovery was not an accident in the accepted meaning of the word.

For X-rays to be produced, the first requirement is for an electrical current of sufficiently high energy to overcome the high electrical resistance offered by the vacuum in a glass bulb. To obtain this high-tension current, various transformers were designed. The second requirement is the means of producing a vacuum, and this was extraordinarily difficult. We take the glass bulb enclosing a vacuum, such as in the simple incandescent light bulb, so much for granted that we forget what a major technological achievement it was. It was the development of the mercury pump that made possible the manufacture of the vacuum tube and resulted in the production of X-rays. Finally, it was Professor Herbert Jackson (1863–1936) of King's College, London, designed the focus vacuum tube, which enabled the cathode rays to be focused onto a metal target, and produced the first real X-ray tube.

EARLY HISTORY

Electrical phenomena are complex and difficult to understand. Early humans would have observed these strange phenomena, such as lightning or the polar lights (the *aurora polaris*), northern lights (the *aurora borealis*) (Figure 1.1), or southern lights (the *aurora australis*) and then speculated as to their origins. The pages of antique history abound with references to the symbolic associations that were established between myth and legend on the one hand and such elementary phenomena on the other. Humanity was ever ready to attribute the mystery of flash and sound to supernatural powers, so the earliest electrical observations were invested with a sacred or religious character. Even an apparently simple question regarding the nature of light

DOI: 10.1201/9780429325748-1

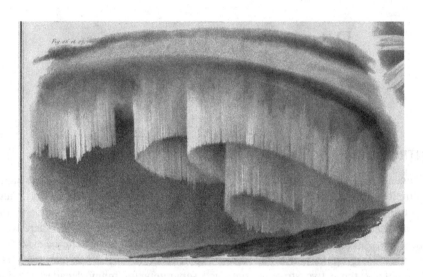

FIGURE 1.1 The Aurora borealis observed from Bossekop, Norway, on 9 January 1834. From Traité de l'électricité et du magnetisme (1840), by Antoine César Becquerel (1788–1878). Public Domain. Author's copy.

was to open fascinating avenues for discovery. In reality, humans do not change, and whilst we may smile at attributing thunder and lightning to the actions of the gods, many contemporary writers, including Fritjof Capra[1] and Rob Bell,[2] now derive a spirituality from the universe of quantum physics.

Thales of Miletus (born c.626BC) a pre-Socratic philosopher, noted interesting properties of amber. Amber when rubbed would attract to itself certain light particles, and this fascinated natural philosophers for many centuries. Pliny the Elder (AD 23–79) and other writers also found that curative powers were bound up with electricity. One of the remedies in use among the Romans was the electric eel of the order Torpediniformes, which could be placed in a bath with a gouty patient (Figure 1.2). Pliny in his *Natural History* (*Naturalis Historia*) implies that it was well known as a therapeutic agent.[3]

The 16th century can be viewed as the dawn of systematic knowledge of natural sciences, and the preeminent name is William Gilbert (1540–1603), Physician-in-Ordinary to Queen Elizabeth the First and President of the Royal College of Physicians. Gilbert first introduced the word electricity, while the appearance of his great work *De Magnete* in 1590 established the basis of electrical science.[4] Gilbert's major conclusion in his book, that magnets worked because the earth itself was a magnet, was a source of wonder to the readers of that day, who still, in the mood of the ancients, considered that such discoveries were direct emanations from a spiritual world. The sense of wonder does persist and for many the boundaries between physics and metaphysics are thin.

In 1643 in Italy, Evangelista Torricelli (1608–1647), a friend and pupil of Galileo Galilei (1564–1642), demonstrated the vacuum above a barometric column of

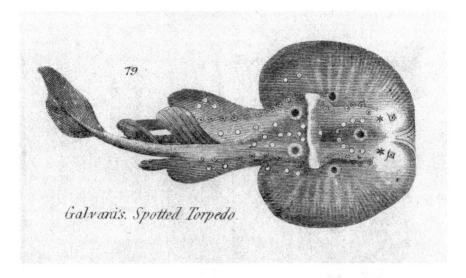

FIGURE 1.2   Galvani's spotted torpedo. From the Museum of Natural History (1860) by William Dallas. Public Domain. Author's copy.

mercury and paved the way for the introduction of the mercury pump. This observation led to the design of various forms of mercury pumps. Almost simultaneously, a German investigator, Otto von Guericke of Magdeburg (1602–1686), was conducting experiments in the hope of obtaining an empty space, in keeping with his theory that the stars maintained their motion by passing through such a negative atmosphere. He attempted to produce a vacuum by emptying a sealed water cask with the help of a piston water pump. Repeated failures led him to construct a mechanical air pump, which was similar to the instrument he had first used, while another of his devices was the first machine for generating frictional electricity. This invention, which was to have some significance for the production of a high-tension current, was a crude method for rubbing a rotating sulphur ball using the hand.

Some ten years later, the development of another pump, improving on von Guericke's principle, was undertaken in about 1660 by Robert Boyle (1627–1691), who carried out experiments on the weight and spring of the air.

If mercury in a vacuum in a barometer is agitated in darkness, then flashes of light will be observed, and this was seen in 1675 by the French astronomer Jean Picard (1620–1682) as the phenomenon of barometric light. In 1705, Francis Hauksbee (1660–1713), who was Curator of Experiments to the Royal Society in London, was investigating mercury in a vacuum. Hauksbee described the phosphorescence of mercury globules, and did not initially see it as resulting from an electrical phenomenon.[5] He developed his apparatus and used a spindle to rub one substance against another in an evacuated bell-jar (Figure 1.3). He was able to rub wool and amber beads together and again observed the pale light flashes. He repeated the experiment in the open air and noted "very little light did ensue in comparison to the appearance of it in vacuo". He continued his work using a variety of materials, and when

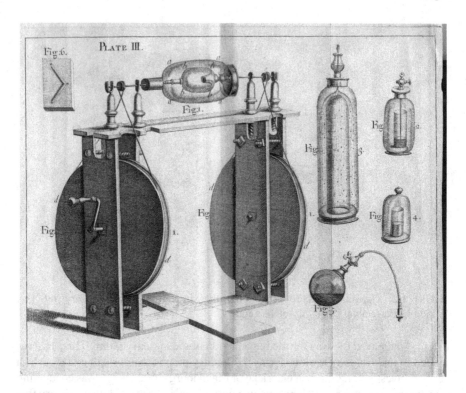

FIGURE 1.3 Apparatus of Francis Hauksbee. From Hauksbee, 1709, see Note 5. Public Domain. Author's copy.

he used glass and wool, he saw a fine purple light when he rotated the spindle. This research carried out by Hauksbee is remarkable, and mark the earliest observation of the results associated with an electric discharge through a vacuum. Hauksbee's work shows a definite commencement of what would become radiological research. The rubbing together of specific bodies is the oldest means of generating electricity, and Hauksbee's purple light is a typical electric discharge, and was regularly seen by early radiologists when the gas X-ray tube had an incomplete vacuum. The next important date is 1729, when Stephen Gray (d.1736), who was a Charterhouse pensioner, demonstrated that some materials would conduct electrical properties for a distance and that some would not. Importantly, he showed that that metal wires conducted electricity.

Throughout the 18th century, the study of electric phenomena was popular in royal courts, including the French court. The leader was Abbé Nollet (1700–1770), who held the Chair of Physics at the College of Navarre and was Preceptor in Natural Philosophy to the Royal Family. His apparatus was his famous electric egg, which at the time was perhaps more an object of curiosity or parlour trick than being of scientific value. In the next century, its significance was appreciated.

The electric egg was a strongly made oval glass vessel, similar to an electric bulb, and when exhausted by an air pump and subjected to the transmission of an electric current produced by means of a frictional machine, it gave rise to a number of startling and colourful effects. These were not unlike the phenomena eventually produced in an X-ray tube of low vacuum. Nollet obtained his results by connecting his electrical generator to the electric egg by wires, which carried the discharges through the bulb (Figure 1.4). His observations were published in Paris in 1753.[6] One day, out of curiosity, his fellow-worker, Charles François du Fay (1698–1739) suspended himself by a silken cord, and Nollet treated him to a charge of electricity from his frictional machine. He then touched du Fay's hand, and a spark, as bright as it was surprising, was seen to pass between the two men. It was in 1734 that du Fay had noted that electricity was in two forms, which he called resinous and vitreous, now named as negative and positive terminals.

In 1745, Pieter van Musschenbroek (1692–1761) at the University of Leiden (Leyden) in the Netherlands developed the Leyden jar. The properties of this electrostatic condenser produced repercussions throughout Europe. The aim was to electrify water by rotating a glass bottle containing water and a central rod which was attached to an electrostatic generator. The Leyden jar is essentially a device to store static electricity and is a high-voltage device with a high-storage capability. A Leyden jar was sent to Benjamin Franklin (1706–1790), the American statesman and philosopher.[7] By far, the best-known experiment in the development of static or high-tension electricity

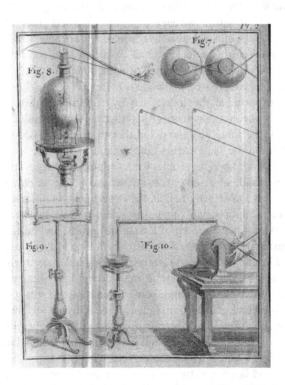

FIGURE 1.4 Abbé Nollet's use of his electric egg. From Nollet, 1746, see Note 6. Public Domain. Author's copy.

was that conducted by Franklin, and at that time only static electricity was known. His belief that the electricity of the earth and in the air were the same essential phenomenon was ridiculed, which Franklin overcame in his famous experiment. He made a silk kite which held an iron point. Fastened to the kite was a hemp string which, continuing as a silken cord, had an iron key attached to its end. The experiment was made in Philadelphia on a rainy day when Franklin released his curious apparatus in the wind. When it was made wet by the rain, the hemp string suddenly became a conductor, and Franklin then touched the key. A spark was immediately created, and this technique, by bringing lightning down to the earth, demonstrated its electrical nature at the same time as it proved his theory. Franklin noted that "thereby the *Sameness* of the Electric Matter with that of lightning is completely demonstrated".[8] Franklin also differentiated between positive and negative electricity, and between conductors and non-conductors of electricity, originating both terms.

Benjamin Franklin was resident in London and his house near Trafalgar Square, currently a museum, is his only surviving residence.[9] During his time in London, Franklin met William Morgan (1750–1833). Franklin and Morgan both shared an interest in electricity, and were both sympathetic to the ideals of the French Revolution. William Morgan was a Welshman from Bridgend in Glamorgan, and was an apothecary and an early actuary.[10] Morgan's experiments were described in a paper read to the Royal Society of London on 24 February 1785.[11] Morgan's experiments were carried out with a mercury gage, or evacuated glass tube, with a piece of tinfoil fastened at the closed end as a terminal. His purpose was to ascertain "the non-conducting power of a perfect vacuum", which was a controversial topic. His paper implies that the various phenomena that accompanied an electric discharge through a vacuum tube were well known at that time.

In Morgan's experiment, he found that by carefully boiling the mercury he obtained so high a vacuum that an electrical discharge was unable to overcome the high resistance of the tube. Morgan noted that the success of this depended upon the boiling, while on the least particle of air being admitted, an electric light, of the usual green colour, became visible. Under repeated charges, the tube at length cracked, when blue and purple colours were obtained. Morgan noted that if the mercury in the gage was imperfectly boiled, the experiment would not succeed; but the colour of the electric light which, in an air rarefied by an exhauster, is always violet or purple appears in this case to be of a beautiful green; and he further noted that, what is very curious, the degree of the air rarefaction may be nearly determined by this means. Morgan had known instances during his experiments when a small particle of air found its way into the tube, the electric light became visible, and was usually a green colour; but the charge being often repeated, the gage had at length cracked at its sealed end, and, in consequence, the external air being admitted into the inside had gradually produced a change in the electric light from green to blue, from blue to indigo, and so on to violet and purple till the medium had at last become so dense that it was no longer a conductor of electricity.

Now Morgan had inadvertently advanced beyond his object, which was only to demonstrate the non-conducting power of a barometric gage from which the mercury had been driven. Morgan had succeeded in producing a high vacuum such as would later be found in the Coolidge tube, which is almost a non-conductor of current. It is not apparent what potential difference Morgan's static machine was able to produce; however, static machines are able to produce enough voltage to generate X-rays. It would seem that the mercury surface would act as an anode and the tinfoil and glass at the other end of Morgan's gage would serve as a cathode. When air was let into his gage, the gas was ionised and resulted in the phenomena that were seen in the early gas or ion X-ray tubes. Essentially, what Morgan demonstrated was that the colour of the light varied with the degree of vacuum in the tube, and this should therefore be called Morgan's phenomenon.[12] The resulting ions would bombard the walls of the tube, producing a fluorescence of the glass and would have resulted in a weak production of X-rays. Certainly, when Morgan described an electrical discharge in an incomplete vacuum, this is similar to a Crookes' tube but without having a defined cathode. However, in this time before the systematic study of fluorescence or the development of photography, there is no way by which Morgan would have detected the invisible Röntgen light.

In 1821–1822, Sir Humphry Davy (1778–1808) at the Royal Institution in London continued the study of electric discharges through a vacuum; however, he was not able to produce a very high vacuum. His work was continued by Michael Faraday (1791–1867), who discovered electromagnetic induction in 1831. Faraday had been Davy's assistant, and was one of the greatest scientists of all time. In 1836, Faraday conducted the first systematic experiments on the discharge of electricity through low-pressure gases. He discovered that when a high-tension current was passed through an exhausted tube into which two electrodes had been sealed, the glow produced was divided by a dark space occurring near the cathode or negative pole. This became known as the Faraday dark space. Also, at about this time, Heinrich Geissler (1814–1879), who was a glassblower and maker of scientific apparatus in Bonn, introduced sealed-in terminals of platinum, as a means of preventing the vacuum tube from being perforated by constant use. The Geissler tube was a low-pressure gas discharge tube, and when a current is passed through it, a beautiful light is observed (Figure 1.5). In 1855, Geissler devised his own mercury pump.

In 1843, Abria, from Bordeaux in France, obtained a vacuum of a still lower pressure, his apparatus essentially being the electric egg, and also the newly invented Ruhmkorff induction coil. Abria observed non-moving striations in the gas discharge; however, little notice was taken of this observation at the time until the work of William Robert Grove (1911–1896) in 1852. In 1842, Grove had developed the first fuel cell, which he termed the "gas voltaic battery".

Heinrich Daniel Ruhmkorff (1803–1877) was a German instrument maker who commercialised the induction coil. This coil, which greatly advanced the production of the necessary high-tension current, may be described as having an "inner" or "primary" coil consisting of a bundle of stout wire, bound together by thick copper wire for the induced low-tension current, while surrounding this is a "secondary" or "outer" coil of fine wire for the induced high-tension current (Figure 1.6).

In 1859, Julius Plücker (1801–1868) demonstrated a green fluorescence which took place opposite the negative electrode in a vacuum tube. Ten years later in 1869, Johann Hittorf (1824–1914) published the results of his studies on the conduction of

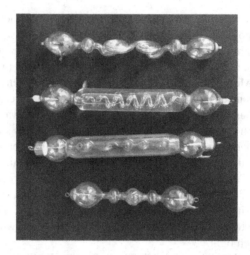

FIGURE 1.5   Collection of Geissler tubes. Author's collection.

FIGURE 1.6   Ruhmkorff induction coil. Author's collection.

electricity through highly rarefied gases. Continuing to observe the fluorescence of the glass vacuum tube, which was caused by rays proceeding from the cathode when air was withdrawn, he noticed that any object placed between the negative electrode and the glass of the tube cast a clearly defined shadow, demonstrating that the newly discovered stream of radiation travelled in straight lines. Hittorf must have produced X-rays in his experiments.

The investigation of the cathode stream of rays was a direct precursor to the discovery of the X-rays; and many physicists became interested in these curious phenomena. An extensive study of the phenomena was now made by Sir William Crookes (1832–1919), who explained his findings before the British Association in

1879. Certain modifications in the glow that accompanied the electric discharge had been noted when the air pressure was still further reduced. The light, attending the discharge, proceeded from the glass wall that was under bombardment by the cathode stream, which fact, later on, was found to be responsible for the production of X-rays. Crookes was a remarkable man and was interested in many areas of scientific research, and like Sir Oliver Lodge (1851–1940) he was interested in spiritualism and became president of the Society for Psychical Research.[13] When the Röntgen Society (the forerunner of the British Institute of Radiology) was founded in 1897, Crookes and Röntgen were the first two honorary members. Crookes is known for his radiometer or light-mill which is depicted on his bookplate (Figure 1.7).

Up to this point the following facts had been found concerning the cathodal radiation from the negative pole of the evacuated glass bulb. These were that an object, placed in its path, would fluoresce certain stones and minerals, producing a brilliant effect that differed according to their quality; that the portion of the tube that was screened by the object from the light of the rays was still in darkness; that the cathode rays heated the object on which they struck, and they could also be deflected by a magnet.

Crookes formed the opinion that the cathode rays were negatively charged moving particles, and he came very close to discovering X-rays. During his work, it happened that a sealed packet of photographic plates was lying close to a vacuum tube which, unknown to him, was emitting X-rays. And when used, the plates were found to be fogged. Crookes attributed this to faulty manufacturing. It can be seen that many physicists might have been the discoverer of X-rays.

In 1892, Heinrich Hertz (1857–1894) investigated cathode rays, and demonstrated their power of passing through certain substances that were placed inside the vacuum

FIGURE 1.7 *Ex Libris* – the bookplate of Sir William Crookes, DSc, FRS. The radiometer is shown centrally at the base of the shield. Author's collection.

tube. Hertz found that cathode rays would pass through thin metal foils such as gold and aluminium, while mica remained impervious to the rays. Unfortunately Hertz died at the tragically young age of 36 following surgery. In his short life, Hertz made major discoveries, including electromagnetic radio waves (initially called Hertzian waves), the photoelectric effect (subsequently explained by Albert Einstein), and in contact mechanics. It is interesting to speculate what Hertz would have achieved had he survived his illness. He is remembered by the SI unit for frequency the hertz (Hz), which is the number of times that a repeated event occurs per second, as established by the International Electrotechnical Commission in 1930.

Further observations were made by Philipp Eduard Anton von Lenard (1862–1947), who was a pupil of Hertz. Lenard inserted a window of fine aluminium opposite the negative pole, and was able to examine the properties of the cathode stream when outside the vacuum. Many substances were caused to fluoresce outside the vacuum tube, similar to when they had been placed inside the tube directly in line with the radiation. Apart from affecting a photographic plate, it was found that the rays which were being produced passed through the hand and certain solid layers to a very slight degree. It was Lenard who first observed that part of this stream issuing from the tube could not be deflected by a magnet, which proved the presence of rays other than the cathode. Lenard was therefore demonstrating X-rays.

Herbert Jackson (1863–1936), who invented the focus X-ray tube, also worked with a tube in which a window had been constructed following Lenard's model. Jackson also observed the fluorescence of objects outside the vacuum, and unfortunately concluded that this was due to ultraviolet radiation. Jackson followed this line of enquiry and so lost his opportunity to discover the X-rays. Many continued investigating the cathode stream when drawn outside the vacuum.

The nature of the cathode stream of rays was still a mystery. William Crookes maintained that they consisted of minute particles of matter. A number of German physicists, led by Gustav Heinrich Wiedemann (1826–1899), supported the view that the cathode rays were similar to the oscillating ultraviolet light, and were caused by vibrations in the ether of an exceedingly short wavelength. The fact established by Lenard, of their power to pass through objects that stopped ordinary light, appeared to strengthen this theory. But, in 1897, the research of Sir Joseph John Thomson (1856–1940) finally disproved the German belief, and demonstrated that the cathode rays were negatively charged particles of matter, now known as electrons.[14] Thomson calculated that the electrons must be very much smaller than atoms and have a very large charge-to-mass ratio.

The above is an outline of the developments that were to culminate in Röntgen's epoch-making discovery. The discovery of X-rays should not have come as a surprise and should have been entirely predicted; however, this does in no way diminish Röntgen's achievement.

RÖNTGEN AND THE DISCOVERY OF X-RAYS

Wilhelm Conrad Röntgen (1845–1923) was the only child of a German mercantile family (Figure 1.8). He was born at Lennep, in the Province of the Rhine, on 27

FIGURE 1.8   Wilhelm Conrad Röntgen (1845–1923) depicted on a plaque at the age of 50 when he made his discovery. A similar plaque was hung for many years in the hall of the old house of the British Institute of Radiology in Portland Place in London. Author's photograph.

March 1845.[15] The family moved to the Dutch town of Apeldoorn, where the young Wilhelm began his schooling. His early life gave little promise of his mature eminence, though he soon revealed a certain talent in the constructing of mechanical devices. Most of his leisure was spent in the open air rather than in academic study, and this love of nature remained a characteristic of his, even during the busiest periods of his life.

At the age of 17, Röntgen was registered as a pupil at the Utrecht Technical School, after which it was his plan to enter the University in Utrecht. But this prospect received a setback as the result of a boyish prank, for which Röntgen was expelled from the School. A caricature of one of the masters was drawn upon a fire screen, and Röntgen, when asked to divulge the name of the pupil who had been responsible, would not speak. He had a series of misfortunes and without a high school diploma was forced to find a new way to university education. His disappointments were probably responsible for his somewhat negative attitude towards examinations, which he saw as necessary evils and an unreliable test of a student's capacity for a given subject.

In 1865, Röntgen was able to enter the Federal Polytechnic Institute in Zurich, where the preliminary regulations were less stringent and he passed the entrance examination. The Professor of Physics at Zurich was August Kundt (1839–1894), who eventually made Röntgen his assistant. In 1869, Röntgen obtained his PhD from the University of Zurich. Röntgen's mode of living was quiet, and in keeping with the modest retirement that always distinguished him, he is described as showing "a dislike of organised pleasure and gaiety". The company of a few friends, a long walk over the mountains, and photography were his most unchanging recreations.

At the end of the day's studies, Röntgen and his friends would gather about the tables of a certain café in the town that was kept by a German refugee named Ludwig. In time, he became closely acquainted with one of his daughters, Anna

Bertha (1839–1919), some 6 years older than him, and who, as a young woman, showed signs of a delicate constitution that gradually became more pronounced as years went by. They were married on 19 January 1872 somewhat to the disappointment of Röntgen's father, who was more materially ambitious than his son. The union was happy, although childless, and after 5 years they adopted Bertha's young niece.

In the spring of every year, the Röntgens visited the Italian Lakes, while autumn found them at Pontresina in the Engadin Mountains in Switzerland. The scientist's love of nature continued to be supplemented by the use of a camera, which he employed during his mountaineering excursion to the Piz Bernina and other summits. It is curious that despite his experimental instinct, he was not sympathetic towards modern travel conveniences, and preferred a horse-drawn carriage even when the train was available.

Röntgen was a favourite student of August Kundt, and in 1872 he followed him to the newly founded German Kaiser-Wilhelms-Universität in Strasbourg. This was a time of general jubilation in Germany, following the recent Franco-Prussian War, and the founding of a German University at Strasbourg was an event of national importance. The bombardment and Siege of Strasbourg was a major event during the Franco-Prussian War, and the French surrender of the fortress took place on 28 September 1870. At Strasbourg, Röntgen was made Privatdozent, and he published a number of high-quality papers.

His work covered investigations on the discharges of electricity under certain conditions; on the conductivity of heat in crystals; on the determination of the ratio of specific heats for air and various gases; on the problem of elasticity, and on capillarity; while he also took part in demonstrating the existence of the plane of polarisation, and the fact that it was subject to quantitative measurement.

Röntgen had a series of appointments, including the Chair of Physics and Mathematics at the Hohenheim Agricultural Academy in Württemberg, and the Professorship of Physics at Giessen University, in Hessen. He continued his study of crystalline phenomena, demonstrated the absorbance of heat radiation by water vapour, dealt with the viscosity of certain fluids, and various theories of compression; while the emergence of an electrical effect, which became known as the Röntgen current, resulted from his important work on moving dielectrics. Röntgen maintained that the basis of judgement was experiment, no matter what hypothesis was supported or cast aside in the process.

Röntgen had various offers that he disregarded, until the end of 1888, when at the age of 43 he became Professor of Physics and Director of the Physical Institute of the University of Würzburg. He had worked there under his old teacher, August Kundt, 17 years prior to his return there as professor.

The greater part of his laboratory work at Würzburg was carried on without an assistant. The apparatus he used was relatively basic, and much of it was self-constructed. His early papers from Würzburg dealt with such subjects as the different physical properties of liquids and solids; the electrodynamic effects of moving dielectrics; the thickness of coherent oil layers on a fluid surface; and the compressibility of liquids and alcohols. In common with other scientific workers all over the world, Röntgen's research gradually became directed towards the properties of the cathode rays, with which he felt many unknown phenomena were associated.

On the evening of Friday, 8 November 1895, Röntgen was working alone in his small and unpretentious laboratory at the Physical Institute (Figure 1.9). His apparatus consisted of a Ruhmkorff coil to create the high-tension current and a Hittorf vacuum tube, through which this high-tension current was passed (Figure 1.10). It had been found previously that the action of the invisible ultraviolet light acting on crystals of barium platino-cyanide caused a most brilliant fluorescence, and it so happened that a card coated with that compound was lying on a table some distance from the actual spot where Röntgen was examining the phenomena that accompanied the passing of a high-tension electric current through a vacuum tube.

The room was darkened, while the tube itself was fitted with a black, light-tight cardboard cover. Röntgen proceeded to pass a high-tension current through the tube, and was surprised to observe that, in spite of its light-tight covering, a sudden illumination played over the screen of crystals, which stood on a table some little distance away. This phenomenon only occurred when the tube was activated, and this apparently simple fact led him on to eight weeks of almost unbroken labour. Sometimes he ate and slept in the laboratory to avoid distraction, while the nervous strain to which he was subject made him even more gruff and irritable to those around him.

FIGURE 1.9 In diesem hause entdeckte W.C. Röntgen im Jahre 1895 die nach ihm benannten strahlen: In 1895, W.C. Röntgen discovered the rays named after him in this house (the Physical Institute at Würzburg). Author's photograph.

FIGURE 1.10 Röntgen's laboratory at the Physical Institute at Würzburg set up as it might have appeared in his day. Author's photograph.

Several incorrect versions of the discovery were circulated in the newspapers at the time. By continuing his research, Röntgen found that he had detected a new type of radiation, differing from the cathode particles which caused the glass walls of the vacuum tube to fluoresce. The new rays arose wherever the cathode electrons were brought to rest by acting on the glass wall of the tube, which acted to transform the energy. It was also found that they passed through various objects and liquids that otherwise repelled light, such as a thick book, a double pack of cards, blocks of wood, rubber, and aluminium, whilst lead and platinum retained their opaqueness. The power of the rays in effecting transparency was seen to be closely dependent upon the compactness or atomic weight of the object concerned.

A further result was observed when Röntgen placed his hand between the activated tube and the screen of platino-cyanide crystals. The image appearing on the screen clearly revealed the bones of his hand, while the softer tissues gave a shadow of much less density. He next substituted a photographic plate for the screen, and again interposed his hand between it and the vacuum tube. After development, an outline of his hand, including a silhouette of the bones, was visible on the plate. It is not known when this first radiograph was made, however it was sometime before 22 December 1895 when the famous image of his wife Bertha's hand showing her ring was made.

Röntgen gave the new radiation the name of "X-rays" ("X" denoting an unknown quantity in Algebra), thus acknowledging that the essential nature of the rays was unknown to him. They are invisible to the eye, only the fluorescence they excite in certain substances being observable. In addition to the latter property, the rays would pass through materials that were opaque to ordinary light, and could be recorded on a photographic plate.

Röntgen proceeded with the utmost caution, and made his discovery known only after a rigid and carefully controlled investigation. A report of the discovery was made to the President of the Physical Medical Society of Würzburg on 28 December 1895 in a document that was remarkable for both brevity, taking only 15 minutes to read, and a straightforward assembling of facts. Röntgen insisted that no mistake was possible, since the greater part his observations rested upon the visual evidence of photography.

The month of January 1896 found Röntgen demonstrating his discovery before Kaiser Wilhelm II in Berlin; and, on the 23 January he spoke at a meeting of the Würzburg Physical Medical Society. To an enthusiastic audience, who were shown the photographs, he explained the likelihood of securing images of the more complicated parts of the body by means of the new rays, and of his own readiness to help in experiments that might prove of medical value. It was at this meeting that the title of Röntgen rays was suggested for the discovery.

His sudden emergence as a world-famous figure made Röntgen even more modest and reticent, except for his happiness in the recognition that the new radiation was of untold value to science and medicine. He continued working without interruption, whilst both the German and foreign scientific community honoured him. The University of Würzburg awarded him the honorary degree of Doctor of Medicine; he was made a corresponding member of the Academies of Science in Berlin and Munich; other awards being the Royal Bavarian Order of the Crown, and the honorary citizenship of his birthplace, Lennep. Röntgen was honoured in popular culture with many depictions on postcards and trade cards, and is feted as a celebrity on a Portuguese trade card given away with Claus & Schweder soap in the late 1890s (Figure 1.11).

Röntgen made a second communication on 9 March 1896, while a subsequent report followed a year later. Röntgen continued his work on crystals, and dealt with their conductive power and reactions under the new rays. In April 1900, now 55 years old, he accepted the post of Professor of Physics at the University of Munich, while the year following saw him receive the first Nobel Prize for Physics. It was customary for Röntgen to avoid, whenever possible, making a public appearance, but on this occasion he made an exception and journeyed to Stockholm for the presentation.

Many offers were made for his services, which he declined, and continued working in the Bavarian capital. Röntgen was very much concerned about the dangers that he saw as invariably accompanying the popular presentations of science, and deplored the reception of superficial and erroneous conclusions by the ordinary public. He saw this state as worse than absolute ignorance.

A personal anecdote illustrating the consistency of his attitude towards social niceties is told by Margret Boveri, the daughter of a professor who was Röntgen's friend. At a function, where the University heads were present, the couples were paired-off for dinner in such a fashion as to leave the wives of the professors without

FIGURE 1.11 *Roentgen*, on Sabonette "Celebridades" (celebrities), published by Claus & Schweder Porto, Portugal, soap packet trade card (c. 1895). Author's collection.

escort. Röntgen took in the situation, and abandoned the lady of title whom it was intended he should accompany in favour of his own wife.

On 2 May 1905, the German Röntgen Society was founded, and a later honour being the granting of the title of Excellency to the scientist. The years of the Great War entailed a twofold sorrow for Röntgen, who displayed a natural patriotism; however, he was unprepared for the defeat of his country, while an equal shadow was cast by the growing illness of his wife Bertha.

There was, however, some comfort in the knowledge that the X-rays were being applied medically to help the wounded, a fact the German Government recognised by awarding him the Iron Cross. His wife Bertha died on 31 October 1919, and Röntgen, who had become a representative of an era that had now passed away, must have looked with saddened eyes towards an immediate future that promised little, either for himself or his country.

In 1920, he resigned his professorship, but still retained an active interest in the laboratory. He also continued in office as the Conservatore of the Physical Metronomical Institute of the Munich Academy of Science, living for the most part at Weilheim with occasional journeys to Switzerland. The early days of 1923 found him once more in the laboratory, but it was to be a last effort. All about him were the signs of a great country and a people in eclipse. The Ruhr was occupied by the victorious troops of a foreign power; Strasbourg, his home for so many of his early and happier memories, was returned to France; and in the Weimar Republic the country was experiencing a post-war decadence. Röntgen developed an intestinal complaint and died in Munich on 10 February 1923, aged 78. On 13th February, accompanied by the tributes of many representatives of science, his body was cremated, and some months later the ashes were placed beside those of his wife and parents in the cemetery at Giessen.

Work on the nature of X-rays continued. In his 1895 first communication, Röntgen had thought that the new rays might be due to longitudinal vibrations in the ether. He knew that this explanation would require further corroboration. The elucidation of Röntgen's question was left to later workers.[16] Max von Laue (1879–1960) was a lecturer (or Privatdozent) at the Institute of Theoretical Physics of Munich University. In 1911, Paul Peter Ewald (1888–1985) from Sommerfeld's Institute in Munich was studying the propagation of electromagnetic radiation in a space lattice. Ewald proposed a resonator model of crystals; however, the model could not be tested using visible light, since the wavelength of light was larger than the spacing between the resonators. Max von Laue thought that X-rays might have a wavelength which was of a similar order of magnitude to the spacing in crystals and could therefore be used to test the model. If the wavelength of X-rays were many times shorter than that of light, then that would also explain the previous failures to produce diffraction effects using gratings that were only suitable for visible light. The theories were confirmed by Walter Friedrich (1883–1968)[17] and Paul Knipping (1883–1935).[18] In May 1912, a fine pencil beam of X-rays was passed through a copper sulphate crystal and the diffraction pattern on a photographic plate was recorded. It was Walter Friedrich who was the first person to build an apparatus that successfully demonstrated diffraction of X-rays by a crystal, and it was Friedrich who first observed the diffraction pattern. This discovery simultaneously demonstrated the wave-like nature of X-ray radiation and the translational symmetry that defined the nature of crystals. The resulting photographic plate showed a large number of well-defined spots which were arranged in intersecting circles around the central beam. Max von Laue then went on to develop a law that connected the scattering angles and the size and orientation of the spacings in the crystal. Although Max von Laue was awarded the Nobel Prize for Physics in 1914, he always recognised the contributions of Frederick and Knipping, and all three should have received the award. The story of the early days of X-ray crystallography is complex and interesting and has been comprehensively reviewed by André Authier.[19]

Thanks to diffraction studies, it became possible to measure the wavelength of X-rays and also study the inner structure of materials. The work of von Laue was taken up by William Henry Bragg (1862–1942) (father) and William Lawrence Bragg (1890–1971) (son) from Adelaide in South Australia.[20] Working in Leeds in England, they both did important work on X-ray crystallography. In 1912–1913, William Lawrence Bragg developed Bragg's law, which connected the observed scattering with reflections from evenly spaced planes within a crystal. The Braggs both shared the 1915 Nobel Prize for Physics for their work on crystallography. The earliest structures to be examined were simple in nature and showed a one-dimensional symmetry, and the structure of common table salt was determined in 1914. As computational and experimental methods improved over the following decades, it became possible to examine ever more complex material. This work resulted in the study of protein structure and then spectacularly to the determination of the double-helical structure of DNA. Dorothy Crowfoot Hodgkin (1910–1994) should be particularly remembered for developing X-ray crystallography to examine biological molecules, and she determined the structures of cholesterol in 1937, vitamin B12 in 1945, and penicillin in 1954.[21] Dorothy Hodgkin was awarded the Nobel Prize for Chemistry in 1964, and in 1969 she determined the structure of insulin.

Perhaps one of the best-known uses of X-ray crystallography was to understand the nature of the DNA and RNA molecules that are central to life. The British biophysicist Rosalind Elsie Franklin (1920–1958) made crystallographic studies of DNA, RNA, and carbon compounds (coal and graphite).[22] Franklin had started working at King's College London in 1951 where Maurice Wilkins (1916–2004) was working using fairly crude apparatus. Maurice Wilkins and Raymond Gosling (1926–2015) had been working on DNA before Rosalind Franklin. In the summer of 1950 using a moistened sample of DNA fibres, they obtained X-ray diffraction images of DNA using a modified X-ray apparatus. The famous experiment of Franklin and Gosling stretched a strand of DNA across a paperclip and set it on a piece of cork. A fine beam of X-rays was passed through the strand of DNA and the diffracted paths were recorded on photographic paper as Photo 51. The resultant image proved the helical shape of DNA. A colleague Alec Stokes (1919–2003) looked at the patterns and made the suggestion that the diffraction pattern could be interpreted as the diffraction pattern of a helix, that is a spiral ladder, and that the DNA molecule could be helical.

James D. Watson (1928–) heard Maurice Wilkins talk about his work on DNA using X-ray diffraction at a meeting in Italy. Watson worked with Francis Crick (1916–2004) and they were interested in X-ray diffraction and the molecular structure of DNA. They were aware of Franklin's work, and to construct their model of DNA, Watson and Crick made use of information from Franklin and Gosling, including unpublished X-ray diffraction images of Franklin's including Photo 51, which were shared by Gosling and Wilkins. In March 1953, Watson and Crick were able to determine the double-helical structure of DNA. This was first announced by Sir Lawrence Bragg who was the director of the Cavendish Laboratory at a Solvay conference on proteins that took place in Belgium on 8 April 1953. Between their periods of work, there were visits by Watson and Crick to pubs in Cambridge, including "The Eagle", where there is a commemorative plaque placed outside (Figure 1.12). The Nobel Prize for Physiology or Medicine for 1962 was awarded jointly to Crick, Watson, and Wilkins. In their Nobel speeches, only Wilkins mentioned Franklin as being, along with Alec Stokes, two of those at King's College London "who made valuable contributions to X-ray analysis". In his book *The Double Helix*, Watson does acknowledge the work of Rosalind Franklin, although the book had a mixed reception.[23] It was Aaron Klug (1926–2018) who had been Franklin's colleague, who won the 1982 Nobel Prize for Chemistry for his work on crystallographic electron microscopy and nucleic acid–protein complexes, who would finally mention Franklin when he gave his Nobel address in Stockholm.[24] Klug was a long-term defender of Franklin's contributions, and had continued her work. The work of Rosalind Franklin continues

FIGURE 1.12 The Eagle Pub, Benet Street, Cambridge, where Watson and Crick worked over lunch to discuss the structure of DNA. The pub, which opened in the 17th century and originally called "The Eagle and Child", was a regular haunt for airmen in the Second World War. Author's photograph.

to interest and cause controversy, and is celebrated in a remarkable play by Anna Ziegler with a memorable performance by Nicole Kidman as Rosalind Franklin, and performed at the Noël Coward Theatre in London in 2015.[25] At the end of the play, Ziegler has Franklin, who died prematurely at the age of 37 from metastatic ovarian cancer, say "I think there must be some point in life when you realize you *can't* begin again. That you've made the decisions you've made and then you live with them or you spend your whole life in regret".

The work of Max von Laue on diffraction gave solid evidence that X-rays were waves of electromagnetic radiation; however, X-rays also behave like particles because they can ionise gases. Indeed, it was this property of X-rays to ionising gases that caused William Henry Bragg to argue in 1907 that X-rays were not electromagnetic radiation at all. We now know that X-rays are photons and as such show characteristics of both particles and waves. The idea of the photon had been proposed by Albert Einstein (1879–1955) in 1905; however, it was not until 1922 when Arthur Holly Compton (1892–1962) demonstrated the scattering of X-rays from electrons that the photon theory was completely accepted. Compton won the Nobel Prize for Physics in 1927 for his discovery of the Compton effect, also known as Compton Scattering, which demonstrated the particle nature of electromagnetic radiation.[26]

There have been many Nobel Prizes related to X-rays, some have been controversial and others have not. Most Nobel Prizes are deserved, and perhaps some are not. It is certain that the race and rivalry to be the first in science is a reality, and scientists are not as dispassionate and detached as they are sometimes presented.[27] It is not immediately obvious why we have made the Nobel Prize such an exemplar of human excellence.[28] We do not need the award of the Nobel Prize in 1901 for us to be aware of the importance of the discovery and Röntgen's greatness. The 1901 award produced significant resentment on the part of Phillip Lenard who presented himself as responsible for the discovery since he had performed all of the groundwork that enabled the discovery to take place.[29] Röntgen himself was unassuming and retiring by nature, and the sensation caused by his discovery was viewed as an unnecessary interruption to his work.

The discovery by Röntgen of his new rays initiated a profound change in physics, and in many other areas, and we are still feeling the reverberations.

NOTES

1. Capra, F. 1992. *The Tao of Physics*. London: Flamingo.
2. Bell, R. 2020. *Everything Is Spiritual: Who We Are and What We're Doing Here*. St. Martins Essentials.
3. Colwell, A.A. 1922. *An Essay on the History of Electrotherapy and Diagnosis*. London: William Heinemann.
4. Gilbert, W. 1958. *De Magnete*. Trans. P. Fleury Mottelay. New York: Dover Publications Inc.
5. Hauksbee, F. 1709. *Physico-Mechanical Experiments on Various Subjects Containing an Account of Several Surprising Phenomena Touching Light and Electricity*. London: R. Brugis.
6. Nollet, J.A. 1746. *Essai sur l'Electricité des Corps*. Paris: Les Freres Guerin.
7. Morgan, E.S. 2002. *Benjamin Franklin*. New Haven: Yale University Press.
8. Franklin, B. 1986. *The Autobiography and Other Writings*. Ed. K. Silverman. Harmondsworth: Penguin Books.
9. https://benjaminfranklinhouse.org (accessed 3 February 2021).
10. Bennetts, N.B. 2020. *William Morgan*. Cardiff: University of Wales Press.
11. Morgan, W. 1785. Electrical experiments made in order to ascertain the non-conductor power of a perfect vacuum. *Philosophical Transactions of the Royal Society*, 75, 272–278.
12. Underwood, E. A. 1957. Wilhelm Conrad Röntgen (1845–1923) and the early development of radiology. In: *Sidelights on the History of Medicine*. Ed. Z. Cope. London: Butterworth & Co.
13. D'Albe, E.E.F. 1924. *The Life of Sir William Crookes*. New York: D Appleton & Company.
14. Thomson, J.J. 1936. *Recollections and Reflections*. London: G. Bell and Sons. Ltd.
15. Glasser, O. *Wilhelm Conrad Röntgen and the Early History of the X-Rays*. Bale, Sons and Danielsson.
16. Segrè, E. 1980. *From X-Rays to Quarks, Modern Physicists and Their Discoveries*. New York: WH Freeman & Company.
17. Heaney, P.J., Kaliwoda, M. 2020. Walter Friedrich unplugged: his 1963 interview in East Berlin. https://www.iucr.org/news/newsletter/volume-28/number-3/walter-friedrich-unplugged-his-1963-interview-in-east-berlin (accessed 26 July 2021).

18. Erwald, P.P. 1999. Paul Knipping 1883–1935. https://www.iucr.org/publ/50yearsofxr aydiffraction/full-text/knipping (accessed 26 July 2021).
19. Authier, A. 2013. *Early Days of X-Ray Crystallography*. Oxford: Oxford University Press.
20. Jenkin, J. 2008. *William and Laurence Bragg, Father and Son*. Oxford: Oxford University Press.
21. Ferry, G. 1998. *Dorothy Hodgkin: A Life*. London: Granta Books.
22. Glynn, J. 2012. *My Sister Rosalind Franklin*. Oxford: Oxford University Press.
23. Watson, J.D. 1970. *The Double Helix: A Personal Account of the Discovery and Structure of DNA*. Harmondsworth: Penguin Books.
24. Maddox, B. 2003. *Rosalind Franklin: The Dark Lady of DNA*. London: Harper Collins.
25. Ziegler, A. 2015. *Photograph 51*. London: Oberon Books.
26. Shankland, R.S. (Ed.). 1973. *Scientific Papers of Arthur Holly Compton, X-Ray and Other Studies*. Chicago: The University of Chicago Press.
27. Myers, M.A. 2012. *Prize Fight: The Race and the Rivalry to the First in Science*. New York: Palgrave Macmillan.
28. Thompson, G. (Ed.). 2012. *Nobel Prizes that Changed Medicine*. London: Imperial College Press.
29. Weissmann, G. 2010. X-ray politics: Lenard vs. Röntgen and Einstein. *The FASEB Journal*, 24, 1631–1634.

2 The Early Radiology Departments and the Problems They Faced

There is a certain sameness about the introduction of X-rays into use throughout the world. This is partly related to the ready availability of the Crookes' tubes which enabled practitioners to repeat Röntgen's findings. Röntgen had reprints made of his *First Communication* and he sent copies to 80 scientists. To approximately 12 eminent scientists, Röntgen, in addition to the copy of the reprint, sent a collection of positive prints of selected radiographs. Of these 12 sets, only 2 full collections are known to exist. Frans W. Zonneveld has investigated the set that was sent to the Dutch theoretical physicist Hendrik Antoon Lorentz (1853–1928).[1] In early 1896, Röntgen sent sets to Franz-Serafin Exner (1849–1926) in Vienna, to Lord Kelvin (William Thomson [1824–1907]) in Glasgow, to Emil Gabriel Warburg (1846–1931) in Berlin, to Arthur Schuster (1851–1934) in Manchester, to Henri Poincaré (1854–1912) in Paris, and Lorentz in Leiden.

ARTHUR SCHUSTER AND EARLY MANCHESTER RADIOLOGY

Arthur Schuster (1851–1934) was professor of physics in Owens College in Manchester.[2] Towards the end of 1895, he was visiting Pontresina where the Röntgens were staying and called to see them in their hotel. Wilhelm Röntgen was out; however, he was greeted in a friendly manner by his wife Bertha. In 1896, he returned to his laboratory and amongst his accumulated correspondence he found a flat envelope with a set of photographs. There was no accompanying explanation and Schuster found the images to be unintelligible. This is perhaps surprising since what they showed might seem to us to be obvious anatomy. One of the photographs was of a hand with the bones marked and labelled. That there was no letter is interesting, but in an "insignificant wrapper" was a thin pamphlet by Röntgen called *Ueber eine neue Art von Strahlen*.[3] Schuster was immediately fascinated, and his daughter Norah related that he kept his pretty wife waiting with the cabman and his horse in the cold winter's evening whilst he read the pamphlet.[4] Schuster was able to repeat Röntgen's experiments and was immediately successful in obtaining radiographs. He wrote to *Nature* on 23 January 1896 and discussed the nature of the rays. That the rays could neither be deflected in a magnetic field nor diffracted suggested that they were not cathode rays or related to light. Schuster differentiated between invisible light, visible light, and Röntgen rays.[5] Schuster's personal assistant Arthur Stanton translated Röntgen's paper for *Nature* and this appeared later in the same issue, but

DOI: 10.1201/9780429325748-2

he never returned the original to Schuster![6] On 7 January 1896, Schuster had the photographs shown at the Manchester Literary and Philosophical Society, and on 8 January, he had a letter in the Manchester Guardian giving a clear account of the discovery and offering to show the radiographs to anyone who was interested. His first clinical radiograph was taken of the foot of a dancing girl from a local pantomime, and showed a retained needle. Schuster kept the print of his for 38 years until his death. The dancer had a septic foot and had been seen at Guy's Hospital in London where she had been told that no treatment was possible. Schuster's radiograph clearly showed why the foot was not healing. Another radiograph showed a fracture-dislocation of the elbow. On 2 March, Schuster gave a lecture at Owens College, and during the lecture took a photograph of the foot of his 6-year-old son Leonard. The exposure time was 5 minutes, and Leonard was anxious about keeping still for such a long exposure. Both Norah and Leonard attended the lectures, and Norah did not fidget like her brother. The consequence of Schuster's success was that his laboratory was inundated with requests from Manchester's doctors for radiography. Norah recalled that her father was not particularly interested in medicine, and later in 1896 Schuster was trying to persuade the Manchester Medical Society to set up its own hospital service. This was not to happen until some years later. It is noteworthy that the service to the Piccadilly Infirmary, which became the Manchester Royal Infirmary, was provided until 1908 by Mottershead & Co., a Manchester-based chemist (pharmacist). Pharmaceutical chemists provided a photographic service to the public, and it is not surprising that they would be interested in the "new photography". Another example from this period is Chas. E.J. Eynon, a pharmaceutical chemist with premises in James Street in Harrogate, and who provided a radiographic service with high-quality prints (Figure 2.1). The pioneer radiologist in Manchester was Alfred E. Barclay (1876–1949) (Figure 2.2) who took up practice in 1906. Norah described being radiographed by Barclay when she was 18. She had worn a new skirt and felt a pain in her leg. She rubbed her leg which became painful, and a deep venous thrombosis or foreign body was suspected. A radiograph showed a needle close to the femur, and she was told that it could never be removed. Schuster had a powerful electromagnet in the physics department and her leg was held against it for an hour. The needle moved, and finally came close to the skin and was removed.

Barclay went on to develop radiology in Manchester and has described the early days of the developing profession.[7] When he started in Manchester there were no radiological appointments at any of the hospitals; however, he was asked to become the Honorary Radiologist to Ancoats Hospital. Although an honorary radiologist was a full member of the staff, the post was offered because the hospital was poor and could not afford even a token honorarium. Barclay noted that to take up radiology as a career in the early days required both courage and faith; there was need of courage to face the fact that clinical colleagues would merely regard the X-ray doctor as a new type of photographer, and faith that there would be a future that might offer a scope of practice beyond that of a mere technical accomplishment. The financial position of the early radiologist was far from secure. Could the doctor make a living by the practice of radiology when, for practical purposes, there was only fracture work that would provide an income? It was difficult even in London where most

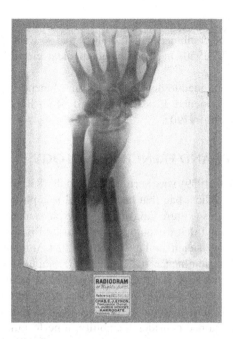

FIGURE 2.1 Radiogram of right forearm, prone, showing a displaced fracture of the radius and dislocated distal radio-ulnar joint. Taken by Chas. E.J. Eynon, Pharmaceutical Chemist from Harrogate on 23 January 1905. Author's collection.

FIGURE 2.2 Alfred E. Barclay (1876–1949). A painting by A.M. Rackow, Assistant Director of the Diagnostic Radiology Department at King's College Hospital, London, and the Belgrave Hospital for Children. Author's collection.

of the specialist work of the country was still centred at that time; Barclay realised that other work must be combined with radiology and so went into partnership with William Bythell (1872–1950) in a combined medical electrology with radiology service.

Arthur Schuster continued to develop physics in Manchester, and in 1900 a new physics laboratory was opened. Ernest Rutherford (1871–1937) succeeded Schuster as Langworthy Professor in 1907.

ANTOINE BÉCLÈRE AND FRENCH RADIOLOGY

Antoine Béclère (1856–1939) was born on 17 March 1856 in Paris, and as a boy was fascinated by a medical bag that had belonged to a physician who had served in the Napoleonic armies (Figure 2.3a and b).[8] The physician had told the boy that he would be given it, and that it would be useful to him in his practice. His father Claude Béclère was a physician, and it was natural that Antoine entered the Hôpital Lariboisère in 1873 to commence his studies. The young student was fascinated by medicine, and even when on holiday would visit the local hospital every day. This was at a time of major changes in medicine with the introduction of Listerian antiseptic practices. Béclère qualified as an intern in 1877 and developed an interest in paediatrics and he gained considerable ability in performing emergency tracheostomy, which was important for cases of diphtheria. Béclère set up a practice in paediatrics in Paris, and he was called out on many occasions for emergency tracheostomy in children. In 1887, he married and was the first of his generation to enjoy the newly invented bicycle.

Béclère had wide interests in medicine, including clinical applications of the work of Louis Pasteur (1822–1895) and in the endocrinological discoveries of

FIGURE 2.3 (a) French medal celebrating Antoine Béclère (1856–1939), engraved by H. Dropsy in 1936. Author's collection. (b) The reverse notes that the medal was presented by Béclère's colleagues, students, and friends on 17 March 1936, and celebrates his life and work. Author's collection.

Charles-Édouard Brown-Séquard (1817–1894). He treated a woman suffering from myxoedema using an extract of sheep thyroid gland, and a patient with Addison's disease using calf adrenal glands. He went to the slaughterhouse at La Vilette to personally obtain fresh material. Béclère was therefore a pioneer of modern endocrinology.

Following the discovery of X-rays in 1895, the first radiographs in France were obtained by Toussaint Barthélémy (1850–1906) and Paul Oudin (1851–1923). The physicist Henri Poincaré (1854–912) presented the first radiograph in their names to the Academy of Sciences on 20 January 1896.[9] At this time, Béclère's main interests were paediatrics and immunological research. In early 1896, Béclère visited the home of Paul Oudin for a demonstration of the radiological findings of Oudin and Barthélémy. Béclère arrived late, having had to attend an emergency; however, he was completely fascinated by the fluoroscopic examination of the heart and lungs and immediately understood the value of the technique. He said, "This path appeared to me like the road to the Promised Land, and I shall make use of it", and he proposed the term radiologie for the new specialty. In 1896, he started his studies in radiology and installed apparatus in his own home at 5, rue Scribe. He made a radiograph of a hydro-pneumothorax and also studied pulmonary tuberculosis. He was appointed Chef de Service at Hôpital Tenon on 1 January 1897, and in his General Medicine Department, and at his own expense, he opened the first Hospital Department of Radiology. For precedence, this department rivals that of the famous New Electrical Pavilion that John Macintyre opened in 1897 at the Glasgow Royal Infirmary. On 5 February 1897, and with his friends Oudin and Barthélémy, Antoine Béclère presented a paper to the Medal Society of Paris Hospitals on the use of radiography in examining aortic aneurysm showing tracings of fluoroscopic examinations. He always kept a record of examinations, either the hard copy of the radiograph or tracings of images. He made particular studies of pulmonary radiography. Béclère's book on chest radiology was published in 1901,[10] the same year that Guido Holzknecht in Vienna published his classic book.

In 1897, and using his principle that "in order to know a subject well, one must teach it", he organised at the Hôpital Tenon regular instructions in medical radiology. This medical radiology teaching was the first in the world and he continued his teaching for the next 30 years wherever he worked, including San Antoine, at the Val-de-Grâce Military Hospital in the Great War, and after the war at the Curie Foundation until 1927. This is a remarkable teaching record.

Antoine Béclère believed that hospital radiology departments should be under the control of a physician. This was in contrast to what had happened at the Necker Hospital where a Radiography Department had been opened in 1898 by a non-physician Gaston Contremoulins (1869–1950).[11] Béclère declared in 1899 that "each hospital would possess a radiographic and radioscopic installation where the Chiefs of Service could send their patients for investigation, and which would be directed only by a qualified doctor".

In 1899, Béclère moved to the St Antoine Hospital. He continued to progress radiology and also developed radiotherapy making many advances. He published widely and presented many courses of lectures.

At the start of the Great War, Béclère was 58 years old and therefore did not have to undertake any military service. He applied for a position where he could serve his country, and was appointed Director of Radiology and Physiotherapy at the Val-de-Grâce Military Hospital. In a similar manner to Marie Curie, he developed mobile radiography vehicles. The numbers of radiologists were found to be insufficient, and he gave courses for nurses, medical students, and radiologists, as did Marie Curie.

After the war, Béclère devoted his time to the practice of radiotherapy. He attended the First International Congress of Radiology held in London in 1925 under the Presidency of Thurstan Holland (Figure 2.4), and also the Second International Congress of Radiology held in Stockholm in 1928 under the Presidency of Gösta Forssell (1876–1950) (Figure 2.5). Antoine Béclère was appointed President of the Third International Congress of Radiology, which was held at the Sorbonne in Paris during 26–31 July 1931. Marie Curie was the Honorary President of the Congress. This was a major task for Béclère; however, the Congress was a huge success, largely owing to his prodigious efforts. The book of the Congress was itself an achievement and is interesting to read, providing a full account of the state of radiology at that period.[12]

The energy of Béclère was prodigious. He continued his work after the 1931 Congress. He was still attending meetings in 1939. He developed a bad cold in

FIGURE 2.4 Cartoon portrait of Antoine Béclère made by Sebastian Gilbert Scott at the First International Congress of Radiology held in London in 1925. Author's collection.

FIGURE 2.5   Cartoon portrait of Antoine Béclère made by Sebastian Gilbert Scott at the Second International Congress of Radiology held in Stockholm in 1928. Author's collection.

January and whilst recovering suffered a fatal heart attack on 24 February 1939. He was modest and did not harbour malice. He loved direct contact with colleagues and was not attracted to honours. In 1957, he was commemorated by a French stamp. It has been said that his sole desire was to serve, for "One has Eternity to repose".

GUIDO HOLZKNECHT AND VIENNESE RADIOLOGY

The dates of Guido Holzknecht (1872–1931) almost completely match those of his fellow Viennese and pioneer of modernist architecture Adolf Loos (1870–1933). Both men transformed their respective disciplines, and both made us look at things differently.

Holzknecht was introduced to X-rays by Vienna's first radiologist Gustav Kaiser (1871–1954), and in 1899 Holzknecht was offered a position working in the department of Hermann Nothnagel (1841–1905), a professor of medicine in Vienna. Holzknecht set to work, and he particularly devoted himself to the study of the chest. His first major observation was on bronchial obstruction when he made the classic observation that the mediastinum shifted on expiration secondary to the resultant air-trapping. He made measurements of the cardiac contour and emphasised the usefulness of the oblique views when looking at the aorta and the oesophagus. On the basis of this work, in 1901 he published the first book devoted to the radiology of the chest.[13] This is an important book and bearing in mind its early date and primitive apparatus, the quality of the images is quite remarkable. It is one of the great books of radiology. The French Radiologist Antoine Béclère considered this work to be on the same level of importance as the *Traité de l'ausculation médiate et des maladies des poumons et du coeur* of René Théophile Hyacinthe Laennec (1781–1826), which described the stethoscope and transformed physical examination of the chest. The pioneer cardiac and MRI radiologist Robert Emil Steiner (1918–2013) said that it is incredible how accurate this particular observer has been in the interpretation of cardiac contours (Figure 2.6).[14] Sadly in 1901, Gustav Kaiser was forced to retire due to his developing radiodermatitis. Kaiser was to live until 1954, and, as we will see, Holzknecht was not to be so fortunate and was to die a radiation martyr.

Holzknecht's book is important not just that it was the first book devoted to chest radiology, but also because of the quality of his work. The book is illustrated with diagrams that show that he was aware of the principles involved in radiography, including cross sections (Figure 2.7). The quality of the images is excellent, and he reproduces both radiographs and also a series of drawings of what he could see (Figure 2.8). Whilst his apparatus looks primitive to us today, it was the cutting-edge high technology of his time (Figure 2.9). We should try to look at old images with the knowledge of that time, although this can be difficult. As can be seen from the image of his apparatus, there is an absence of radiation protection with no shielding around the X-ray tube seen centrally, and also no protection around the fluorescent screen. The operator would be exposed to both primary radiation from the main beam and also secondary radiations from scatter within the patient, and this is why there were so many injuries in this first generation of radiologists. There was also the pernicious

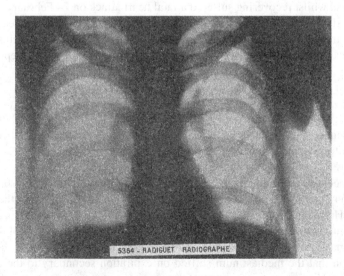

FIGURE 2.6 A normal chest radiograph. The structures can be seen well in this radiograph from 1901. From Holznecht, 1901, see Note 13.

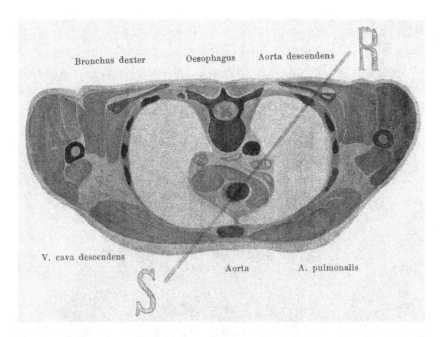

FIGURE 2.7 An axial cross section of the chest. Sagittal and coronal sections are also shown. From Holznecht, 1901, see Note 13.

FIGURE 2.8 A normal lateral chest. This is a drawing taken at fluoroscopy, and Holzknecht identifies the spaces behind the sternum and the heart. From Holznecht, 1901, see Note 13.

FIGURE 2.9 Guido Holzknecht's apparatus for examination of the chest. The Ruhmkorff induction coil is seen to the left, the glass X-ray tube is seen centrally, and the fluorescent screen is suspended from the ceiling and held in the hands. Note the absence of any radiation protection in this image from 1901. From Holznecht, 1901, see Note 13.

habit of using one's own hand to test the quality of the X-ray beam. What is interesting is the significant use of cross-sectional images to illustrate the principles of radiography, and we can imagine how he would have regarded modern CT scanning. Holzknecht also made radiographs of a specimen of the heart to aid his understanding of cardiac anatomy that was shown at radiography. There is an interesting image of a hydro-pneumothorax, which was common at that time and presumably related to the blind puncturing of the chest in cases of suspected empyema or effusion.

A particularly interesting image that Holzknecht depicts is one of a young lady wearing a corset and is presented as showing the harmful effects of the contemporary fashion of tight corsetry for women (Figure 2.10). The radiograph is labelled as showing the subject as having a high moderately strong corset, and described as a ventro-dorsal recording with a deep-lying X-ray tube. Fashion for both men and women has often sparked controversy, and the tightly laced corset for women remains controversial to this day. Writing from a feminist perspective, Summers presents the corset as the perfect vehicle through which to police femininity, with women consciously resisting its restrictions.[15] For Summers, the corset was a crucial element in constructing middle-class women as psychologically submissive subjects. Many modern women have not accepted this narrative, and Sarah Chrisman sees the corset as a tool of empowerment and not of oppression.[16] Chrisman notes that all women of

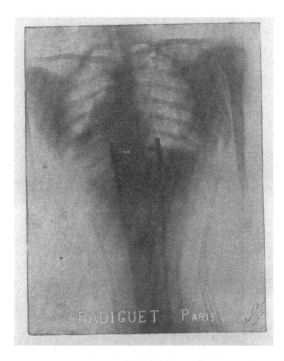

FIGURE 2.10   A radiograph showing a woman wearing a high and moderately strong corset. From Holznecht, 1901, see Note 13.

all social classes wore corsets, and women were able to express themselves through their fashion. It is difficult to find any nineteenth century image where the women are not wearing stays or corsets (Figure 2.11). In the nineteenth century, whilst much of the fashion industry was for women, it was also led by female dressmakers and milliners, and many businesses were owned by women. Alison Matthews has shown that expressions such as "fashion slaves" or "victims" have been used, and that in 1827 the Italian poet Giacomo Leopardi has fashion exclaim that she plays many deadly games "crippling people with tight shoes; cutting off their breath and making their eyes pop out because of their tight corsets".[17] Matthews also noted that the medical profession has encouraged our cultural bias towards blaming women for the health hazards that are the result of larger systemic problems. As the nineteenth century progressed, many doctors wrote regularly about the harmful effects of fashion that resulted in a wide range of health issues with damaged organs and even death from the tightly laced boned corsets. It is noteworthy that there is no skeletal evidence of rib damage in women during this period. The topic of deaths related to corsets was discussed in letters to *The Lancet* in the early 1890s.[18] In the 1890s, women were taking a more active part in public life, and this was causing concern. The advent of radiography, with Holzknecht's dramatic image of the lady wearing the tight corset, was seen as giving objective scientific evidence for the harmful results of tight corsetry. The radiographic studies were continued by the French doctor Ludovic

FIGURE 2.11 A typical late-Victorian woman in her corset. Unknown woman photographed by the photographic studio of E. Darlington, Shortlands, Kent. Author's collection.

O'Followell in his 1908 book *Le Corset*.[19] In his book, O'Followell was probably trying to promote a less severe corset design rather than calling for the abolition of the corset altogether. Holzknecht's image is labelled "RADIGUET, Paris". This is Arthur-Honoré Radiguet (1850–1905) who manufactured X-ray apparatus with his son-in-law Georges Jules Massiot (1875–1962). Sadly, Massiot died in 1905 from the effects of radiation. The firm of Radiguet & Massiot played a pioneering role in radiology. In 1960, the firm became a partial subsidiary of Philips, then becoming a full subsidiary of Philips France as Massiot-Phillips and finally Philips Systemes Medicaux.

As was the case in those days, Holzknecht was interested in both diagnosis and therapy, and in 1902 he described his well-known chromoradiometer which was the first device designed to measure radiation dose. This was a major advance in dosimetry.

Holzknecht was probably the first to suggest that radiology should be a medical specialty in its own right. In 1903, with his Viennese radiology colleague Robert Kienböck (1871–1953), he proposed that there was more to radiology than just simply the use of a helpful technique. He was actively involved in the training of young radiologists and emphasised the need for radiologists in training to be familiar with physiology, anatomy, and pathology as well as the clinical features of disease, in

FIGURE 2.12   The statue of Guido Holzknecht (1872–1931) in the Arne Karlsson Park in Vienna. Author's photograph.

order to make accurate and helpful radiological diagnoses. The Viennese school of radiology became hugely influential and attracted many foreign students. Holzknecht was also the first to describe gastric cancer using radiology and the opaque meal.

There is a statue of Guido Holzknecht in the Arne Karlsson Park in Vienna (Figure 2.12). There is a poignancy about the figure, with his poor damaged hands held in front of him. It was Antoine Béclère who said of Holzknecht that no one brought more passionate ardour and inspiration to the pursuit of a good and high ideal, no one had a deeper zeal, was more indefatigable, had more courage, dedication, and all selflessness.

PRACTICE IN THE EARLY RADIOLOGY DEPARTMENTS

The early X-ray departments resembled each other quite closely. It was relatively straightforward to acquire the basic apparatus; however, obtaining diagnostic radiographs was not easy and the early practitioners required skills in a number of areas, including photography, electricity, and anatomy. There were no courses for the training of the would-be X-ray expert, and the early textbooks, when they did appear, were often inadequate and because of the rapid developments in knowledge, they quickly became out of date. Medical journals were important in the rapid dissemination of information about the developing specialty, and throughout 1896 the *British*

Medical Journal (*BMJ*) ran a series of review articles by Sidney Domville Rowland (1872–1917) with the aim "to investigate the application of Roentgen's discovery and to study practically its applications", and the first report appeared on 8 February 1896.[20] Essentially the *BMJ* became the journal of the new discipline, and played a vital role in developing the new specialty. The articles gave "the new photography" a clinical and scientific credibility. Following these articles, Rowland started a new journal *The Archives of Clinical Skiagraphy*, which continues today as *The British Journal of Radiology*.[21] A skiagram was Rowland's term for a radiographic photograph and is from *skia*, which is the Greek word for a shadow. His *BMJ* series would have put Rowland in contact with all who were interested in the new X-rays, and he was therefore in an ideal position as editor.

Writing on 2 April 1896 in his Preface to the first issue, Rowland commented that the object of this publication was to put on record in permanent form some of the most striking applications of the new photography to the needs of medicine and surgery.[22] Rowland went on to say that the progress of this new art has been so rapid that, although Röntgen's discovery was only a thing of yesterday, it had already taken its place among the approved and accepted aids to diagnosis. This is a remarkable statement to have been made so soon after the discovery. The pioneers experienced many difficulties as expressed by Charles Thurstan Holland (1863–1941), the Liverpool pioneer radiologist and the president of the First International Congress of Radiology that was held in London in 1925. Towards the end of his life in 1936, Holland wrote that "there were no X-ray departments in any of the hospitals. There were no experts. There was no literature. No one knew anything about radiographs of the normal, to say nothing of the abnormal".[23] The radiological journal became essential to both disseminate information and experiences and to give advice about practical aspects of radiography. There was a page in the journal devoted to giving answers to questions sent in by correspondents, and there were also book reviews and advertisements. The role of advertisements in journals has never been simply about generating income for the publishers. Advertisements give the readers of the journal information about resources that they will need to develop their clinical practice and describe the photographic plates, X-ray tubes, and the apparatus that will be needed.[24] The rapid progression is well illustrated by reading the articles by Rowland, and even today there is a sense of fascination and excitement. The *BMJ* had a wide circulation throughout the world, and many would have been attracted to the new specialty. In spite of all this, the pioneer Sebastian Gilbert Scott (1879–1941), who was at that time working at King's College Hospital in London, commented that the medical profession as a whole remained sceptical about the whole business of X-rays. He also observed that even the president of the British Physical Society was at that time among the doubters, saying in public that "I admit that the discovery is very interesting, but I do not see at the present time how it can lead to results of any significance". Scott noted that, as might have been expected, a number of surgeons not only in England but also in Germany were also sceptical about the value of X-rays in surgery. The German surgeons M. Breitung and G. von Bergmann warned their colleagues against unwarranted expectations from the new photography.

The number of doctors specialising in radiology was small, and they were usually too busy to give up their time for teaching. The doctors who wished to learn the subject were advised to attach themselves to a department and so pick up a knowledge of radiology. There was no trained technical assistance, and the radiologist did most of the radiographic work by themselves, possibly assisted by a porter, a dispenser (pharmacist), or a nurse. In Manchester, a dispenser helped Barclay in his hospital work, but also competed for the private patients. The technique of that time was no "press-the-button" business but was a very complex combination of procedures. The skills required by the early practitioners were many and varied, and taking the early plain films was as complex and difficult as any modern imaging. The X-ray department had few amenities and was usually housed in a basement. The apparatus was feeble, inconstant in its output, and quite unreliable. Barclay noted that, looking back, it seemed remarkable that any results could be obtained with such makeshift and unreliable apparatus, and still more remarkable was the range of examinations attempted and their comparative success. The radiologist's main concern was of necessity focused on obtaining results, and there was little time given to the interpretation of what was seen. What was seen was "often uncharted virgin soil in which there were as yet no tracks or signposts". The exposures were long, and sandbags were used to keep the patients still. It was not until about the year 1902 that general hospitals really began to think that there might be some diagnostic value in the examination and proceeded to install the necessary apparatus.

FLORENCE STONEY AND THE ROYAL FREE HOSPITAL

It was in 1902 that Florence Stoney (1870–1932) installed the X-ray apparatus at Royal Free Hospital (RFH) and also at the New Hospital for Women (NHW – later renamed the Elizabeth Garrett Anderson Hospital).[25] On 4 December 1901, the Weekly Board meeting of the Royal Free Hospital opened a fund "to defray the cost (about £100) of a Röntgen Ray apparatus". The Electrical Department at the RFH was opened in April 1902, and the medical school's magazine in May 1902 stated that

> an advance has been made ... by the organisation of an electrical department. An X-ray apparatus is now fitted up in Elizabeth Ward .... Miss Stoney is appointed as electrician [with] two students as assistants, who will be initiated into the mysteries ... . Miss Stoney attends on Wednesdays at 12.30 p.m. and on Fridays at 9 a.m.[26]

The involvement of medical students in the introduction of what was cutting-edge technology is interesting and to us seems more than a little curious. However, the same was the case at King's College Hospital when Sebastian Gilbert Scott and a fellow student named Williams were chosen to work out the possibilities of this new method of medical diagnosis. In addition, Sidney Rowland was a medical student when he was writing his classic series of articles for the *British Medical Journal*. The medical profession had what now seems to us to be a remarkably apathetic attitude towards the introduction of the X-rays. The department at the RFH had taken some time to get itself into good working order, and in January 1903 Florence wrote, "but now the X-rays are available twice a week, … and … patients are treated by electricity with the constant or interrupted current as required".

The rooms that were provided were badly ventilated, with no separate room for the X-ray work. This should not be seen as unusual and many early departments had unsatisfactory locations. The apparatus was only available for use when the room was not needed for other purposes, which is curious. In the beginning, the room was left open so that everyone could make use of it, including the house surgeons for the casualty work, but the room had to be locked after the equipment was damaged after only a week from its opening. Florence noted that with "a delicate electrical instrument connected with an electric main with over 200 volts it does not blindly do to turn one handle after another and observe results". The electrical power was obtained from the mains electricity supply at nearby St Pancras. During this period, the radiologist, commonly called a radiographer, was not a member of the hospital medical staff and as a result Florence was not part of the committee that discussed the work of her department. Following an appeal for improved apparatus, a piece of second-hand equipment had been bought, but unfortunately there had been no enquiry about its suitability for use with the other equipment. In 1902, Florence started the X-ray Department at the NHW in Euston Road, and was given the title "Rayologist". The NHW was only a short distance from the Royal Free Hospital. Using her experience gained at the RFH, Florence proposed several changes and arranged for the X-ray department to be enlarged. Florence saw patients on one or two mornings a week, and the entry for the Röntgen Ray Department in the 1905 Hospital Report shows 59 diagnostic cases and 22 therapy cases for the whole year. This seems a small number; however, no statistics had been recorded before Florence was appointed. Most of the diagnostic examinations were for suspected renal calculi. Florence was concerned to promote the work of her department, which steadily grew under her care and the 1906 report showed Florence as the head of the Electrical Department and she was included as one of the medical staff. The financial challenge that the hospital faced by equipping a modern X-ray department became clearer, and a special fund was created for the X-ray department, with Florence contributing some of her own money.

By the year 1910, the X-ray department needed a new room in the hospital, since it was "almost dangerous in its old quarters". In 1911, Florence was able to obtain £123/3s/6d for X-ray equipment for the new room, this being the single most expensive item purchased by the Hospital Guild. The X-ray department did not have a formal representation within the hospital's committee structure. This was changed in March 1911when the Pathology Committee enlarged its remit becoming the *Pathology, X-ray and Mechanical Therapy Committee*. All of these changes enabled Florence to put forward a case that the X-ray service should be available for two full days each week, rather than only two mornings as was previously the case. By 1914, the diagnostic workload had doubled which would indicate that this extension was approved. Florence also suggested recommended charges: 2 shillings and 6 pence (12.5p) for a diagnostic examination and 7 shillings and 6 pence (37.5p) for radiation treatment for a month. These charges would only be for the outpatients, and Florence argued against cross-charging for in-patients, believing that this should be covered by the overall hospital costs. Florence recommended that a darkroom technician to carry out all the photographic development should be appointed, a task which she had previously done by herself. There are stories of Florence taking plates to her home to be developed.

Florence recommended that an X-ray nurse be appointed at an enhanced nurse's salary. The not unexpected response was that there were no spare nurses. This could not have been easy for Florence, since caring for patients without nursing assistance, particularly for therapy patients, would have been difficult. When the department at the Glasgow Royal Infirmary had opened in 1897, there were X-ray nurses from the beginning, as well as two medical officers and an unqualified assistant, although this was a considerably bigger hospital than the NHW. More costly requests followed. On 2 December 1913, the Management Committee meeting of the NHW discussed a new coil costing more than £42. At that time, the management was becoming concerned about the rising costs of providing an effective X-ray service, and the decision was referred back to the hospital's Medical Council with a request to "go thoroughly into the matter with Miss Stoney". On 30 January 1914, a request was made for £300 for new X-ray tubes. The X-ray tubes for therapeutic use lasted for about three months, and for diagnostic use for about six months reflecting their different uses. By 11 June that year, there was "an urgent need to enlarge the X-ray department", and that request was turned down by the Management Committee meeting in July since "Enlarging the X-Ray accommodation would amount to approximately £250", and therefore nothing could be done in the present state of hospital finances. The NHW was part of the voluntary hospital movement and received no state funding, monies being derived from fees, fundraising, and bequests. Funding expensive radiology equipment has always been problematic, no matter what financial model is used.

War broke out in August 1914, and Florence left the NHW for her war service, leaving a robust and respected department, which the hospital continued to support.

There were many challenges faced by the first generation of pioneers. The early apparatus was quite basic, and was only after 1905 that more powerful apparatus was developed, and abdominal and chest radiography became easier to perform. The period up to 1914 was one of rapid progress and steady development of apparatus and technique, and also accompanied by a large increase in the volume of work. In Barclay's department, the number of cases that were radiographed in 1914 was approximately three times that of 1909. In the early period, the predominant work was the examination of fractures, but special types of examination, including those for renal and gallstones, were gradually introduced. By 1912, the bismuth opaque meal examination was becoming routine, and nearly all surgical gastrointestinal cases were submitted for radiological examination.

So much was to change during the First World War, and the post-war world was different in many ways. As Vincent Cirillo has demonstrated, the war was important for the development of the medical specialty of radiology, and this was certainly the case in in the United States.[27] The military surgeons had become accustomed to working as a team with radiologists, and this continued after the war. The number of radiologists increased exponentially, and the returning soldiers, who had experienced radiography overseas, now expected radiography when they were confronted with illness when back home. The introduction of the Coolidge hot filament X-ray tube and the replacement of glass plates by X-ray film made the process of radiography significantly easier. The value of chest radiography was only more slowly appreciated. Certainly, chest radiography had been employed by Guido Holzknecht in Vienna, and Hugh Walsham and Harrison Orton had published the first book in English on chest radiography in 1906.[28] Whilst Walsham and Orton commented on the increasing number of cases sent to the X-ray department for examination of the chest, the use of radiography in chest pathology was much less common than in skeletal trauma. As an illustration, in 1919 the famous physician Sir William Osler (1849–1919) developed pleural complications following a chest infection. An empyema or a lung abscess was suspected, however the physicians caring for him saw no need for chest radiography and relied on physical examination. Osler himself did not request a radiograph, although in the 1916 edition of his textbook *The Principles and Practice of Medicine* he had noted the value of chest radiography in the diagnosis of pleurisy.[29] The value of radiography of the chest following trauma in the Casualty Clearing Station and Stationary hospitals in the war was demonstrated by Rea.[30] Rea said that the surgeons who insisted on radiography of the chest prior to surgery attested to its extreme value. If any doubt remained about the value of radiography of the chest, then the influenza epidemics of 1918 and 1919 demonstrated beyond doubt the great value of chest radiography.[31] Sequential chest radiography in lobar pneumonia gave new insights into the pathology and natural history of the condition, and LeRoy Sante (1890–1964) who was professor of radiology at St. Louis University School of Medicine found that his results disagreed with the previously accepted views. Sante's study of lobar pneumonia made many primary observations on the natural history of the condition.

Following the Great War, radiology entered a fruitful period, and classical radiology gave increasing insights into anatomy and pathology.

NOTES

1. Zonneveld, F.W. 2020. Spectacular rediscovery of the original prints of radiographs Roentgen sent to Lorentz in 1896. *Insights into Imaging*, 11, 46–53.
2. Thomas, A.M.K., Banerjee, A.K. 2013. *The History of Radiology*. Oxford: Oxford University Press.
3. Schuster, A. 1932. *Biographical Fragments*. London: Macmillan & Co.
4. Schuster, N.H. 1962. Early days of Roentgen photography in Britain. *British Medical Journal*, ii, 1162–1166.
5. Schuster, A. 1896. On Röntgen's rays. *Nature*, 53, 274.
6. Röntgen, W.C. 1895. On a new kind of rays. Trans: A.T.Stanton, from: Sitsungsberichte der Würzburger Physik-medic. Gesellschaft, 1895. *Nature*, 53, 274–276.
7. Barclay, A.E. 1949. The old order changes. *The British Journal of Radiology*, 22, 300–308.
8. Béclère, A. 1973. *Antoine Béclère*. Paris: J B Baillière.
9. Pallardy, G., Pallardy, M.-J., Wackenheim, A. 1989. *Histoire Illustrée de la Radiologie*. Paris: Les Éditions Roger Dacosta.

10. Béclère, A. 1901. *Les Rayons de Röntgen et le Diagnostic des Affections thoraciques (non tuberculeuses)*. Paris: Libraire J-B Baillière et fils.
11. Mornet, P. 2019. *Gaston Contremoulins, 1869–1950, Visionary Pioneer of Radiology, Forgotten Heritage*. EDP Sciences.
12. *Le IIIme Congrès international de Radiologie*. 1931. Paris: Masson ed Cie.
13. Holznecht, G. 1901. *Die Röntgenologische Diagnostik der Erkrankungen der Brusteingeweide*. Hamburg: Lucas Grafe und Sillen.
14. Steiner, R.E. 1973. The impact of radiology on cardiology. *British Journal of Radiology*, 46, 741–753.
15. Summers, L. 2021. *Bound to Please: A History of the Victorian Corset*. 2nd ed. Berg 3PL.
16. Chrisman, S.A. 2013. *Victorian Secrets: What a Corset Taught Me about the Past, the Present, and Myself*. Skyhorse.
17. Matthews, A. 2015. *Fashion Victims: The Dangers of Dress Past and Present*. London: Bloomsbury Visual Arts.
18. Isaac, S. 2017. The dangers of tight lacing: the effects of the corset. https://www.rcseng .ac.uk/library-and-publications/library/blog/effects-of-the-corset/ (accessed 20 May 2021).
19. https://publicdomainreview.org/collection/the-corset-x-rays-of-dr-ludovic-o-followell -1908 (accessed 20 May 2021).
20. Rowland, S. 1896. Report on the application of the new photography to medicine and surgery. *British Medical Journal*, ii, 1676–1677.
21. Thomas, A. 2020. 125 years of radiological research: BJR's history is radiology's history. *British Journal of Radiology*, 93, 20209002.
22. Rowland, S. 1896. Preface. *Archives of Clinical Skiagraphy*, 1, 3–4.
23. Holland, C.T. 1938. X-rays in 1896. *British Journal of Radiology*, 11, 1–24.
24. Thomas, A.M.K., Anderton, S. 2018. History of the *British Journal of Radiology*. *Medical Physics International Journal*, 6, 242–245.
25. Thomas, A.M.K., Duck, F.A. 2019. *Edith and Florence Stoney, Sisters in Radiology* (Springer Biographies). Switzerland, Cham: Springer Nature.
26. McIntyre, N. 2014. *How British Women Became Doctors*. Woodford Green: Wenrowave Press.
27. Cirillo, V.J. 2000. The Spanish–American war and military radiology. *American Journal of Radiology*, 174, 1233–1239.
28. Walsham, H., Orton, G.H. 1906. *The Röntgen Rays in the Diagnosis of Disease of the Chest*. London: H.K. Lewis.
29. Bliss, M. 1999. *William Osler: A Life in Medicine*. Oxford: Oxford University Press.
30. Rea, R.L. 1919. *Chest Radiography at a Casualty Clearing Station*. London: H.K. Lewis.
31. Sante, L.R. 1927. *Lobar Pneumonia: A Roentgenological Study*. New York: Paul B. Hoeber, Inc.

3 Radiology and Culture

There was a huge and immediate popular interest in Röntgen's remarkable discovery. When the publications of that era are read, one can sense the excitement that was in the air, and the sense of excitement is almost palpable. That Röntgen could produce images of the inside of the human body was greeted with incredulity by all. There is a myriad of articles, stories, and poems from the early period. There was a cultural impact following the discovery of X-rays in 1895. The "New Photography" as it was called had its influences felt in all aspects of culture.

WRITTEN LITERATURE

Within a short time, accounts of the new Röntgen rays were appearing in print. Since that November in 1895, many stories and novels have been written. In some of the stories the presence of the X-rays may not be central to the narrative, and in others the X-rays are of primary significance.

George Griffith (1857–1906) wrote a remarkable story *A Photograph of the Invisible* that appeared in the popular *Pearson's Magazine* of April 1896,[1] and this would seem to be the first fiction published with an X-ray theme. Griffith wrote visionary stories and fantasies and was a popular author. His story is based on the Röntgen rays and appeared only five months after their discovery. It's remarkable that such fiction appeared earlier in the same issue of the magazine that reproduced Dam's famous interview with Röntgen.[2] Dam had been asked by the editor of the magazine to interview Röntgen, and the editor had shown Dam the text of Griffith's story. Dam noted that not long ago this story would have been read with utter incredulity, and possibly not unmixed with ridicule. Science fiction and science fact interact in a complex and fertile manner, and there are many examples of the developing genre of science fiction prefiguring science facts. For the reader of science as fiction, it will be the case that science as fact will be made more accessible. The fiction may prefigure the fact, and so as an example space flight to the moon was shown in fiction long before it became a reality. Tattersdill[3] has reviewed the relationships between science and fiction in the period of the fin de siècle, at the end of the 19th century and the turn of the 20th century. This period was a time of major developments in science and technology, and if it was difficult for the scientists to understand the changes that were taking place, how much more difficult must it have been for the members of the public? Stories with a scientific theme appeared in popular magazines such as *Pearson's Magazine* and *The Strand Magazine* and laid the foundation for what became known as science fiction. However, to define science fiction and its limits is not easy, and the writer Damon Knight somewhat frustratingly said that "science fiction is what we point to when we say it".[4] Science fiction has covered important topics and philosophical issues and should not be dismissed simply as trivial pulp fiction.

DOI: 10.1201/9780429325748-3

In the story by Griffiths, Denton, who is a jilted lover, wants revenge upon an inconstant lady. He approaches a professor who knew of Röntgen's recent discovery. A plan is formulated to wound the beautiful lady where it would hurt the most, that is in her vanity. The professor takes a radiograph of Denton's hand which he looks at with horror (Figure 3.1). The professor then arranges to take a photograph of the lady using the Röntgen photography and sends the print to her and her new husband. The couple are horrified by the likeness for above the dress

> were the face and hair not of a living woman, but of a ghost, and, beneath all, sharp in outline and perfect in every hideous detail, a fleshless skull – her own skull … grinned at her through the transparent veil of flesh and seemed to stare at her out of the sockets in which two ghostly eyes seemed to float (Figure 3.2).

Denton had his revenge, and the lady was admitted to a private lunatic asylum. She imagined that she was now a skeleton, and that her clothing and skin and flesh were nothing more than transparent shadows which everyone could see through. She was forced to live in a darkroom lest she saw her flesh. The fact that we are walking and animated skeletons covered in flesh is now self-evident, it is difficult for us now to place ourselves in the mindset of the 1890s.

A story by L.T. Mead and Clifford Halifax appearing in the *Strand Magazine* of July 1896 was the first in a series about "The Adventures of a Man of Science". The

"Read that, and tell me if you don't think it's about enough to knock a man over."

FIGURE 3.1 Denton and the professor: "Read that, and tell me if you don't think it's about enough to knock a man over". Drawing by George Henry Grenville Manton (1855–1932). From Griffiths, 1896, see Note 1. Public domain.

The blood died out of her face till it was grey and white and ghastly.

FIGURE 3.2   The inconstant lady and her husband examine the radiograph: "The blood died out of her face till it was grey and white and ghastly". Drawing by George Henry Grenville Manton (1855–1932). From Griffiths, 1896, see Note 1. Public domain.

story is centred on a famous diamond "The Snake's Eye" that is stolen.[5] A servant was taken violently ill and approached Gilchrist, the main character, who determined to find out the source of the problem. Gilchrist lived in a flat in Bloomsbury in London and had fitted himself up with a Röntgen ray laboratory. Gilchrist noted that the new discovery was the craze of the scientific world and he was "of course, bitten with it" and had purchased several vacuum tubes. The discovery was seen in the story as one which would make rapid advances and would be of immense importance to medical science. Gilchrist and the sick man went to the laboratory where Gilchrist thought that the Röntgen rays might be really tested, and he surmised that the rays could be used to discover the crime. Having checked that the Crookes' vacuum tube was working well, the suspect was placed in front of it. The cap was taken off the camera and with an exposure lasting from seven to ten minutes a photograph was obtained. The plate was developed, and when Gilchrist saw what the mysterious X-rays had produced, he could scarcely restrain a loud and joyful exclamation. The suspects skeleton was shown, and below the ileo-caecal valve, he could see a foreign substance about the size of the swallowed jewel. Gilchrist thought that he was seeing the gold in the jewel since the diamond itself was "probably not impervious to the X-rays". Gilchrist showed the plate to a doctor who diagnosed peritonitis and recommended removal of the swallowed jewel. Removal was not possible, and the criminal died with the stolen jewel recovered only after death. Whilst the story is fanciful in some respects, the authors give a reasonably accurate account of radiography of the period. However, abdominal radiography was difficult in this time with the low-powered tubes then in use. However, the use of radiography in diagnosing abdominal disease is certainly anticipated; and in fact forensic uses of radiography had early applications in 1896.[6] Since these early stories, X-rays have featured in innumerable stories and novels.

The detective novel and murder mystery were very popular in the 20th century and many famous writers produced their own detective novels. *The X-Ray Murders* was an interesting addition to the genre and was popular having American editions in 1942[7] and 1944, with a British edition in 1945. Milton Scott Michel (1916–1992) was an American crime fiction writer and playwright, and was active from the 1940s to the 1960s. Michel worked as radiography technician in New York City from 1940 to 1944, following which he became a self-employed and successful writer. *The X-Ray Murders* is a typical example of detective mystery fiction, and as the *Winston Salem Journal* says of the first edition, it is "one of the best of its class, furnishes the reader with all the clues, yet keeps him guessing to the last". During the story, the reader is invited to work out the identity of the killer with a classic denouement at the end. The hero is a "gumshoe" or private eye who has turned down an academic post in psychology to work as a detective. The victim is a radiologist who dies of aplastic anaemia, under the impression that he has prostate cancer. The radiographer's daughter enlists the assistance of the detective, who works out that the radiologist has been slowly murdered by removing the lead protection in the wall between his office and the radiotherapy room. Technical knowledge of radiography was obviously needed to write the book which is an exciting read and is warmly recommended. The reader does not need previous technical knowledge and any pertinent facts are explained clearly. The book is of its period, and true to the genre. I will not reveal more of the plot in case anyone is moved to read the book. Other popular novels where radiological knowledge was needed are *X-Ray* by Charlie Hope[8] and *Team Twenty Ten* by George Korankye.[9] In the novel *X-Ray*, the story is set in a hospital where a plot is discovered in an X-ray department. A radiographer is changing the names on the radiographs as an insurance scam so that patients with abnormal chest radiographs are given apparently normal examinations in order to reduce insurance premiums. The theme is fanciful; however, it does introduce young people to hospital life. The lead characters are healthcare students, and it is seldom that a student radiographer features in fiction. *Team Twenty Ten* by George Korankye is more interesting in its themes. Like Scott Michel, Korankye is a radiographer with specialist knowledge, and part of his reason for writing the novel was to raise the profile of radiographers. The hero is a superintendent radiographer at the time of a global pandemic when the virus RAD-121 is infecting the world population, and which when it mutates becomes undetectable. The only means of diagnosis of RAD-121 involves radiographers. This is prescient of the role of radiography in the COVID-19 epidemic which took place 10 years after the novel was published, and in which an understanding of the imaging features is necessary to guide management, to produce national protocols, and to assist doctors to accurately identify and diagnose episodes of COVID-19 infection.[10] Fiction again prefigures real events.

One of the greatest novels ever written, and also one that features X-rays, is *The Magic Mountain* (Der Zauberberg) of 1924 by Thomas Mann (1875–1955).[11] The novel won the Nobel Prize for Literature in 1929 and has been the subject of much critical analysis. *The Magic Mountain* is set in an exclusive tuberculosis sanatorium located in the Swiss Alps. Mann had visited such a sanatorium in Davos when his wife had been a patient with a pulmonary condition. The sanatorium can be seen as a fictional microcosm for the state of Europe in the period leading up to the First World War. The thin air of the alps was beneficial to those with pulmonary tuberculosis. There are many themes in the novel of which the X-rays are one. Not unsurprisingly, the X-ray department or laboratory was located in the cellar, as it usually was in the early years of radiology, and as such has been likened to the Hades of Greek mythology. José van Dijck has examined X-ray vision in *The Magic Mountain*[12] in some detail. Van Dijck notes that radiography was presented by Mann as giving both an objective verification and an indisputable proof of reality. Radiography also signalled a dominance of the visual over other senses such as sound and touch. This has continued with the increasing importance of medical tests, with patients often being sent for imaging before a physical examination has taken place. The X-rays, or the new photography, were a sort of super photography that might prove "the existence of immaterial substances, the materiality of things heretofore unseen". van Dijck presents The *Magic Mountain* as demonstrating three invisible aspects of the body: the verification of disease, the visualisation of intimate feelings such as love, and the ultimate proof of the spiritual self after death.[13] In all these three areas of medical, psychological, and the metaphysical, the X-ray became almost a magical instrument that could render both the body and soul transparent. The radiograph has a superficial objectivity, and its introduction became part of the transformation of medicine from an art into a science, and to the development of modern technological medicine. In the 1920s, Mann was concerned about the growing division between the two cultures of art and science, and the radiographic image became a symbol of medical objectivity. The idea that our society, in both its educational and intellectual life, was characterised by a split between the two cultures of the arts or humanities became of increasing concern as the 20th century progressed. It was C.P. Snow's Rede lecture of 1959 and subsequent book[14] that stimulated the public debate that is still unresolved today. However, as van Dijck has noted, the interpretation of the radiograph has an observer variability, and medical image interpretation has elements of an art. It should also be remembered that it was not uncommon for the early practitioners to "touch up" the negatives, the hard copy equivalent of the digital Photoshop, so that the image showed what was believed to be present. The interpretation of an image is not objective, and Paul Goddard and colleagues have shown that the report of a CT scan will vary depending on the presence or absence of clinical information.[15] The current fascination with artificial intelligence (AI) in radiology continues the quest for an objective perception; however, is even AI-assisted perception objective? In modern radiology, the radiologist no longer looks at the patient since the patient is replaced by the image of the patient. In AI, the radiologist may be replaced by the computer, and neither the doctor nor the patient need be present. Indeed, we have a state when neither the doctor nor the patient need be in the same room, nor even on the same continent.

The relationship between patient and the X-ray is developed in a most interesting book *The Doctor Is Sick* written by Anthony Burgess (1917–1993) in 1960.[16] Edwin Spindrift, the central character of the novel, is a lecturer who collapses whilst teaching abroad. He is repatriated to England where he is investigated in the neurosurgical department of a London hospital for a suspected brain tumour. Edwin is a linguist and there is much discussion of words and their origins in the book. The book is semi-autobiographical since Burgess himself collapsed whilst teaching history in Malaya, and doctors diagnosed him with an inoperable brain tumour. Burgess was invalided home in 1959 and was investigated on a neurological ward of a London hospital. The character of Doctor Eddie Railton, who looks after Edwin in the novel, is based on Sir Roger Bannister (1929–2018), the London neurologist who looked after Burgess, and was also the first man to run a mile in under four minutes which he achieved on 6 May 1954, shortly before the novel was written. Edwin describes the patient's experience of both having a medical neurological examination and then of the radiology department. Spindrift is the spray that is blown from the crests of waves during a gale, and Burgess may have chosen the name to indicate Edwin's passivity during his illness, investigation and treatment.

Edwin's skull X-rays were performed by one of the "crisply permed, white-coated young women who were jauntily self assured" and who deferred to no one. Edwin's grinning skull was recorded, and he commented to the radiographer that the poet Webster "saw the skull beneath the skin". These words were from *Whispers of Immortality* by T.S. Eliot (1885–1965), who wrote: "Webster was much possessed by death, And saw the skull beneath the skin; And breastless creatures under ground, Leaned backward with a lipless grin". The young woman thought it better to be a radiographer than a poet, since radiographers "save lives, don't we?" Edwin wondered what was the purpose of saving lives, and what did she want people to live for? The radiographer replied that it was no concern of hers, and that anyway "That didn't come into my course". She is presented as living a world of sterile technology, and of ignoring the dictum of Socrates that the unexamined life is not worth living. The medical machines may not be able to give any overarching meaning to our existence, and yet what meaning can medicine provide? The doctor might save lives, but what is then the purpose of life for the saved individual? The character of Alex in the novel *A Clockwork Orange* that Burgess published in 1962 is brought to mind. Alex is a violent and atomised individual who has little relation to the wider society, and yet what meaning can the wider culture give to Alex? Does the postmodern culture with the state as the only authority have any more value or meaning than does the violent subculture that Alex inhabits? The answers must surely lie elsewhere.

Following his plain films, Edwin had tests that required more than a single white-coated operator, and he realised that greater opportunities now presented themselves for treating him "as a thing". The tests required a passivity on Edwin's part, and he saw that when he was particularly docile and plastic, he was elevated to a pet's level and patted. So, the docile patient is perhaps the ideal that the fictional Ludovico behavioural modification technique that Alex receives in *A Clockwork Orange*. The prison authorities are pleased with Alex in his quiet and passive state, and he is released from captivity. The hospital staff are similarly pleased with Edwin's passivity.

The next investigation is a cerebral arteriogram, performed by a direct injection into both carotid arteries. Edwin "saw faces, upside down, peering at him incuriously" and he notes how horrible is the inverted human face "far more monstrous than any monster from outer space", this presumably reflecting the experiences of Burgess. The staff are not curious about him and chat about their own inconsequential matters ignoring Edwin. The doctor greets him with a "Hiya, Mister" and since he has a PhD, Edwin corrected him saying "Doctor". The doctor completely ignores him replying "That's right, I'm the doctor". Next follows a description of a carotid arteriogram with direct needle puncture using traditional ionic contrast media. Edwin describes a pain that seemed green in colour and tasted of silver iodide. The pain showed "by some synaesthetic miracle" what the momentarily tortured nerves looked like. Edwin described the pain as shooting down his face, gouging his eyes out, and extracting teeth with cold pliers. The pain was likened to a tree shouting out, and the arteries are indeed a tree, the arterial tree. The test is a ritual, and the punctured artery is likened to a snake that is caught, tamed, and force fed. We are reminded of the experience of Torsten Almén[17] (1931–2016) performing head and neck angiography in the days of classical ionic high osmolar contrast agents and with their attendant severe symptoms when the ionic contrast is injected.

Edwin was too active after the procedure, and Dr. Railton had to tell him to lie still since he would need every ounce of stamina that he can find before they are finished with him. The next procedure was the lumbar air encephalogram, in which the spine is punctured with a needle, and the removed spinal fluid replaced with air. The procedure was performed by a psychiatrist which was not unusual at that time. Commonly invasive procedures were performed by a clinician and were later interpreted by a radiologist. During the lumbar puncture, Edwin felt his vertebrae collapse as before, and that his life juice was splattered everywhere. In describing the procedure, he noted that the air entered coyly, easing its way up the bony chimney, splitting up into "quiet crocodiles tramping corridors they had never seen before". Edwin thought that the staff resented his body and saw it like a potato that they were trying to roll around. He thought that it should be possible for the head to be removed and examined, and then afterwards reattached using an epoxy glue. In this period, shortly before the introduction of the head CT scanner, the patient would be examined in a complex chair with the apparatus being driven by a motor around the patient (Figure 3.3). The procedure was complex and required an organised team.

The results in Edwin's case, although not for Burgess, proved positive and as was common at the time, his wife was told the diagnosis and consented to the neurosurgery. The remainder of the book is spent giving an account of Edwin's adventures in London, and it is uncertain as to whether the outrageous events described in detail are taking place in real life or in an anaesthesia-induced dream world.

The relationship of the patient to the image of the patient, and to how the image is obtained is a complex one. The description by Burgess of being on the receiving

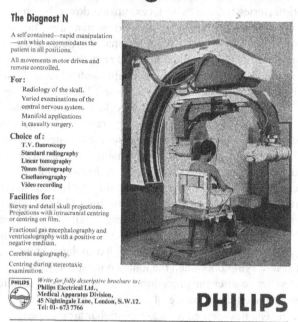

**Philips
Neuro-Diagnostic Unit**

The Diagnost N

A self contained—rapid manipulation—unit which accommodates the patient in all positions.

All movements motor driven and remote controlled.

For:
   Radiology of the skull.
   Varied examinations of the central nervous system.
   Manifold applications in casualty surgery.

Choice of:
   T.V. fluoroscopy
   Standard radiography
   Linear tomography
   70mm fluorography
   Cinefluorography
   Video recording

Facilities for:
Survey and detail skull projections. Projections with intracranial centring or centring on film.

Fractional gas encephalography and ventriculography with a positive or negative medium.

Cerebral angiography.

Centring during stereotaxic examination.

Write for fully descriptive brochure to:
Philips Electrical Ltd.,
Medical Apparatus Division,
45 Nightingale Lane, London, S.W.12.
Tel: 01- 673 7766

PHILIPS

FIGURE 3.3 Advertisement from the *British Journal of Radiology* from the late 1960s for the Philips Neuro-Diagnostic Unit for examination of the head. Courtesy of Philips Medical Systems and the *British Journal of Radiology*.

end of medical imaging is a fascinating account by a writer of the first order, and to obtain the correct balance between efficiently obtaining a diagnostic result of a complex study and the patient not being passive and as simply an object is not an easy one, either then or now.

FILMS AND THEATRE

The use of medical imaging in films is common, although this is usually as part of the story when the central character needs a medical investigation. Such examples are *The Theory of Everything* of 2014 and *The Danish Girl* of 2015, both directed by James Marsh and both starring Eddie Redmayne. In *The Miracle on 42nd Street* of 1947, directed by George Seaton, the X-ray equipment is used as an example of a charitable donation to a children's hospital. *Casualty 1900s* (released in the United States as *London Hospital*) is a television series set in the London Hospital before the Great War and released in 2006. It was directed by Bryn Higgins and Mark

Brozel. The radiation martyr and pioneer radiographer Edward Wilson is depicted in his department and with authentic apparatus.

The film *Fantastic Voyage* of 1966 explores interesting ideas. The screenplay was by Harry Kleiner with an accompanying novel by Isaac Asimov.[18] The book was promoted as combining the best of science and fiction and as being accurate and convincing. The film prefigures the use of nanotechnology and also of interventional neuroradiology. A blood clot has to be removed from the brain and a novel technique is used. A team is miniaturised in a submarine and are guided around the circulation for an interventional procedure. The science was as accurate as then possible, and extracts from the film were subsequently shown in lectures for medical students in the 1970s. The submarine and crew act like a nanobot. Nanorobotics is an emerging technology using robots near the size of a nanometre. In the film, the submarine and crew are reduced to the size of a microbe or molecule for their fantastic adventure.

Perhaps the most curious production related to radiography is *It Happened in Key West* by Jill Santoriello and Jason Huza.[19] It was performed on the London stage in 2018 as musical theatre. The musical retells the story of Carl Tanzler or Count Carl von Cosel (1877–1952), who was employed as a radiology technologist at the Marine-Hospital Service in Key West, Florida. Tanzler fell deeply in love with his patient Elena Milagro de Hoyos (1909–1931). Sadly, she died following complications from pulmonary tuberculosis. In 1933, and almost two years following Hoyos' death, Tanzler removed her corpse from the grave, and lived at his home with the body for seven years until he was found by Hoyos' relatives and authorities in 1940. Tanzler lived with the body in apparent happiness and dressed the body in appropriate clothes. The story is a curious one and has been retold many times.[20] Perhaps somewhat surprisingly, the production was popular and made for an enjoyable evening at the theatre.

It is perhaps in the genre of the horror film that X-rays are seen more frequently, such as in *Teeth* of 2007 by Mitchell Lichtenstein, when a named abdominal radiograph of the main character is depicted with the threatening words "It's what's inside that counts". In *The Breeder* of 2011 directed by Till Hastreiter, a remotely controlled X-ray room becomes an image of an unspecified but impersonal and malevolent medical investigation.

The film *X-Ray* of 1981 directed by Boaz Davidson, also released as *Hospital Massacre*, starred Barbi Benton[21] who was *Playboy*'s Hugh Heffner's girlfriend, and has developed a certain cult status. The central character has her radiographs altered in order to make it appear that she has a terminal illness and keep her in the hospital. The film poster has a certain lurid appeal declaring "X-ray, you have nothing to fear until they operate" or "Hospital Massacre, There's no recovery room at all". As a slasher film, *X-Ray* may not appeal to many, and perhaps it should better be viewed as a surreal art film.

VISUAL MEDIA

The relationship of radiography to the visual media is very fertile and many artists are working in this field. This was shown in 2009 when the British Institute of Radiology put on a successful exhibition *The Art of Medicine*. The exhibition brought together professionals involved in the business of radiology, with artists who draw inspiration from radiological techniques and images. Such art should provoke people to think about the link between science and art, and also contribute to the ongoing exchange of ideas between two worlds. Art and science are not, and never should be, two separate cultures. Art and Medicine have been influencing each other for centuries, and the new photography that was the X-ray was highly influential on contemporary artists.

Cubism was an influential avant-garde art movement in the early 20th century. The Cubist pioneer Jean Metzinger (1883–1953) in his influential *Note sur la peinture*,[22] noted that in Cubist artwork, objects are analysed, and then broken up and reassembled in an abstracted form.[23] And so instead of depicting objects from a single viewpoint as in conventional painting, the Cubist artist depicts the subject from a multitude of viewpoints to represent the subject in a greater context. This raises the question as to whether we could have had Cubism without radiography since looking at the subject from a multitude of viewpoints is the essence of radiography.

From the 1930s onwards the artist Francis Bacon (1909–1992) became interested in medical images. Sir Michael Sadler had sent Bacon an X-ray photograph of his own skull and had asked that it be incorporated into a painting. It became part of "Crucifixion" in 1933; however, the painting has not survived. The often shocking images in the paintings of Bacon reflect his wide use of medical images. When his studio was examined after his death, a significant number of medical textbooks were found.[24]

Bacon had visited medical bookshops in London and owned the book *Positioning in Radiography* of 1939 by Kathleen Clark.[25] When the book was published, Bacon was undertaking very little painting, and its illustrations of patients being radiographed were of considerable generic significance to him.[26] Bacon always acknowledged the influence that the book had on him, and many motifs in *Positioning in Radiography* can be seen in the paintings of Francis Bacon. He cut out many images from the book and used them in a number of ways. In one instance he cut out two plates from the book and then used a safety pin to hold together an image of a male torso and female legs. Bacon's portraits have a multi-layered effect and show deeper structures as in radiography. It is as if the many layers of the body are shown at once in an almost radiographic manner. The inside of the body is now seen and a portrait simply displays the exterior. Artists have traditionally learnt anatomy, and the influence of radiography encouraged artists to display deeper structures.

Radiography and Francis Bacon both have influenced many artists. The Polish artist Urszula Zajkowska is currently working in London.[27] She trained at the Silesian University in Poland and in painting under the Polish painter Anna Rybka. Her inspirations are texture and light, including X-rays. Zajkowska commented that she knows Bacon's art by heart, and he is her great inspiration. The primal fascination of insight into the human body and X-ray films she owed to Francis Bacon. Her painting of a shoulder is most striking (Figure 3.4). It is a black and white X-ray painting on paper, and made with a mixed technique, including gauche, oil pastels, and coloured pencils. It is inspired by an X-ray image as were some earlier paintings,

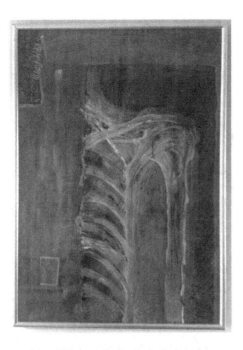

FIGURE 3.4 *Untitled*, X-ray inspired painting by Urszula Zajkowska. Author's collection.

and this painting is an interpretation of a shoulder radiograph. The painting is very moving, and reflects the incurable bone disease of one of her family members.

The artist Susan Aldworth (b. 1955) has a background in philosophy, with an interest in the human mind, especially consciousness and our sense of self.[28] In 1999, following an episode of illness, she experienced a cerebral angiogram in a similar manner to Anthony Burgess. However, technology had advanced since Burgess described Edwin's angiogram in 1960 in *The Doctor Is Sick*. Aldworth's experience of observing her brain live on a computer monitor during the angiogram triggered in her an ongoing fascination with the relationship between the physical brain and the sense of self.[29] The real-time visualisation of the living body in diagnostic imaging produces a sense of wonder in many people. Aldworth talks of "scribing the soul", and yet Bacon seems to see only meat. X-rays may show death, and yet they also show life. Aldworth's artistic representation of angiography is interesting and again blurs the boundaries of the arts and the sciences. In *Transience*, Aldworth describes how struck she was by human frailty, and in particular the dependence of "Self" on the physical brain.[30] The modern brain scan revealed to Aldworth the extraordinary anatomical landscape inside the living brain, and the scans can promise to show us something of who we are. Prior to her brain scan, Aldworth had seen herself as an external surface, and her scan was a revelation. Where was the "me" in the physical structure of the brain? The image *Elisabeth* (Figure 3.5) is a detail from an installation of nine monoprints, and was first shown at the National Portrait Gallery London, in 2013.[31] Each of the nine pictures in this large-scale portrait visually references the

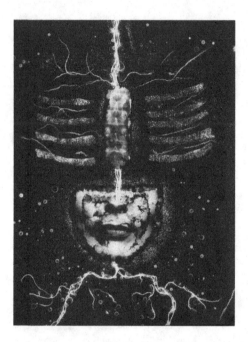

FIGURE 3.5 "Elisabeth", detail from an installation of nine monoprints by Susan Aldworth, 70 × 50 cm each. Susan Aldworth, 2013. First shown at the National Portrait Gallery London, in 2013. Image courtesy of the artist.

many different narratives which make Elisabeth who she is. This allowed Aldworth the freedom to explore the internal person as part of this portrait by including images and information revealed by neuroscientific techniques (brain scans and EEG data) as well as the external appearance and personal narratives. In this image, Elizabeth's face can be seen in her womb; she was living with epilepsy and worried about having a baby.

The question "What is art?" will always result in a variety of responses; however, the question "What is science?" raises its own variety of responses as well. Is the artistic representation of the angiogram any less accurate than the original angiogram? The London neurologist Peter "PK" Thomas (1926–2008) was a pioneer of electron microscopy of the peripheral nerve.[32] Thomas gave a lecture on peripheral nerve injury to the University of London Audio-Visual Centre in 1972.[33] He used a wide range of visual aids to show the anatomy and pathology of peripheral nerves, and the talk was illustrated with detailed electron micrograph studies. The electron micrograph studies were hard copies mounted on grey cards to allow them to be photographed by the television camera. His neighbour, who had no scientific background, was most taken with the photographs and asked if he might have one. The image was then framed and placed on a wall in his house, and presumably the medical image now became "art". It should be remembered that the term "art" is derived from the Latin word "ars"; which means an art, skill, or craft. Both the radiologist and the artist practice their respective crafts. Is there then a meaningful difference between the artisan and the artist, or is it simply a question of context? Diffusion MRI images showing the white matter tracts of the brain, the so-called tractography, appear very much like art created by an artist. The artist working with medical images blurs the art and science distinction and gives us new insights into ourselves. In this blurring the two cultures of art and science no longer seem so divided as they were to C.P. Snow in 1965.[34]

Many photographers have worked with the new photography since the earliest days of radiology. The pioneer Birmingham radiologist John Hall-Edwards (1858–1926) was a painter and a photographer. He produced artistic radiographs of flowers, and also painted. Sadly, he had amputations for radiation injuries and was able to paint holding a brush in a prosthetic hand. His paintings have a melancholic ambience with a sad foreground, and yet there is a light in the distance promising a hope in the future (Figure 3.6). Others have developed artistic radiography and the images of Nick Veasey have a fascination, with many of his creations resonating with early images in radiological journals.[35] Hugh Turvey is an X-ray artist and a photographer, and his work involves bridging the gap between the arts and sciences, and between graphic design and pure photography.[36] In 2009, Hugh was appointed as the first Artist in Residence to the British Institute of Radiology and his images have a fascination for the public, as shown in his exhibition *X-posé: Material and Surface* at London's Oxo Gallery in 2014.[37] In his work *Fabrication* of 2018, Turvey continues his interest in the "material" world following his 2014 exhibition. The image of the common object, the jacket, may be viewed as positive (Figure 3.7) and negative prints (Figure 3.8), and combines light and X-ray photography. The relationship between the surface structures and the hidden interior produces a creative tension. What is interesting about the work of Veasey and Turvey is the absence of the sense of the macabre. The modern radiologically based images produce a fascinated delight, but the sight of our own bones no longer shocks, and perhaps this is

FIGURE 3.6 Untitled watercolour landscape by John Hall-Edwards. Author's collection.

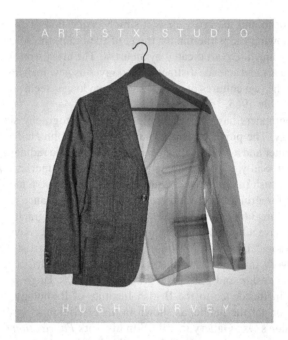

FIGURE 3.7　Fabrication 2018 by Hugh Turvey (positive image). Image courtesy of the artist.

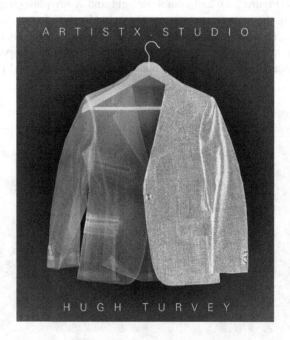

FIGURE 3.8　Fabrication 2018 by Hugh Turvey (negative image). Image courtesy of the artist.

because the modern eye has become used to images of our interior anatomy. Another approach is taken by the Winchester artist Andrew Carnie who produces images that can disturb the superimposed radiological images and photographs of the body.[38] Marius Kwint reflects on the work of Carnie and the dreadful fascination of seeing one's innermost parts brought to life as anatomy.[39] Perhaps it is the artist who has to reinterpret the images to make them seem dreadful since most will view normal anatomy with pleasure. Radiology has given us an internal aesthetic to match the aesthetic of surface anatomy.

X-RAY VISION

To see inside the body directly has long been a popular theme in both apparent fact and fancy. Prior to radiology, it was claimed that a clairvoyant mesmeric trance could produce an internal vison, although this does not seem to have been a common claim.[40] However, following Röntgen's discovery, there were a series of claims of X-ray vision which were taken seriously. In 1865, John Tyndall (1820–1893) noted that "Different nerves are suited to the perception of different impressions".[41] We do not see with the ear or hear with the eye, and Tyndall noted that the optic nerve "was limited to the apprehension of the phenomenon of radiation". John Tyndall worked on diamagetism, infrared radiation, and the greenhouse effect. Tyndall noted that the eye could not see "the entire range even of radiation". These rays that could not be seen Tyndall called invisible or obscure rays, and he produced a spectrum of electric light in 1865 (Figure 3.9). Since there were rays outside of the visible range, it would not seem unimaginable that these rays might by some technique or skill be perceived. Röntgen, in his Third Communication that he made in 1899 to the Royal Prussian Academy of Sciences, discussed the topic, noting that "If our eyes were as sensitive to X rays as they are to light rays, a discharge tube would appear to us like a light burning in a room uniformly filled with tobacco-smoke".[42] The eyes are

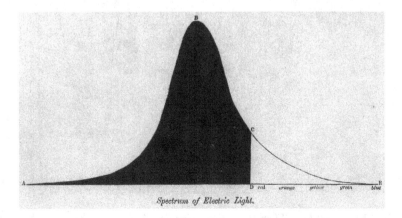

Spectrum of Electric Light.

FIGURE 3.9 Tyndall on Electric Spectrum 1865: Spectrum of Electric Light. From John Tyndall, "On Radiation", Longman, London. Author's copy. Public domain.

indeed sensitive to X-rays, and Howard Pirie from Montreal writing in 1934 commented that "The power of X rays to stimulate the retina has been known for many years but no practical use has been made of it".[43] Pirie made an apparatus for reading with closed eyes and he observed that patients could read letters and locate foreign bodies. For his work Pirie was awarded the gold medal by the American Roentgen Ray Society. Whilst the retina might be sensitive to X-rays,[44] there is all the difference in the world between seeing a "blue-gray mist" and the clear X-ray vision of popular fancy.

In 1899, the case of the 11-year-old Boston boy Alfey Lionel Brett was discussed in detail in The New York Press (Figure 3.10).[45] The boy's father was a physician from South Braintree, MA, and the powers were revealed when Alfey exclaimed: "Papa, I can see your bones". Dr. Brett had been hypnotising his son and the boy's vision was associated with a hypnotic state. It was said that the boy could see things

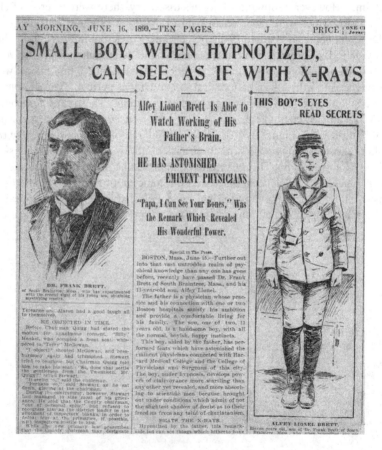

FIGURE 3.10 *Small Boy, When Hypnotised, Can See, As If with X-Rays.* The New York Press (New Jersey edition), New York, 16 June 1899. Public domain. Author's collection. No longer in press and ceased before the Second World War.

that "hitherto have been revealed to the material optics of the X-ray machine". However, the boy's vision was presented as superior to the X-rays since the fluoroscope reveals only "darkened outlines, the boy sees all the colors". The X-rays and what they revealed were a subject of major popular fascination at that time. Many eminent doctors were apparently astonished that Alfey could see things only shown by X-rays. Interestingly, in another experiment, Alfey could see "brain flashes of the sensory and motor nerves from the brain centres to the arm and back again" much to the astonishment of his audience. As a confirmation, the boy could point to the spot on the brain where the flashes originated, and which scientists knew contained the centres in the brain that controlled the upper arm. Alfey would seem to be anticipating modern functional imaging. Dr. Brett also reported that Alfey could see the workings of his brain. Whilst the boy's gift is described as X-ray vision, the phenomenon is presented as one of clairvoyance, and the hypnotic trance of the boy seems similar to the mesmeric trance. The newspaper took the boy's story seriously, and certainly many, and at that time in particular, took psychical phenomena seriously. Perhaps the best example was the physicist Oliver Lodge (1851–1940) who made many contributions to physics, including early X-ray studies, whilst at the same time advancing psychical research.[46] Lodge viewed his psychical research with as much seriousness as his scientific work. The borders of science are of necessity ill-defined.

A story appeared in *The Sketch* newspaper of 7 March 1906 describing the activities of a Frenchwoman Louise Bar from St. Quentin, described as "The Lady with the X-ray Eyes".[47] In a similar manner to Alfey Brett, Mlle. Barr was hypnotised by her father and could "see into the human frame and diagnose illness of any kind". When Louise holds the patient's hand, she described the disease and prescribed the remedy which was written down and the prescription was signed by a Dr. Hamand. The French authorities rather wisely seem to have taken a pragmatic approach, and whilst not apparently commenting on the truth or falsehood of her claims, prosecuted her for the illegal practice of medicine.

In 1923, two men who claimed to be able to see through a brick wall were tested before the king and queen of Spain.[48] They could see a watch through bandaged eyes even when it was behind a lead screen and they could read newspapers in a sealed box. The experiments were conducted by an engineer, Madriline Manuel Maluquer, and "precautions against trickery were taken". The phenomenon was ridiculed by British scientists, including Professor Rankine from Imperial College in London.

There is obviously a continuing attraction to the idea of X-ray vision. A recent example is the Russian woman Natalya "Natasha" Nikolayevna Demkina (1987–) from Saransk, called "the girl with the X-ray eyes". Natasha apparently became aware of her gift when she was aged 10. She suddenly was aware that she could see inside her mother's body, and then started describing the organs that she could visualise. Natasha was able to switch between her regular vision and what she called her medical vision. In a similar manner to Alfey Brett, she saw a colourful image of the organs, and then started to analyse what she viewed.[49] The fees that Natasha charged for her consultations were used to pay for her medical school tuition fees. In 2004, Natasha came to London and appeared on the Discovery Channel. An experiment was filmed which had been formulated by scientists from the Committee for the

Scientific Investigation of Claims of the Paranormal (CSICOP). Even though she did not pass the test, this did not stop the Discovery Channel from advertising the film by saying that Natasha "still manages to astound even the most hardened critics". Many remained impressed by Natasha's skills, and many seem to accept what they want to believe and ignore what does not suit their perspective. We believe as true that which fits our worldview.

The continuing interest in X-ray vision is curious. We seem to be unhappy simply being ourselves, with our normal and limited senses and abilities, and desire augmentations. So many modern novels, comic books, and films have superheroes with special powers as the main characters, and there are increasingly themes of transhumanism. The archetypic superhero is the man of steel or the Superman. Superman was probably the bestselling superhero in American comic books until the 1980s. Superman was created by the writer Jerry Siegel and the artist Joe Shuster, and first appeared in print in 1938 and then later in film.[50] Superman was a humanoid alien from the planet Krypton and has superhuman powers, including the iconic X-ray vision. The X-ray vision allows Superman to see through walls to find the criminal, and also seem to work in a transmission mode allowing him to melt metals. As an alien, the X-ray vision that Superman possesses has no philosophical or existential implications, and seems entirely natural to the character. This is probably because until about the 1990s, comics were aimed at children who were interested in action and excitement and not in deeper and philosophical themes.

Perhaps of more interest than the Superman stories is the short story "The Man Who Saw Everything" (also known as "The Man with the X-Ray Eyes") which appeared in the book *The Horror on the Asteroid and Other Tales of Planetary Horror* of 1936 written by Edmond Hamilton (1904–1977) who was one of the developers of science fiction. "The Man Who Saw Everything" was considered to be one of Hamilton's best stories and has been reprinted several times.[51] In the story, David Winn is a newspaper reporter in New York who wants to marry his girlfriend, but his finances are not adequate. Winn learns that a scientist Dr. Jackson Homer has developed a process to alter the eyes to make the retina more sensitive to radiation, thereby allowing vision through inorganic materials. Homer needed a human subject following his animal tests, and David Winn seemed ideal. Winn believed that with his new vision, he would become the reporter who could see everything, and then as the world's best reporter, would become the best paid reporter and could marry his girlfriend. The process is a complete success and Winn can see through the inorganic walls and view only animal subjects. Since the experiment was a success, Winn returned to his work as a reporter. However, it does not turn out quite as Winn expected. Politicians behind closed doors are seen to be corrupt and he is shocked. He goes to see his fiancée and travelled through a slum, past a prison, and then past a hospital. Since Winn can see through walls and doors and floors, he observes every type of human evil, misery, and misfortune. He is exposed to the grim reality of human life and determines that following marriage, the couple will move to the country where there are fewer people and walls. However, as he approaches the house, he observes his fiancée and her mother talking about him and finds out that both think he has nothing and never will have anything. His fiancée only agreed to the marriage because she realised that she must marry someone, and that Winn was the best that she could get. Winn rushes off and in anguish drowns himself in the neighbouring river. His body was found with one hand in front of his eyes, and Dr. Homer said that Winn was trying to keep himself from seeing everything. Winn saw everything just as he had wanted to, and it was too much for him. Dr. Homer exclaimed "God keep us blind in this world! Prevent us from the horror of doing what he did, of seeing too well".

Perhaps the greatest science fiction film of all times is *X: The Man with the X-Ray Eyes* of 1963. The film was produced and directed by Roger Corman (b. 1926), and featured Ray Milland (1907–1986) in the title role of Dr. James Xavier. Although Corman stated that the idea behind the film was his, since he had made a series of science fiction films before 1963, it would seem unlikely that he was unaware of Hamilton's story. Corman described the film as a low-budget Greek tragedy which seems quite appropriate. The film received critical acclaim winning the Silver Spaceship (Astronave D'argento) prize at the first International Festival of Science Fiction Film (Festival internazionale del film di fantascienza) held in Trieste, Italy, later that year. The film was subsequently made into a comic book and a novel.[52] Dr. Xavier is a medical research doctor who develops eyedrops to increase the range of his vision beyond the visible into the range of X-rays and still further. Before he starts the experiment, he has a colleague test his vision and is assured that his eyes are perfect. Xavier exclaims that his eyes are not perfect since he is blind to all by a tenth of the universe. However, his colleague warns Xavier that only the gods see everything. The drops prove successful, and Xavier is able to see a young patient's thoracic tumour and also to see beneath clothing. He keeps on using more drops and after a series of events finds it increasingly difficult to comprehend what he is now seeing. In the final scene, Dr. Xavier drives into the desert and visits a religious tent revival where he tells the preacher that he has come to "tell you what I see". Xavier tells the pastor that "there are great darknesses … and beyond the darkness is a light that glows … and in the centre of the universe the eye that sees us all". The pastor is horrified and tells Xavier that "If thine eye offends thee … pluck it out!".[53] Xavier tears his eyes out and the film ends. In an interesting alternate ending, suggested by the writer Stephen King, after having torn out his eyes, Xavier exclaims "I can still see!".[54]

The film raises many interesting questions, including those about the limits to human perception. Is it worth seeing more if we cannot understand what we see? Is there a limit to what we can understand? Both David Winn and Dr. Xavier see more than they can cope with and are driven mad. At some point will science show us a depth of reality that is no longer comprehensible? Technology has developed since 1963 and the visions of Winn and Xavier can easily be replicated with modern medical imaging. Small and easily hidden cameras and microphones for domestic use are readily purchased and raise their own ethical issues. The diagnosis of the young patient by Xavier would now be straightforward using modern medical imaging. However, how to emotionally cope with excessive knowledge remains as much a problem now as then. As Roger Corman said, "Can you see God? What are the limits of seeing through & through & through?" Does an increasing knowledge give one any deeper knowledge of reality in any meaningful way, and is a deep philosophical study of epistemology any better than a naive realism? This topic is discussed by Leo Tolstoy (1828–1910) in his short story *The Coffee House of Surat (After Bernardin de Saint-Pierre)*.[55] In the story, there is a discussion in which the sun is used as a metaphor for the divine. A philosopher is described who became blind from staring at the sun whilst trying to decide what it is. The philosopher concluded that since the sun is neither solid nor liquid nor gas, it must not exist at all. He then uses his own blindness as evidence for his theory. So, what are the borders of physics and meta-physics, and is there any value in making unverifiable theories? Will physics have its end, either when we cannot understand what is observed, or when there are no more discoveries to be made?

There have always been toys and puzzles associated with X-rays, and Röntgen himself had a magician's box at Weilheim containing magical tricks and puzzles.[56] Cards that have an aperture containing a grid give the illusion of X-ray vision when looked through. Such cards were sold from an early time as X-ray viewers. An Austrian X-ray Postcard of 1904 (Figure 3.11a and b) is labelled "Origenell! Reflex = Röntgen-Strahlen Postkarte". The aperture with the grid can be seen in the centre of the star. In the 1960s, the idea of placing such grids into spectacles was thought of, and these were sold in comic books. When looking at a person through such a grid, an illusion of seeing bones is produced. Such X-ray spectacles have long been a part of popular culture as X-ray "specs" or "gogs".

In the Punk Rock world, a South London band was formed in 1976 calling them-selves X-Ray Spex.[57] Their music has a raw and invigorating quality, and featured the singer Poly Styrene, who was born as Marion Joan Elliott-Said. X-Ray Spex gave their debut performance at London's Roxy nightclub in London's Covent Garden, having

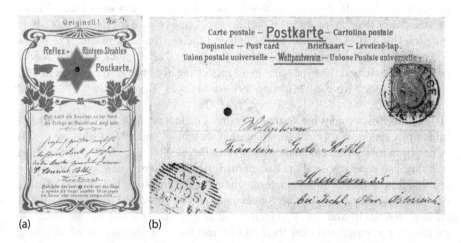

(a)   (b)

FIGURE 3.11 (a) Austrian X-ray Postcard: "Origenell! Reflex = Röntgen-Strahlen Postkarte". The aperture with grid can be seen in the centre of the star. Author's collection. No publisher listed. (b) Austrian X-ray Postcard posted 14 August 1904. Address side, aper-ture with grid can be seen to the left side. Author's collection. No publisher listed.

had only six rehearsals in their manager's front room. The Roxy was the centre of London's developing punk music scene. As their debut album *Germfree Adolescents* of 1978 showed, their music was highly energetic, but perhaps a little shambolic. X-Ray Spex appeared under the EMI label, as did other punk rock groups, including Johnny Rotten and the Sex Pistols. Johnny Rotten was the stage name of John Lydon. It was the Sex Pistols that was to cause EMI considerable anguish in November 1976 with an infamous interview on an EMI-controlled Thames Television programme. The British nation was both scandalised and fascinated at the same time, and EMI deleted their single *Anarchy in the UK* from its catalogue. The EMI management had to engage in damage limitation, and the Sex Pistols were sacked.[58] Dropping the Sex Pistols in 1977 was a financially unwise move since they moved to Virgin Records where they sold more than a million records.[59] Management attention was unfortunately diverted from EMI Medical Electronics and the EMI scanner which was having its own problems at that time.

A children's film *The Kid with the X-Ray Eyes* directed by Fred Olen Ray was released in 1999. In the film, the 12-year-old Bobby finds a special pair of glasses that gives him X-ray vision, and has a series of exciting adventures. These ideas obviously have an appeal to children and a comic strip *X-Ray Spec* appeared in the British children's comic *Beezer* and in the 1990s, featuring a schoolboy and his escapades with the glasses that gave him X-ray vision. His X-ray spectacles resulted in many adventures and saved him from many scrapes. The boundaries between the culture of children and young adults have become increasingly blurred since the 1990s.

CONCLUSIONS

The discovery of the new rays by Wilhelm Röntgen in 1895 has had a pronounced impact on almost all aspects of our culture that shows little signs of diminishing.

NOTES

1. Griffiths, G. 1896. A photograph of the invisible. *Pearson's Magazine*, 1 (April 1896), 376–380.
2. Dam, H.J.W. 1896. A wizard of today. *Pearson's Magazine*, 1 (April 1896), 413–419.
3. Tattersdill, W. 2016. *Science, Fiction, and the Fin-de-Siècle Periodical Press* (Cambridge Studies in Nineteenth-Century Literature and Culture Book 105). Cambridge: Cambridge University Press.
4. Knight, D. 1967. *In Search of Wonder: Essays on Modern Science Fiction*. Chicago: Advent Publishers.
5. Meade, L.T., Halifax, C. 1896. The adventures of a man of science. 1. The snake's eye. Told by Paul Gilchrist. *The Strand Magazine*, 12, 57–68.
6. Brogdon, B.G., Lichtenstein, J.E. 1998. Forensic radiology in historical perspective. In: *Forensic Radiology*. Ed. B.G. Brogdon. Boston: CRC Press, 13–34.
7. Michel, M.S. 1942. *The X-Ray Murders*. New York: Coward-McCann Publishing Company.
8. Hope, C. 2000. *X-Ray*. No. 3 in Heart Rate Series (reprint). London: Collins.
9. Korankye, G. 2010. *Team Twenty Ten*. Basingstoke: McTaggart Publishing.

10. British Society for Thoracic Imaging. 2020. COVID-19 imaging database. https://www.bsti.org.uk/training-and-education/covid-19-bsti-imaging-database/ (accessed 5 April 2021).
11. Mann, T. 2020. *The Magic Mountain*. Trans. Helen Tracy (first published 1924, S. Fisher Verlag, Berlin). London: Woolf Haus Publishing.
12. van Dijck, J. 2011. *The Transparent Body: A Cultural Analysis of Medical Imaging*. Seattle: University of Washington Press.
13. van Dijck, J. 2000. Röntgenstralen tussen wetenschap en kunst in De Toverberg [X-rays between art and science in The Magic Mountain] (Dutch). *Gewina*, 23(2), 91–106.
14. Snow, C.P. 2012. *The Two Cultures (Canto Classics)*. Cambridge: Cambridge University Press; Reissue edition.
15. Leslie, A., Jones, A.J., Goddard, P.R. 2000. The influence of clinical information on the reporting of CT by radiologists. *The British Journal of Radiology*, 73, 1052–1055.
16. Burgess, A. 1960. *The Doctor Is Sick*. London: Heinemann.
17. Almén, T. 1985. Development of non-ionic contrast media. *Investigative Radiology*, 20, 2–9.
18. Asimov, A. 1966. *Fantastic Voyage*. London: Dennis Dobson.
19. https://britishtheatre.com/review-it-happened-in-key-west-charing-cross-theatre/ (accessed 26 April 2021).
20. Harrison, B. 2009. *Undying Love: The True Story of a Passion that Defied Death*. Marathon, Florida: Recent Work. Ed. A. Carnie The Ketch & Yawl Press.
21. *Barbi Doll: Pictorial*. 1970. *Playboy*, 17(3), 141–149.
22. Metzinger, J. 2008. Note sur la peinture, Pan (Paris), October–November 1910, 49–52; in Antliff, M., Leighten, P., *A Cubism Reader, Documents and Criticism, 1906–1914*. Chicago: The University of Chicago Press, 75–83.
23. https://www.wetcanvas.com/forums/topic/jean-metzinger-artist-of-the-month/ (accessed 26 April 2021).
24. Cappock, M. 2005. *Francis Bacon's Studio*. London: Merrell.
25. Clark, K.C. 1939. *Positioning in Radiography*. London: Messrs. Ilford: W. Heinemann.
26. Harrison, M. 2005. *In Camera, Francis Bacon. Photography, Film and the Practice of Painting*. London: Thames & Hudson.
27. https://www.artmajeur.com/en/abstractshinytexture/artworks (accessed 26 April 2021).
28. https://susanaldworth.com/biography-and-statement-2020/ (accessed 26 April 2021).
29. Aldworth, S. 2008. *Scribing the Soul*. Susan Aldworth.
30. Aldworth, S. 2013. *Transience*. London: Cassland Books.
31. Aldworth, S. 2013. *The Portrait Anatomised*. London: Cassland Books.
32. Thomas, P.K. 1963. The connective tissue of peripheral nerve: an electron microscope study. *Journal of the Anatomical Society (London)*, 97, 1, 35–44.
33. Wellcome Collection. https://wellcomecollection.org/works/tzufkc5s (accessed 14 August 2021).
34. Snow, C.P. 1965. *The Two Cultures*. Cambridge: Cambridge University Press.
35. Veasey, N. 2008. *X-Ray, See through the World Around You*. London: Viking Studio.
36. https://www.x-rayartist.com/about.html (accessed 3 May 2021).
37. https://www.xogram.com/store/c3/INSTALLATIONS.html (accessed 3 May 2021).
38. https://www.andrewcarnie.uk/andrewcarnie (accessed 3 May 2021).
39. Kwint, M. 2008. Animated mediations on mortality: the work of Andrew Carnie. In: *Time Will Tell: Andrew Carnie, Recent Work*. Ed. A. Carnie. Winchester: Winchester Gallery.
40. Miller, J. 1983. A Gower street scandal. *Journal of the Royal College of Physicians of London*, 17, 181–191.
41. Tyndall, J. 1865. *On Radiation*, The Rede Lecture delivered before the University of Cambridge. London: Longman, Green, Longman, Roberts & Green.

42. Roentgen, W.C. 1899. Further observations on the properties of X-rays. *Archives of the Roentgen Rays*, 3, 80–88.
43. Pirie, A.H. 1934. An apparatus for reading with closed eyes. *British Journal of Radiology*, 7, 111–116.
44. Reid, J.A. 1972. The "visibility" of X-rays. *Oral Surgery, Oral Medicine, Oral Pathology*, 34, 330–334.
45. *Small Boy, When Hypnotised, Can See, as If with X-Rays*. New York: The New York Press (New Jersey edition), June 16, 1899, 1–2.
46. Lodge, O. 1932. *Past Years: An Autobiography*. New York: Charles Scribner's Sons.
47. Thomas, A.M.K. 2017. The lady with the X-ray eyes: the invisible light. *The Journal of The British Society for the History of Radiology*, 41, 19.
48. X-Ray Eye Man: Demonstration at the Spanish Court. *Daily Express*, 2 June 1923.
49. Sample, I. 2004. Visionary or fortune teller? Why scientists find diagnoses of "X-ray" girl hard to stomach. *The Guardian*. https://www.theguardian.com/uk/2004/sep/25/russia.health (accessed 3 April 2021).
50. Created by: J. Siegel, J. Shuster. 1999. *Superman: The Dailies. Volume III: Strips 673–966, 1941–1942*. New York: DC Comics.
51. Hamilton, E. 1946. The man with X-ray eyes. *Startling Stories* (British Edition), 3, 47–53.
52. Sundak, E. 1963. *X*. New York: Lancer Books.
53. Book of Matthew 5:29-30. King James Version of the Bible.
54. Phipps, K. 1977. Roger Corman reflects on his long, legendary career: but he isn't finished yet. https://uproxx.com/movies/interview-roger-corman/ (accessed 4 April 2021).
55. Tolstoy, L.N. 1885. *The Coffee-House of Surat* (After Bernardin de Saint-Pierre). Trans. L.A. Maude. In *What Men Live By, and Other Tales*, Many editions.
56. Glasser, O. 1933. *Wilhelm Conrad Röntgen and the Early History of the Roentgen Rays*. London: John Bale, Sons & Danielsson, Ltd.
57. https://www.x-rayspex.com/xrayspex (accessed 25 April 2021).
58. Cavendish, W. 2017. *His Masters Voice, Sir Joseph Lockwood and Me*. London: Unicorn.
59. Southall, B. 2009. *The Rise and Fall of EMI Records*. London: Omnibus Press.

4 Radiology and Anatomy

The radiographic images are primarily anatomical in nature, although functional information may be derived. In 1945, the pioneer paediatric radiologist John Caffey wrote in the first edition of his classic book that shadows are but dark holes in radiant streams, to be seen as twisted rifts beyond the substance, and therefore meaningless in themselves. Caffey realised that if we wish to comprehend Röntgen's pallid shades, we need always to know well the solid matrix from which they spring. The physician therefore needs to know intimately each living patient through which the racing black light darts. The black or invisible light produces cold, silent, and empty shadows. It is the role of the radiologist to demonstrate the warm, lively, and fleshly humanity and their story is root and key to the shadows.[1] Radiography shows us ourselves.

The discovery of X-rays in 1895 by Wilhelm Conrad Röntgen transformed our understanding of both the physical world and ourselves. Traditional anatomy, as demonstrated by anatomists such as Andreas Vesalius (1514-1564), was learnt from the dissection of the supine deceased body. Radiology showed anatomy in the living in a manner previously not possible, and has transformed our anatomical understanding, particularly of human growth and variation. The knowledge of radiological anatomy is central, since without knowing the normal, we cannot interpret the abnormal.

Andreas Vesalius may be seen as the father of our modern understanding of anatomy. Vesalius performed his dissections in the standard manner, examining bodies lying supine on a bench in an anatomical theatre. However, in his *De humani corporis fabrica* (On the Fabric of the Human Body) of 1543, Vesalius presents human anatomy in both an artistic and scientific manner.[2] His dissections are shown as animated figures in a landscape, and the figures resemble the living. This brings to mind the European Dance of Death, otherwise known as the Danse Macabre or Totentanz. The Dance of Death presented the dead dancers as very lively and agile, making the impression that they were actually dancing, whereas their living dancing partners looked clumsy and passive in comparison.

TRADITIONAL ANATOMY

Traditionally, internal human anatomy is encountered in defined locations. These are the anatomy theatre, in operative surgery, the graveyard, and the battlefield. The discovery of the X-rays in 1895 by Wilhelm Conrad Röntgen was to transform our understanding of both the physical universe and ourselves, and internal anatomy could now be seen in the living. The famous radiograph that Röntgen made of his wife Bertha's hand was taken on 22 December 1895, and Röntgen then presented his preliminary report *Über eine neue Art von Strahlen* (On a New Kind of Rays)

DOI: 10.1201/9780429325748-4

FIGURE 4.1 An early postcard of figures dancing on a beach, *The Strand-Idyll á la Röntgen*: the happy bathers on a beach become dancing skeletons under the influence of the X-rays, resembling the Danse Macabre. Author's collection.

to the Physical-Medical Society of Würzburg on 28 December 1895.[3] The discovery caused a worldwide sensation, and the public had to be reassured that this was an authentic discovery by a respected scientist. This early postcard of figures dancing on a beach, the Strand-Idyll á la Röntgen as shown by the X-rays, resembles both the Danse Macabre and the anatomical figures of Vesalius in a landscape (Figure 4.1).

CHARLES THURSTAN HOLLAND

The impact of Röntgen's discovery is shown clearly in the work of the Liverpool pioneer Charles Thurstan Holland (1863–1941).[4] Holland became a leader of world radiology, and in 1925 was President of the First International Congress of Radiology.

Holland was working as a general practitioner in Liverpool when on 7 February 1896 he saw some of the early X-ray work of Sir Oliver Lodge (1851–1940) at Liverpool's University College. Lodge was able to take radiographs of a boy who had shot himself in the hand. A successful radiograph was obtained, and writing in 1937,[5] Holland says that one cannot today imagine the excitement in the department when the plate was brought out into the daylight and the shadow of the bullet was demonstrated. A more recent equivalent is the excitement generated by early CT or MRI images. By the end of May 1896, Holland had acquired an X-ray apparatus, and as he later recalled, "There were no X-ray departments in any of the hospitals. There were no experts. There was no literature. No one knew anything about radiographs of the normal, to say nothing of the abnormal".

On 1 September 1896, Holland was able to examine a full-term stillbirth (Figure 4.2). He was fascinated to see the bones, particularly of the hands and feet. He noted the absence of ossification of the epiphyses at the ends of the bones. Holland immediately realised the role that X-rays could have for anatomical studies and in observing skeletal growth. On 17 September 1896, Holland examined the hand of a child at the age of 1 year and was again fascinated to see the ossifications

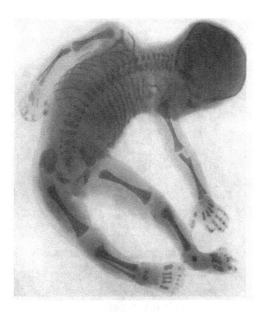

FIGURE 4.2 A full-term stillborn child examined on 1 September 1896 by Charles Thurstan Holland showing epiphyseal detail. From Holland, 1923, see Note 4. Courtesy of the *British Journal of Radiology*.

of the developing skeleton. He observed the bones with a clarity that had not been previously possible. He started collecting radiographs at different ages of development and later that year he showed them at the British Association Meeting held in Liverpool that September.

JOHN POLAND AND THE DEVELOPMENT OF STANDARDS FOR BONE MATURATION

This early work on bone age by Holland was developed by the surgeon John Poland (1855–1937) from the Miller Hospital in Greenwich near London. Poland described the eponymous Poland's syndrome and was influential in development of orthopaedics. His bone age atlas[6] was published in 1898, and in it Poland pointed out that the development of the ossification centres differed quite considerably from that which had been previously described.

For the hand radiograph of a boy aged 17 years, John Poland commented that "In this instance the epiphyses of the metacarpal bones and phalanges of finger and thumb, though fully developed, have not, as in the two preceding skiagrams (radiographs), joined their respective shafts" (Figure 4.3). So, the older boy had relatively younger bone age, and this showed the need for further work. Poland's population sample was small; however, this remarkable book demonstrated future possibilities. This information on normal skeletal development would be essential with the establishment of well-baby clinics, school health programmes, and the routine health

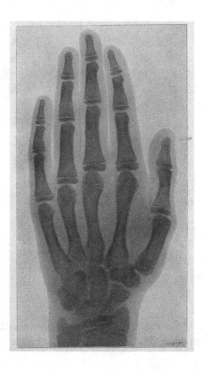

FIGURE 4.3   A hand radiograph of a boy aged 17 years by John Poland. From Poland, 1898, see Note 6. Public domain.

examination of children, which developed in the first half of the 20th century and which was being increasingly requested by parents.

In 1929, Prof. T. Wingate Todd from Cleveland, Ohio, initiated his long-term studies of human growth and development. In 1931, 3-month-old children were enrolled into the programme, and in subsequent years children from aged 3 months to 14 years were added into the study, which finished in the summer of 1942. The children were selected since they were "free from gross physical and mental defects". Since they were admitted on the application of a paediatrician, they came from families of above average economic and educational status. Todd published his *Atlas of Skeletal Maturation of the Hand* in 1937[7] based on his X-ray studies. This ground-breaking book used data from the study group and also included children from public schools (these are US state–run schools) and from various social agencies. Todd found that there was a measurable difference between these two groups, attributable to the less privileged background of the children not in the research group.

William Walter Greulich (1899–1986) and Sarah Idell Pyle (1895–1987), who were both anatomists from Stanford, using the extensive radiographic material from Todd's research series on child growth and development, published their *Radiographic Atlas of Skeletal Development of the Hand and Wrist* in 1950, with a second edition appearing in 1959[8] and reprinted in 1988. That this book remains the standard over 60 years later shows its enduring value. The alternative Tanner-Whitehouse (TW) method[9] was developed using radiographs from average-class children in the UK between 1950 and 1960, with an update TW3 (Tanner-Whitehouse 3) published in 2001. The TW method was never as popular as the Greulich and Pyle method and is currently out of print.

However, to what extent do the standards of 1950 based on the growth patterns of Caucasian upper socio-economic class American children living in Cleveland from 1931 to 1942 apply to contemporary populations that are nutritionally different and more culturally diverse? The standards apply only to the American population studied, and other populations require a different standard. In 2019, Sumit Patil published his *Assessment of Bone Ages in Indian Children: Applicability of Greulich and Pyle Standards to Indian Children.*[10] This study was designed to determine bone ages (skeletal ages) of Indian children in Maharashtra and the applicability of the Greulich and Pyle American standards to Indian children. Patil found that male children skeletally lag behind the American standard in all age groups. Female children also lag behind the American standard in all age groups except the 12–13-year age group in which in which they are accelerated. The assessment of bone age is important and Patil notes the incomplete registration of birth in India, and since parents may conceal a true age, an objective method of assessing true age is important. Bone age assessment has also been used by authorities where foreigners cannot provide evidence of their date of birth, and also in cross-border migrations such as between India and Pakistan, Bangladesh, and Nepal. Chronological age is also important for thresholds for legal responsibility in many countries and may vary from 14 to 21 years of age. Bone age assessment in relation to migration raises many ethical and legal issues and the topic is controversial.

JOHN POLAND AND EPIPHYSEAL ANATOMY
AND CLASSIFICATION OF FRACTURES

John Poland had a particular interest in epiphyseal anatomy and was a pioneer of the X-rays. In a talk given to the West Kent Medico-Chirurgical Society in 1886, before X-rays were discovered, Poland noted that whilst epiphyseal fractures in children were a daily occurrence, they were seldom accurately diagnosed and that the subject was rarely dealt with well in surgical textbooks. It is therefore not surprising that considerable effort was made in the late 19th century and early 20th century in understanding the anatomy of the developing epiphyses and radiographic anatomy. Eugene Corson, from Savannah, GA, wrote to John Poland on 21 November 1900, admiring Poland's book *The Traumatic Separation of the Epiphyses*[11] and enclosing some reprints and radiographic prints. Corson enclosed an article *A Skiagraphic Study of the Normal Membral Epiphyses at the Thirteenth Year* which had appeared in the November 1900 *Annals of Surgery*. Corson wrote that "The X-ray will prove to be a valuable aid in the study of many points of normal anatomy" and that "The bone relationships in joints, the various joint movements, and the different steps in bone development can all be studied in a striking way by the X-ray" (Figure 4.4). It was the clarity of radiography that impressed Colson, who commented that "the

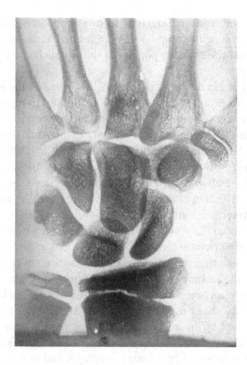

FIGURE 4.4   A detailed wrist radiograph sent by Eugene Corson, from Savannah, GA, to John Poland on 21 November 1900. Author's collection.

discovery of Röntgen, a discovery which makes possible and easy and an absolutely correct diagnosis where previously uncertainty and error outweighed definite knowledge". Corson knew that before the abnormal could be interpreted, there should be a thorough knowledge of the normal. The work of Poland remained influential for many decades, and was still being quoted in the 1970s. The assessment of epiphyseal trauma in children remains challenging, and the utilisation of CT may now be beneficial in addition to the plain films.

WILLIAM J. MORTON

William James Morton (1845–1920) was "Professor of Diseases of the Mind and Nervous System and Electro Therapeutics" in the New York Post Graduate Medical School and Hospital. His book *The X-Ray or Photography of the Invisible*[12] is undated; however, the preface is dated 11 September 1896. Morton's co-author was Edwin Hammer who was an electrical engineer. This book was written following the huge worldwide interest that had taken place following the discovery of X-rays by Wilhelm Conrad Röntgen and summarised the current state of knowledge. This book is important because it is the first book on radiology that was written by a physician, and Morton made interesting speculations about potential future uses for the new rays. Morton makes the pertinent observation that in teaching the anatomy of the blood vessels, the X-rays open out a new and feasible method. Morton said that the arteries and veins of dead bodies may be injected with a substance opaque to the X-ray, and thus their distributions may be more accurately followed than by any possible dissection. The feasibility of this method applies equally well to the study of other structures and organs of the dead body. So, Morton believed that to a certain extent, X-ray photography could replace both dissection and vivisection, and in the living body the location and size of a hollow organ, for instance the stomach, may be ascertained by causing the subject to drink a harmless fluid, more or less opaque to the X-ray, or an effervescent mixture which will cause distension, and then taking the picture. Morton's words are most interesting. This book was written less than a year following the discovery of X-rays, and Morton is not only predicting contrast gastrointestinal studies, but also the use of radiology in the equivalent of modern virtual autopsy, or virtopsy. The pioneers so often realise the exact significance and importance of their observations. Morton had immediately seen that the radiological examination of the body, either living or dead, could produce more information than could be found in either the operating theatre or the pathology department.

GROWTH OF THE RADIOLOGICAL ATLAS

It was because of the difficulties in image interpretation, even of the normal, that many radiographic atlases were published as the 20th century progressed. In 1907, William Ironside Bruce (1876–1921) published his *System of Radiography with an Atlas of the Normal*.[13] The early atlases showed two-dimensional images, and the three-dimensional structure of the body was difficult to appreciate. So, when we consider plain skeletal radiographs of today, they seem superficially straightforward to interpret; however, this was far from the case and early radiographs were as difficult to interpret then as complex MRI scans are today. Part of the reason was the low power of the apparatus and the large focal spot resulting in a blurred image. It was therefore not uncommon for the radiologist to touch up the negative so as to make it clearer; however, the resultant image might bear limited resemblance to normal anatomy. As the technical quality of the X-ray photograph was supposed to indicate the value of the examination, various photographic tricks could be applied to improve the photographic results and make the radiographs more impressive. Even a superficial examination of the radiographs published in the *British Medical Journal* in 1896 reveal that many radiographs have been modified. However, improvement in apparatus gradually did away with the necessity of "touching up" the radiograph.

In addition to the radiographic image quality, a problem for the pioneers was that inadequate attention was given to the relationship between the X-ray source and the part that was being radiographed. There were striking distortions produced by differing relationships between the X-ray source, patient position, and film position. There was no agreement as to the best location for the X-ray tube and positioning was left to the individual operator and the tube was placed in an indeterminate position based on individual experience. To solve this problem was part of the motivation of Ironside Bruce in writing his 1907 book. Bruce also reassured his readers that none of his radiographs were retouched and were obtained under ordinary conditions.

Bruce realised that radiographic interpretation would be greatly facilitated if every radiograph reproduced an exactly similar picture of the normal parts examined. To this end, his book was illustrated with standard projections and standard radiographs. Before his early death of excess radiation exposure, he was working on an X-ray couch that would assist in the accurate standardisation of radiographs.[14] Most of the early textbooks of radiology had a section on radiographic positioning with illustrations of positions and typical radiographs, with an example being the influential book of 1903 by William Pusey and Eugene Caldwell.[15] In time specific manuals for radiographers were written such as by the US Army in 1918.[16] One of the best-known radiography books is *Positioning in Radiography* by Kathleen Clara Clark (1898–1968).[17] It was published in 1939 and sold for 3 guineas (£3.15p), which was a significant sum. It had 500 pages and 1,400 illustrations and diagrams, and by 1942 had sold about 6,500 copies. The book became the most important book on radiography ever written and was hugely influential. The great theme throughout Clark's life was the standardisation of radiographic projections and techniques. Whilst standardisation might be seen in a negative manner for some occasions, for Clark standardisation was about promoting high quality and on advising as to the best technique that should be used to obtain optimal results. The book became the standard work of reference for radiographers and has been through many editions.[18]

In modern times, the equivalent of the anatomical atlas was to reach its peak in the Visible Human Project (VHP).[19] The first images of the VHP were made available by the National Library of Medicine in the United States in November 1994. The images of the first person to become a "Visible Human" was Joseph Paul Jernigan from Texas.[20] Jernigan had been convicted of burglary and murder in 1981 and he was executed by lethal injection in August 1993 and as Waldby declared, "The choice of Jennigan's body by the project team provided media coverage with a set of stock narratives and an appealing moral economy of criminal transgression, punishment, sacrifice, and redemption". The use of Jennigan's body makes a logical connection with early anatomical studies. The dissection and then imaging of the body in modern times produced accurate anatomical information and this could be compared to classical Galenic texts which emphasised the importance of anatomy. In former times, many of the bodies dissected were those of criminals, and it is salutatory to consider that the modern VHP marks a return to this tradition. Whilst Waldby notes that the team chose the body, it would seem that it was also the career criminal and murderer that was chosen. As Vivian Nutton noted, when annual university lectures were introduced in the 14th century in Italy, the corpses used were those of criminals and those on the margins of society.[21] Nutton describes the dissections as being as much a spectacle as instruction. This point has also been made by Roy Porter who wrote that early anatomical demonstrations were public occasions, almost spectacles, for the purpose not of research but for instruction, and allowed the professor to parade his proficiency.[22] The VHP data sets comprise MRI, CT, and anatomical images. Visible Woman data sets were added to the Visible Man in 1995. The VHP data sets were to be a reference for the study of human anatomy, to act as public domain data for the testing of medical imaging algorithms, and to be a test bed and model for the construction of network-accessible image libraries.[23] As might have been anticipated, the VHP caused a public sensation with huge media attention. The VHP revealed known knowledge and as a public spectacle which was instructive and displayed proficiency was similar to the 14th-century anatomical university dissections in outcomes.

ANGIOGRAPHY AND VASCULAR ANATOMY

Morton had indicated that radiography performed with the addition of opaque material (what we would call contrast media) would show vascular anatomy. In fact, the first angiographic demonstration of anatomy was undertaken in Vienna in January 1896. Edward Haschek (1875–1947) who was Professor of Experimental Physics at the University of Vienna and his medical colleague Otto Theodor Lindenthal (1872–1947) injected a calcium carbonate emulsion (Teichmann's mixture) into the severed arm from a cadaver.[24] The arteriogram exposure was for 57 minutes, which was not unreasonable when one remembers the low power of the apparatus that was then available, and showed the vessels well. This procedure was performed in Vienna, and the radiograph can be seen at the Museum in the Josephinum in Vienna.

The angiographic work in Vienna was almost immediately followed by work of the group in Sheffield in England. Prof. Hicks, who was the Principal of Firth College in Sheffield, and Dr. Addison achieved both a renal and a hand arteriogram. Radiological work had been started at Firth College in Sheffield on 1 February 1896. Hicks and Addison injected specimens that were available in the medical school and their results were published later that month in the *British Medical Journal* of 22 February 1896 in an article entitled *The New Photography in Sheffield*.[25] The apparatus was simple and consisted of an ordinary battery of cells with an induction coil and a Crookes' tube. The apparatus was of low power and the current was never above the strength of one that would give a 3-inch spark. In their earlier experiments, the exposure time varied between 20 and 30 minutes; however, they stated that more recently they had obtained good shadows of the bones of the fingers using an exposure of a minute and a half. The Crookes' tube was used with a glass shield, with a window in it of about three-fourths of an inch in diameter. The vascular injections had been performed by Dr. Addison in the medical school and on 6 February he had injected samples using the ordinary red lead mass which was used in the dissecting rooms showing radiographic images of the arteries in the hand and kidney. The hand was nailed to a half inch wooden board and injected, whereas the kidney was simply laid on a wrapped photographic plate. The delicate branching pattern of the arteries in the kidney and hand were shown in a similar manner to those that have been demonstrated in Vienna a few weeks earlier.

The first full X-ray atlas of the arteries of the body was written by Herbert Charles Orrin (1878–1963), and was published in 1920 as *The X-Ray Atlas of the Systemic Arteries of the Body*.[26] Orrin is described as a civil surgeon who was attached to the 3rd London General Hospital RAMC(T) located in Wandsworth. The book is beautifully illustrated with beautiful radiographs (Figure 4.5a and b).

The book was designed to be used by students of anatomy, surgical anatomy, and operative surgery. It was intended to provide a series of natural illustrations of the

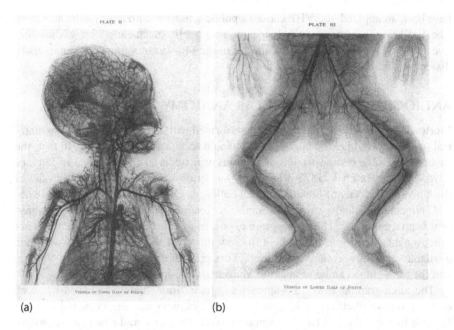

(a)   (b)

FIGURE 4.5   (a) Vessels of the upper half of foetus by H.C. Orrin. (b) Vessels of the lower half of foetus by H.C. Orrin. From Orrin, 1920, see Note 26. Public domain.

systemic arteries in continuity, and precisely as they exist in situ in the undissected body. The aims of the book were therefore purely anatomical in nature. Orrin wrote in his introduction that no matter how well dissection is performed, complete continuity of the vessels; their exact relationship to bones; their finest terminal branches; the series of anastomosis into which they enter are seldom if ever accurately displayed or intelligently appreciated by dissection alone.

Orrin therefore echoes the earlier words of William Morton. The atlas was accompanied by a full set of stereoscopic radiographs, "which provide the only possible means of accurately rendering visible the points and details specified" (Figure 4.6). It is interesting that in 1920, Orrin recognised the value of 3D angiography, which is now shown so well using CT or MRI scanning.

Angiographic atlases continued to develop, and in 1939 Charles Laubry (1872–1941) and others published his important *Radiologie Clinique du Coeur et des Gros Vaisseau*.[27] The book contained a detailed collection of post-mortem angiograms of the heart and great vessels, and this time in the adult. The early atlases were of post-mortem studies and their importance was twofold. Firstly, they demonstrated a degree of anatomical detail that was impossible to obtain by simple dissection and gave a truer image of anatomical relationships. However, they also indicated that with the correct equipment and non-toxic contrast media, such studies might be possible in life. Thanks to the work of Egas Moniz and other Portuguese radiologists, the goal of practical angiography in the living was finally realised[28,29] in 1927.

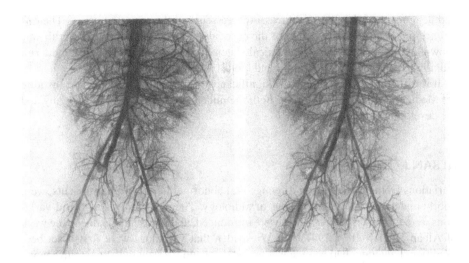

FIGURE 4.6 Stereo-pair of the vessels of the foetal torso by H. C. Orrin. From Orrin, 1920, see Note 26. Public domain.

ANGIOGRAPHY AND THE NUFFIELD INSTITUTE FOR MEDICAL RESEARCH, OXFORD

Angiography proved to be of value in primary anatomical research. Alfred Barclay (1877–1949) developed techniques for cineangiography[30] and during the Second World War studied the foetal circulation using foetal and new-born sheep. Angiography was performed on the foetal lamb with injections into the umbilical or internal jugular vein. Barclay and his colleagues were able to make the first direct radiological demonstration of the foetal circulation. The exact physiology of the foetal circulation had been a matter of dispute for over three centuries and Barclay and the unit at Oxford demonstrated that oxygenated blood from the placenta was diverted from the deoxygenated blood in the superior vena cava enabling it to pass mainly into the left atrium and ventricle, being separated from the venous return by the crista dividens. Their work replaced hypotheses with objective facts. The details of the foetal circulation were accurately determined for the first time, and also the closure of the ductus arteriosus and foramen ovale after birth. The work was published in an influential monograph in 1944.[31] The team went on to perform primary research into the vascular anatomy of the liver in the foetus, and also work on the renal circulation. The work on the renal circulation was performed with Josep Trueta (1897–1977), the Catalan surgeon, and was described as revolutionary and advanced the knowledge of hypertension and the effects of crush injuries.[32] Crush injuries were seen following injuries sustained in bombing during the Second World War and the work was of major importance, all the more remarkable because of the difficult wartime circumstances under which it was performed. Post-war Barclay went on to develop renal microangiography to show the microscopic structure of the renal circulation. Barclay made new discoveries of the nature of the renal circulation and demonstrated the fine structure of the vessels and their anastomoses.[33] The difference in structure and function of the cortical and juxta-medullary glomeruli was shown and the normal and pathological states of the vasa recta. The vasa recta run radially towards the renal pyramid and loops back up to the cortex.

It is difficult to overestimate the significance of the team at the Nuffield Institute for Medical Research in Oxford to both the understanding of vascular anatomy and the development of angiography.

ALBAN KÖHLER

Variations from normal, either as congenital abnormalities or normal variants, were poorly understood in the early days of radiology. The majority of congenital variations were unknown before X-rays were introduced, and it was largely due to the work of Alban Köhler (1874–1947)[34] of Wiesbaden that variations were first described. Traditional anatomy had been learnt on the dead and the new living anatomy shown on radiographs required a new level of appreciation and understanding of anatomy and its variations. Köhler was an initial member of the German Röntgen Society, which was founded in Berlin in 1905, becoming its president in 1912.

Since earliest times, variations from normality had been recognised, particularly in the animal kingdom. Before radiography, the knowledge of human congenital anomalies, apart from gross and visible anomalies, was limited to those found by anatomists at dissection.

Köhler's book the *Lexikon der Grenzen der Normalen und der Anfänge des Pathologischen im Röntgenbilde*[35] was published in 1910, and it has gone through numerous of German editions. For writing the book, Köhler received the highest radiological award in Germany, the Rieder Gold Medal. The book was enormously influential and became an immediate classic. Instead of reproducing radiographs, the book was illustrated using line drawings (Figure 4.7). The book was translated into many languages, and was called the "Bible of Röntgen-Rays". Köhler said that his book "is a guide to the diagnosis of those tissues which exhibit or appear to exhibit slight and not particularly noticeable divergencies from the normal anatomical picture". When Köhler wrote his book in 1910, there was no such work in the radiological literature, and the atlases of pathological anatomy showed only gross abnormalities that were easy to recognise. Köhler emphasised his primary aim was to avoid confusing quite normal findings with pathological ones. So as an example, at that time the fabella behind the knee was being mistaken for an intra-articular body resulting in unnecessary surgery.

The book was translated into English by Arthur Turnbull in 1931 appearing as *Röntgenology, The borderlands of the Normal and Early Pathological in the Skiagram*,[36] with a second edition appearing in 1935.[37] In the preface to the 2nd English edition, the famous American radiologist James Thomas Case (1882–1960) from Chicago indicated the usefulness of the book not only to clinicians, but also "to physicians and lawyers whose work brings them in contact with problems on legal medicine". Case wrote that "how many foolish actions would be avoided and unjust

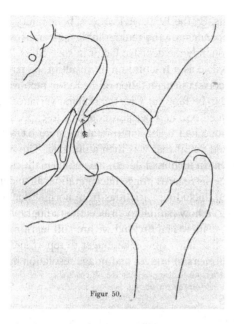

Figur 50.

FIGURE 4.7 The normal anatomy of the hip joint, a line diagram from 1910 by Alban Köhler. From Köhler, 1910, see Note 35. Public domain.

decisions righted by a sufficient dissemination of the knowledge of developmental appearances in the radiogram". He then described a

> ridiculous damage (legal) suit over an alleged fracture of the spine, allowed as a just claim in a high court of law. The deciding testimony was that of a surgeon who declared the radiograph clearly demonstrated a fracture, whereas in reality was a long-standing hypertrophic osteoarthritis with huge osteophytes almost uniting the lumbar vertebrae into one bony mass; and what he interpreted as a fracture was in reality only a small island of calcification just separating two of the opposing bony outgrowths.

In this instance, faulty evidence based on ignorance leads to a serious miscarriage of justice.

SEBASTIAN GILBERT SCOTT

Sebastian Gilbert Scott (1879–1941) was the director of the radiologic department of the London Hospital and was interested in medical jurisprudence and skeletal variations. His influential book *Radiology in Relation to Medical Jurisprudence* was published in 1931.[38] By the 1930s, many claims for compensation were being made in the UK at Common Law or under the Employer's Liability and Workmen's Compensation Acts of 1905 and 1925. Scott held that large sums were being paid out each year that would not have been awarded if there were more awareness of skeletal variations and other conditions that made accurate diagnosis difficult, even for experienced radiologists. By the 1930s, evidence of bone injury in medico-legal cases was almost entirely dependent on the radiographic appearances, and the evidence of the radiologist was frequently the decisive factor in such cases. The misinterpretation of a supernumerary ossicle as a fracture might result in the payment of a large sum in compensation. The correct interpretation of the radiograph was therefore essential, and the radiologist had to be familiar with the skeletal variations that may simulate a fracture. As Scott pointed out, and perhaps contrary to expectations, improvements in radiographic technique had made interpretation more difficult instead of easier because of the increased detail that was then attainable. This was because with the "continued improvement in technical detail, new abnormalities are bound to be met with as time goes on". So as regards image interpretation, an understanding of normal appearances is required, including variations from normal that are not pathological, and also understanding of how pathology affects the radiographic appearances. This knowledge took many years to acquire and we are still learning, and as an example, the images from the CT scanner now take longer to report and are more difficult to interpret as both the numbers of images and image resolution have increased.

THEODORE KEATS

The work of Alban Köhler has been continued by Theodore Eliot Keats (1924–2010) from Charlottesville in Virginia. His *Atlas of Normal Roentgen Variants that May Simulate Disease* first appeared in 1973 and is currently in its ninth edition.[39] The book is a modern classic and its presence in most, if not all, radiology departments is a witness to its value. As each new imaging technique develops, the normal and abnormal appearances need to be learnt afresh. Keats pays tribute to Köhler, and also to the pioneer paediatric radiologist John Caffey (1895–1978), as pioneers in the field of skeletal Röntgen variants. The publication of Caffey's book in 1945 was a landmark[40] and whilst the text illustrated pathology as well as possible bearing in mind the limited techniques of the time, significant space was devoted to discussing normal anatomy and anatomical variations. Caffey felt that the radiological findings were often seriously misleading when the radiologist was not familiar with normal variations and with the limitations of the technique. Caffey recorded his indebtedness to Thomas Morgan Rotch (1849–1914), who was Professor of Pediatrics at Harvard University and a pioneer of neonatal care. Rotch's book[41] on radiography in children published in 1910 is fascinating and gives a brilliant account of living anatomy and pathology with many high-quality radiographs. In this book, Rotch stressed the importance of mastering normal appearances before interpreting the abnormal, and he also carefully correlated clinical findings with the X-ray findings he depicted. The book was based on his experience of Boston Children's Hospital and it is remarkable that by 1910 he had examined more than 2,300 children.

FANTASY SURGERY AND ARBUTHNOT LANE'S DISEASE

If the nature of normality is not appreciated in clinical practice, there is a danger of medical intervention for non-existent conditions. For example, a lack of understanding of the normal anatomy and physiology as shown on radiography may lead, as Ann Dally has indicated, to the "fantasy surgery" for dropped organs (visceroptosis and floating kidneys) and chronic intestinal stasis.[42] Dally states: "This was the surgery of fantasy. Its rationale lay in the mind of the surgeon and the public concerning the origin of symptoms and how to cure them". However, we need to be careful that we are not guilty of what C.S. Lewis called chronological snobbery. Regarding his viewpoints, Lewis exclaimed: "I still had all the chronological snobbery of my period and used the names of earlier periods as terms of abuse".[43] It is all too easy to be blind to our own failings and overly critical of those in the past. There are fashions today as there were in the past, and do we know how posterity will judge us? As imaging becomes increasingly common, the interpretation of more minor abnormalities becomes increasingly difficult.

There is a significant difference between the anatomy as learnt on the cadaver and that seen in the living. To determine the significance of any difference required experience and even by the 1930s such experience was still limited. Dropped organs or visceroptosis associated with a low position of the colon and a variety of symptoms had been described by Frantz Glénard (1848–1920) in 1885, and the introduction of radiology increased interest in the condition. In 1915, William F Braasch (1878–1975) from the Mayo Clinic in Rochester wrote his influential book based on his experience of retrograde pyelography.[44] Braasch improved the cystoscope and developed ureteric catheters. In his book Braasch discusses the movable kidney and noted that "The condition is usually accompanied by functional nervous disturbances which are reflected by a series of subjective symptoms that may render it difficult to identify any actual pain which might result from renal excursion". Braasch illustrates the mobile kidney with many examples and discusses the indications for surgical intervention. His conclusions are quite conservative and guarded; however, he does note that "subjective data may be so distinct as to warranty an operation in selected cases". Almost 25 years later, William Lower and Bernard Nichols from the Cleveland Clinic in Ohio published their definitive account of the radiology of the urinary system.[45] Lower and Nichols illustrated multiple examples of renal ptosis. The cases of ptosis show angulation of the ureter and a variable degree of hydronephrosis. Many of the cases illustrated look essentially normal to modern eyes. The book also illustrates radiographs showing wire sutures that were used to fix the kidneys in position (Figure 4.8). In reality, many surgeons were doubtful about the significance of the findings of renal ptosis and Charles Jennings Marshall (1890–1954), a surgeon from Charing Cross Hospital and Bromley & District Hospital in England, stated in his influential book on chronic abdominal pain that "It is dubious whether symptoms may be attributed to a mobile kidney without positive evidence of obstruction or torsion".[46] Marshall was cautious about the overdiagnosis of visceroptosis generally and commented that "One wonders how many thousands of individuals are wearing an abdominal belt, with no benefit apart from moral support to the serious atrophy of their abdominal musculature, simply because a radiogram has shown a festoon of colon in the pelvis!"

There would seem to have been more evidence for, and belief in, the harmful sequelae of chronic intestinal stasis. The condition was promoted by Sir William

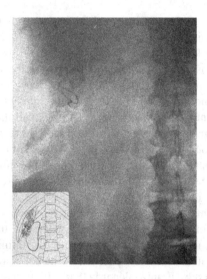

FIGURE 4.8 A plain film showing the right kidney well anchored to the 12th rib by a wire suture. The gallbladder is full of stones. From Lower and Nichols, 1933, see Note 45. Public domain.

Arbuthnot Lane (1856–1943) who promoted treatment by colectomy. Lane was an immensely innovative and important surgeon and promoted abdominal surgery. During the First World War, Lane organised and opened Queen Mary's Hospital in Sidcup, which pioneered reconstructive surgery. As a result of his services to surgery, Lane was made a baronet in 1913. It is overly simplistic to portray Lane as a proponent of "Fantasy Surgery". In the 1920s, Lane founded the New Health Society to promote good health, including making fruits and vegetables as abundant as possible. In the 1920s, Lane was derided by many in the medical protection for his views on autointoxication and his views on the value of colectomy for constipation fell out of favour. Lane recorded his indebtedness to the radiologist Alfred E Jordan as being the first to apply the resources of radiology to the study of stasis. Whilst this is not exactly true, Jordan's book of 1923,[47] *Arbuthnot Lane's Disease* was very influential, with a second edition appearing in 1926.[48] The book is profusely illustrated and makes interesting reading. A chapter on stasis and the prevention of cancer resonates with the work of Dennis Burkitt (1911–1993). Burkitt noted differences in diseases patterns between rural Africa and in the developed West and made a correlation between dietary fibre and health.[49] Burkitt's 1979 book on dietary fibre[50] was an international bestseller and he was a captivating speaker, resembling Arbuthnot Lane in this respect. Both Lane and Burkitt communicated directly to the public. Whilst Lane perhaps made excessive claims for the medical consequences of colonic stasis and was excessive in his recommendations of surgical treatments, there is more to his work than can easily be dismissed as "fantasy surgery".

MODERN ANATOMICAL STUDIES

RADIOLOGICAL ANATOMY

The development of modern cross-sectional imaging since the 1970s has resulted in a flourishing of anatomical knowledge. In 1993, the UK pioneer of CT, Adrian Dixon, with others published their findings on the position of the normal vermiform appendix at computed tomography.[51] The authors noted that the conventional uninflamed appendix may not be most commonly situated in a retro-caecal position, as reported in surgical and post-mortem studies, but was most commonly seen in a retro-ileal position. Indeed, the concept of normal anatomy becomes interesting, and are we referring to classical anatomy as described in anatomy textbooks or are we referring to the living anatomy as shown by radiology? The concept of normal anatomy may indeed be a statistical one, and in a recent study of the venous drainage of the left liver,[52] it was found that only 26.5% of patients had conventional venous drainage. These anatomical variations become important when invasive procedures, either surgical or radiological, are being considered.

In 2006 and 2007, a series of radiological anatomy books were published by Amirsys focusing on the anatomy that is visible on imaging studies. In his foreword to one of the volumes, Michael Huckman[53] recalls his medical student days in the 1950s and learning anatomy from cadavers in the traditional manner using manuals of dissection and having no idea of the importance of what he was learning. In the 1960s and 1970s, the radiology that he experienced was two-dimensional. It was only with the development of the CT and MRI scanners in the 1970s and 1980s that internal anatomy was demonstrated directly without the use of a scalpel, in a detail that was previously undreamed. Huckman's anatomic images in this new book look at human anatomy *in the projections radiologists use* (his emphasis), and this is important "with the bulk of diagnoses today being made in the radiology department". The level of anatomical detail shown in modern imaging is now so detailed that many, including Huckman, have recommended that medical imaging be used for teaching anatomy to medical students. The book discusses the question as to the nature of the boundary between classical anatomy and medical imaging; however, this has been a controversial area since the earliest days of radiology.

Medical imaging is increasingly used for the teaching of anatomy. The replacement of dissection by medical imaging in medical schools has taken place for a variety of reasons. So, the Peninsular Medical School[54,55] teaches that since we encounter anatomy in clinical practice through living and surface anatomy and medical imaging, it would be best to teach students anatomy in these settings right from the beginning of their studies. However, Gunderman and Wilson[56] reviewed arguments about how radiology and anatomical dissection can work synergistically to create a level of understanding that is difficult to achieve by either method alone. Dissection of a body will give an insight into the nature of mortality and the human condition that medical imaging alone cannot provide. Part of the reason that dissection remains valuable is that in order to appreciate medical imaging, a prior knowledge of anatomy is presupposed.[57]

Cadaveric dissection offers an active and hands-on exploration of human structure, and provides deep insights into the meaning of human embodiment and mortality, and can also offer a rite of passage into the medical profession. As Gunderman and Wilson emphasise, despite its important strengths, radiology cannot be simply substituted for cadaveric dissection, and so the best models for teaching gross anatomy will surely use both approaches. The combination of dissection and radiology will therefore allow the student to develop a clinically appropriate and useful three-dimensional image of the body.

NOTES

1. Caffey, J. 1945. *Pediatric X-Ray Diagnosis. A Textbook for Students and Practitioner of Pediatrics, Surgery and Radiology.* Chicago: The Year Book Publishers.
2. Thomas, A.M.K. 2016. Vesalius, Röntgen and the origins of modern anatomy. *Vesalius,* 22, 79–91.
3. Röntgen, W.C. 1896. *On a New Kind of Rays.* Translated by Arthur Stanton (Arthur Schuster) from Sitzungsberichte der Würzburger Physik-medic. Gesellschaft, (1895) *Nature,* 53, 276–274.
4. Holland, C.T. 1923. X-rays and diagnosis. *Journal of the Röntgen Society,* 19, 1–25.
5. Holland, C.T. 1937. X-rays in 1896. *Liverpool Medico-Chirugical Journal,* 65, 61.
6. Poland, J. 1898. *Skiagraphic Atlas Showing the Development of the Bones of the Wrist and Hand.* London: Smith, Elder, & Co.
7. Todd, T.W. 1937. *Atlas of Skeletal Maturation (Hand).* St Louis: C V Mosby.
8. Greulich, W.W., Pyle, S.I. 1959. *Radiographic Atlas of Skeletal Development of the Hand and Wrist.* 2nd ed. California: Stanford University Press.
9. Tanner, J.M., Whitehouse, R.H., et al. 1976. *Assessment of Skeletal Maturity and Prediction of Adult Height: TW2 Method.* 2nd ed. Cambridge: Academic Press Inc.
10. Patil, S.T. 2019. *Assessment of Bone Ages in Indian Children: Applicability of Greulich and Pyle Standards to Indian Children for Determination of Skeletal Age.* Chişinău: Lambert Academic Publishing.
11. Poland, J. 1898. *The Traumatic Separation of the Epiphyses.* London: Smith, Elder, & Co.
12. Morton, W.J. 1896. *The X-Ray or Photography of the Invisible and Its Value in Surgery.* London: Simpkin, Marshall, Hamilton, Kent & Co. Ltd.
13. Ironside Bruce, W. 1907. *A System of Radiography with an Atlas of the Normal.* London: H K Lewis.
14. Ironside Bruce, W. 1924. *A System of Radiography with an Atlas of the Normal.* 2nd ed. Ed. J. Magnus Redding. London: H K Lewis.
15. Pusey, W.A., Caldwell, E.W. 1903. *The Practical Application of the Röntgen Rays in Therapeutics and Diagnosis.* Philadelphia: W. B. Saunders & Co.
16. Division of Roentgenology. 1918. *United States Army X-Ray Manual.* New York: Paul B. Hoebner.
17. Clark, K.C. 1939. *Positioning in Radiography.* London: Messrs. Ilford: W. Heinemann.
18. Thomas, A.M.K. 2020. "Our Katie": Kathleen Clara Clark, MBE, MSR, Hon. FSR, Hon. MNZSR, FRPS (1898–1968). *The Invisible Light,* 47, 26–38.
19. Waldby, C. 2000. *The Visible Human Project, Informatic Bodies and Posthuman Medicine.* London: Routledge.
20. Headsman. 2013. 1993: Joseph Paul Jernigan, visible human project subject. https://www.executedtoday.com/2013/08/05/1993-joseph-paul-jernigan-visible-human-project-subject/ (accessed 17 February 2021).

21. Nutton, V. 1996. The rise of medicine. In: *The Cambridge Illustrated History of Medicine*. Ed. R. Porter. Cambridge: Cambridge University Press, 52–81.
22. Porter, R. 1996. Medical science. In: *The Cambridge Illustrated History of Medicine*. Ed. R. Porter Cambridge: Cambridge University Press, 154–201.
23. National Library of Health. The visible human project. https://www.nlm.nih.gov/research/visible/visible_human.html (accessed 17 February 2021, last reviewed 9 July 2019).
24. Doby, T. 1976. *Development of Angiography and Cardiovascular Catheterisation*. Littleton: Publishing Sciences Group, Inc.
25. Anon. 1896. The new photography in Sheffield. *British Medical Journal*, 1, 495–496.
26. Orrin, H.C. 1920. *The X-Ray Atlas of the Systemic Arteries of the Body*. London: Bailière, Tindall and Cox.
27. Laubry, C., Cottenot, P., Routier, Heim de Balsac R. 1939. *Radiologie Clinique du Coeur et des Gros Vaisseau*. Paris: Masson et Cie.
28. Veiga-Pires, J.A., Grainger, R.G. 1982. *Pioneers in Angiography: The Portuguese School of Angiography*. Lancaster: MTP Press Ltd.
29. Moniz, E. 1931. *Diagnostic des Tumeurs Cérébrales et épreuve de l'Éncephalographie Artérielle*. Paris: Masson et Cie, Éditeurs.
30. Guy, J.M. 1988. A. E. Barclay and angiographic research in Oxford. *British Journal of Radiology*, 61, 110–1114.
31. Barclay, A.E., Franklin, K.J., Prichard, M.M.L. 1944. *The Foetal Circulation and Cardiovascular System, and the Changes that They Undergo at Birth*. Oxford: Blackwell Scientific Publications, Ltd.
32. Trueta, J., Barclay, A.E., Franklin, K.J., Daniel, P.M., Prichard, M.M.L. 1947. *Studies of the Renal Circulation*. Oxford: Blackwell Scientific Publications, Ltd.
33. Barclay, A.E. 1951. *Micro-Arteriography: And Other Radiological Techniques Employed in Biological Research*. Oxford: Blackwell Scientific Publications.
34. Thomas, A.M.K. 2010. Early forensic radiology. In: *Der durchsichtige Tote – Post mortem CT und forensische Radiologie*. Eds. H. Nushida, H. Vogel, K. Püschel, A. Heinmann. Hamburg: Verlag Dr. Kovač, 103–113.
35. Köhler, A. 1910. *Lexikon der Grenzen des Normalen und der Anfänge des Pathologischen im Röntgenbilde*. Hamburg: Lucas Gräfe & Sillem.
36. Köhler, A. 1931. *Röntgenology: The Borderlands of the Normal and Early Pathological in the Skiagram*. London: Ballière, Tindall & Cox.
37. Köhler, A. 1935. *Röntgenology: The Borderlands of the Normal and Early Pathological in the Skiagram*, 2nd English ed. London: Ballière, Tindall & Cox.
38. Scott, S.G. 1931. *Radiology in Relation to Medical Jurisprudence (Employers' Liability and Workmen's Compensation Acts)*. London: Cassell & Co.
39. Keats, K.E., Anderson, M.W. 2012. *Atlas of Normal Roentgen Variants that May Simulate Disease*, 9th ed. Amsterdam: Mosby.
40. Caffey, J. 1945. *Pediatric X-Ray Diagnosis: A Textbook for Students and Practitioner of Pediatrics, Surgery and Radiology*. Chicago: The Year Book Publishers.
41. Rotch, T.M. 1910. *Living Anatomy and Pathology: The Diagnosis of Diseases in Early Life by the Roentgen Method*. Philadelphia: J B Lippincott Company.
42. Dally, A. 1996. *Fantasy Surgery 1880–1939*. Amsterdam: Rodopi.
43. Lewis, C.S. 1955. *Surprised by Joy: The Shape of My Early Life*. London: Geoffrey Bles.
44. Braasch, W.F. 1915. *Pyelography (Pyelo-Ureterography): A Study of the Normal and Pathological Anatomy of the Renal Pelvis and Ureter*. Philadelphia: W. B. Saunders Company.
45. Lower, W.E., Nichols, B.H. 1933. *Roentgenographic Studies of the Urinary System*. London: Henry Kimpton.

46. Marshall, C.J. 1938. *Chronic Diseases of the Abdomen: A Diagnostic System*. London: Chapman and Hall.

47. Jordan, A.C. 1923. *Chronic Intestinal Stasis (Arbuthnot Lane's Disease): A Radiological Study*. London: Henry Frowde.

48. Jordan, A.C. 1926. *Chronic Intestinal Stasis (Arbuthnot Lane's Disease): A Radiological Study*. 2nd ed. London: Henry Frowde.

49. Burkitt, D.P. 1973. Some diseases characteristic of modern western civilization. *British Medical Journal*, 1, 274–278.

50. Burkitt, D.P. 1979. *Don't Forget Fibre in Your Diet: To Help Avoid Many of Our Commonest Diseases*. London: Martin Dunitz Ltd.

51. Picken, G., Ellis, H., Dixon, A.K. 1993. The normal vermiform appendix at computed tomography: visualization and anatomical location. *Clinical Anatomy*, 6, 9–14.

52. Cawick, S.O., Johnson, P., Gardner, M.T., Pearce, N.W., Sinanan, A., Gosein, M., Shah, S. 2020. Venous drainage of the left liver: an evaluation of anatomic variants and their clinical relevance. *Clinical Radiology*, 75, 964.e1–964.e6.

53. Huckman, M.S. 2006. Foreword. In: *Diagnostic and Surgical Imaging Anatomy, Brain, Head & Neck, Spine*. Eds. H. Ric. Harnsberger, A.G. Osborn, et al. Utah: Amirsys.

54. McLachlan, J.C., Bligh, J., Bradley, P., Searle, J. 2004. Teaching anatomy without cadavers. *Medical Education*, 38, 418–424.

55. Mclachlan, J.C. 2004. New path for teaching anatomy: living anatomy and medical imaging vs. dissection. *The Anatomical Record (Part B: New Anatomy)*, 281B, 4–5.

56. Gunderman, R.B., Wilson, P.K. 2005. Viewpoint: exploring the human interior: the roles of cadaver dissection and radiologic imaging in teaching anatomy. *Academic Medicine*, 80(8), 745–749.

57. Dijck, J. van. 2005. *The Transparent Body: A Cultural Analysis of Medical Imaging*. Seattle: University of Washington Press.

5 Dangers in the X-Ray Department

In the excitement following the discovery of X-rays with the promise of so many benefits, that the new rays might be harmful was not considered. The dangers experienced in the new X-ray departments were varied. Those that were specific to radiology were electrical accidents, radiation injuries, and the chemical toxicity of photographic processing. It would be a mistake to imagine that medical practice is ever without risk. For example, the pioneer Sidney Rowland (1872–1917) made many contributions to early radiology in the years after 1895.[1] After his early interest in the new photography, Rowland left the practice of radiology and worked as a bacteriologist for the Lister Institute. There is no record of Rowland having been injured by radiation, presumably because he was never involved in providing a radiological service. Unfortunately, Rowland was to die prematurely at the age of 44 of cerebrospinal fever (meningitis) whilst investigating an outbreak of that disease among the troops in Mesopotamia in March 1917.[2] It was not uncommon for either doctors or nurses to contract infections from patients, either then or now.

ELECTRICAL INJURIES

Many early X-ray departments developed from electrical department with their attendant hazards. The use of electricity has always involved a degree of risk, although this might not be appreciated with modern apparatus and electrical accidents are now almost unknown. The earlier X-ray equipment could be quite daunting to use, and the working conditions were unhealthy, with X-ray departments commonly situated in unventilated and damp basements, and protection for the operator from stray radiation and electrical shocks was basic. At the London Hospital, a special trolley was designed by Ernest Harnack, the senior radiographer.[3] The radiologist Sebastian Gilbert Scott described it as a very large and very heavy contrivance with a marble top and a number of accumulators. On one occasion, an obese alcoholic female had been admitted with a possibly fractured kneecap. An X-ray examination was asked for, and in due course, Harnack and his assistant arrived at her bedside with his awe-inspiring X-ray contraption (Figure 5.1). The patient eyed both Harnack and the machine with considerable suspicion mingled with undoubted fear. After considerable preparation, Harnack switched on the "juice" for a "try out". There was the usual explosion and sparks flew in all directions. This was too much for the patient, who leapt out of bed and was finally caught as she was disappearing through the front door in her nightdress. After this exhibition of activity, Scott commented rather dryly that the X-ray examination was now not considered necessary. The frightening nature of the apparatus with its noises and sparks was also observed by Francis

DOI: 10.1201/9780429325748-5

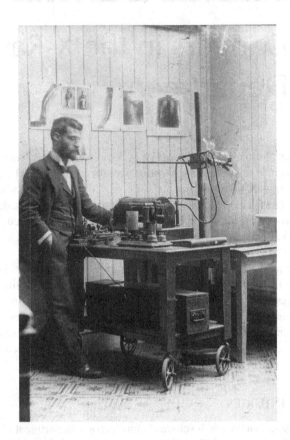

FIGURE 5.1 The "awe-inspiring X-ray contraption" – Ernest Harnack and mobile radiography at the London Hospital. From Sebastian Gilbert Scott's album. Author's collection.

Abbott who was a Red Cross surgeon in the Graeco-Turkish War of 1897. Abbott commented that the local inhabitants looked at the whole affair as the work of "o Διάβολος" (the devil). It made it difficult to take a skiagram (a radiograph) when the subject was constantly crossing himself unless watched.[4] This is in contrast to our modern apparatus when the patient lies on the X-ray couch and nothing perceptively happens when the exposure is made apart from the noise of the rotation of the anode of the X-ray tube. Even as late 1928, Leggett commented[5] regarding X-ray exposures that perhaps the most objectionable feature is the noise and flashing of sparks, which try severely nervous patients and children.

Whilst the early apparatus did involve the presence of exposed electrical cables passing to the X-ray tube, the apparatus itself was of low power. The high-tension current to activate the tube was produced by a 10-inch Ruhmkorff induction coil, activated by a low-tension current from a nest of six Grove cells. Making and breaking of the primary current was accomplished, or was supposed to be accomplished, by the means of a platinum hammer. As already indicated, the X-ray departments at that time were consigned to the region of the cellars or boiler house in the hospital basement. Scott said that high-tension electrical current, which had to be forced through the X-ray vacuum tube, would leak all over the place and would often refuse to pass through the X-ray tube at all. The tubes used were of the early ion or gas variety. This high-tension current caused noisy sparks like miniature lightning, which frightened the patients, and causing electrical shocks to the operator. Many times, particularly on a damp and foggy morning, an hour might have to be spent drying the apparatus and coil with methylated spirit and hot dusters. Dampness was only one of the many trials of the operator in those days. Scott said that the fellow conspirators causing grief to the radiologist were the erratically tempered X-ray tube, the hammer platinum break of the early days which was always sticking, the mercury breaks which exploded periodically, and finally the accumulators. For many years, a nest of Grove's cells, and later on accumulators, were the sole means of supplying the low-tension primary current to the induction coil.

In the absence of electrical and radiation protection, Scott said that it was really a wonder that not more operators were damaged, and that it was probably the weak output of the apparatus at that time that protected them. That this was the case has been demonstrated by Kemerink and others[6] in their review of the topic. They noted that the early apparatus consisted of induction coils, static electricity generators, and high-frequency coils as sources of the high-voltage for the X-ray tube. The static generators of that time delivered currents that were far too weak to threaten human life. The high-frequency coils, whilst they could cause serious burns, would only stimulate muscles and nerves to a limited extent, and this made the induction of ventricular fibrillation highly improbable. In 1919, Gunther from the manufacturers Etablissements Gaiffe-Gallot-Pilon in Paris reviewed various causes of deaths and recommended prudence.[7] He was aware that all users of X-ray apparatus occasionally received electrical shocks, and he noted that although the shocks were painful and that they could result in burns, the longer-term sequelae were minimal.

After 1920, the situation changed and the higher powered transformers now delivered currents above the threshold that was needed to induce ventricular fibrillation. Kemerink was able to document 51 fatal and 62 serious non-fatal electrical accidents, most occurring from 1920 to 1940. The problem was surveyed in 1931 by Wintz for the League of Nations.[8] By 1931, Wintz noted that the high-power transformers in use for diagnostic radiology were deadly dangerous, and that all apparatus then in use was more or less dangerous. Even in the early 1930s, there were still X-ray tubes that were not self-protected from radiation, and were still to be found enclosed in an open protective container (a lead glass bowl), and the unprotected high-tension terminals and cables projected from the ends of the tubes (Figure 5.2). However, even the newer ray-proof tubes that did not need external radiation protection were still dangerous since both ends of the tube were in reach of the patient's hands. Wintz gave an account of various safety devices, including audible and visible signals when the voltage was in use, and also recommended warning notices. It was recommended that specially empowered authorities should carry out regular inspections.

In a typical early 1930s department,[9] the high-tension unit was located in a room adjacent to the diagnostic room and the high-tension leads were brought through large insulators in the wall and coronaless overhead conducting cables passed across

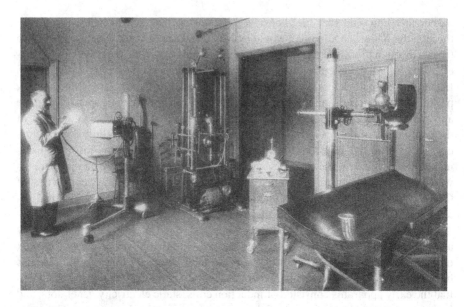

FIGURE 5.2   The X-ray room at St. Elisabeth Gasthuis in Antwerp. The X-ray tube can be seen on the right above the couch and covered with a glass bowl. Old postcard. Public domain. Author's collection.

the room at a height of no less than 9 ft. from the ground (Figure 5.3). The high tension was fed to the tube using spring-loaded reels of insulated cables. High-tension currents would be seen to spray in the form of corona from any sharp points or corners. The corona would cause the nitrogen and oxygen in the air to combine producing nitric oxide, and nitric oxide has complex effects on the body.[10]

There had been a long-term aim to produce entirely shockproof apparatus. The early shockproof apparatus was unwieldy and the radiation output was limited. Shockproof apparatus was developed and gradually introduced in the 1930s and was achieved using shockproof X-ray tubes and cables (Figure 5.4).[11] To achieve a shockproof tube, the ray-proof tube was enclosed in an earthed metal cover. Initially there was air between the shockproof cover and the ray-proof tube insert, which in time was replaced with oil. The fully shockproof tube could only be produced when insulation had been perfected, and shockproof cables could then be run between the transformer and the X-ray tube. The cables were enclosed in oiled paper or rubber and then enclosed in a metallic braid. The braid was then connected to earth. The shockproof cables were then fixed close to the ceiling and passed down to the shockproof tube.

In any review of the newspaper archives for the 1920s and 1930s, many tragic cases of radiological electrical accidents will be found, and even with more modern apparatus care needs to be taken (Figure 5.5). One case was described in 1929, and was particularly tragic since it involved the death of an 8-year-old child.[12] The child was being held by a nurse on one side of the couch and the mother on the other. There was a sudden flash, with the nurse becoming unconscious, the mother thrown into

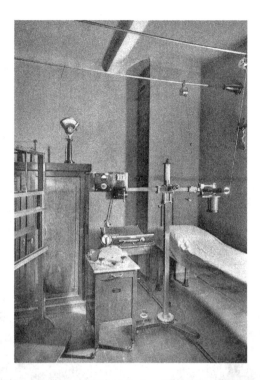

FIGURE 5.3 The X-ray room at the Klinek St-Josef in Esschen, Belgium. The high tension is seen being fed to the tube using spring-loaded reels of insulated cables, from the cables seen passing across the ceiling. Old postcard. Public domain. Author's collection.

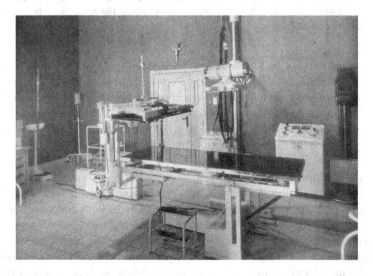

FIGURE 5.4 The X-ray room at the St-Josefkliniek in Bornem, Belgium. The room is now shockproof. Note the tilting X-ray table and simple fluorescent screen in this period before image intensification. Old postcard. Public domain. Author's collection.

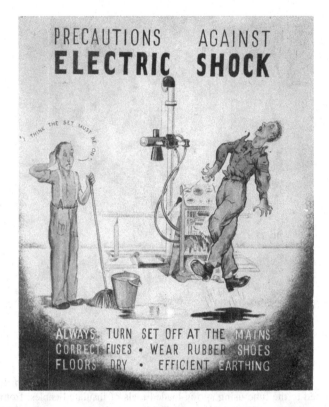

FIGURE 5.5 "Precautions against electric shock. Always turn set off at the mains, correct fuses, wear rubber shows, floor dry, efficient earthing". Early 1940s radiographer educational poster. Author's collection.

the air, and the child being killed. The radiologist thought that the nurse had formed an electric circuit from the high-tension cables through to the child. The nurse had only been in an X-ray room once before and had no idea of any possible risks. The coroner noted that the hospital authorities should consider the wisdom of admitting an inexperienced person into X-ray rooms. It should be remembered that it was not necessary to touch the cables, and at this time before image intensification, the room would have been in complete darkness. The coroner also noted that some manufacturers at that time in their advertising were indicating that excellent radiographs could be produced using their apparatus without any previous training, allowing the practitioner to set up as an expert radiologist. The modern equivalents are MRI accidents with the magnet due to inexperienced staff, and the use of diagnostic ultrasound without adequate training with machines that are easily purchased online.

A further electrical danger of radiology, particularly in the pre-shockproof era, was in the use of radiography in the operating theatre with the risk of explosion. A general anaesthetic could be obtained using the inflammable agent diethyl ether (ether), which could either be administered using a face mask and dropping bottle, or

FIGURE 5.6 "Famous last words. 'You wont mind if I give the Patient more Ether'. Never X-ray a patient when inflammable anaesthetics are being used". Early 1940s radiographer educational poster. Author's collection.

via an endo-tracheal tube. Ether is now obsolete as a general anaesthetic since it is highly inflammable and thereby incompatible with modern surgical and anaesthetic methods, and a risk of explosion (Figure 5.6). Ether is still used as an anaesthetic in some countries because of its low cost and high therapeutic index and also because it has minimal cardiac and respiratory depression.

RADIATION INJURIES

X-rays are invisible light and care needs to be taken (Figure 5.7). In the earliest period after the discovery of X-rays, the possibility of X-ray dermatitis, which is a skin inflammation caused by radiation, was unknown and the sore fingers that radiologists experienced were attributed to the chemicals in the metol-developer. Scott commented that the warning that an overexposure of the skin to X-rays would cause a skin irritation came in time to save any damage to hands. Unfortunately, this was not to be the case. The first person who suffered radiation damage was Emil Herman Grubbé (1875–1960)[13] from Chicago. Towards the end of 1895, Grubbé was working

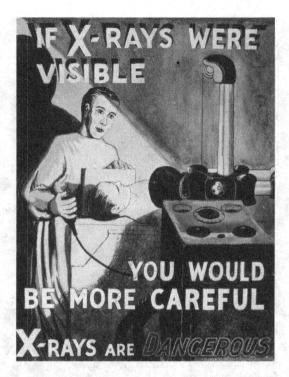

FIGURE 5.7 "If X-rays were visible you would be more careful. X-rays are dangerous". Early 1940s radiographer educational poster. Author's collection.

with a Crookes tube, and by January 1896, he had developed dermatitis of his fingers which he showed to a colleague. These biological effects suggested him a possible therapeutic use and on 29 January 1896, he treated a lady with an advanced breast tumour. Grubbé also used a lead foil to circumscribe the lesion. On 12 December 1896, Wolfram Conrad Fuchs (1865–1908) described his four X-ray burns,[14] noting that they were slow to heal but should be regarded as slight "in comparison with the benefits of this wonderful discovery". He made what were probably the first recommendations ever to prevent future injuries. His suggestions were to shorten exposures to a minimum, to position the tube no less than 12 inches (30.48 cm) from the body, and to rub Vaseline into the skin over the part to be examined. Fuchs made the assertion that X-ray burns were no more dangerous than ordinary burns; however, his later career showed this not to be the case. Fuchs made many contributions to the exciting new speciality of radiology and his importance is shown by his being called to the dying US President William McKinley (1843–1901) after his assassination at Buffalo, New York in 1901, although as it transpired the X-rays were never used. His extensive use of X-rays produced a severe radiodermatitis of the thumbs and fingers which had to be amputated. He suffered from severe pain, and Fuchs was forced to retire in 1905. He developed enlarged axillary lymph nodes which proved to be

FIGURE 5.8   The memorial to the radiology martyrs in the garden of St. Georg's Hospital in Hamburg. Image courtesy of the late Hermann Vogel.

secondary to malignancy, and following multiple operations Fuchs died on 24 April 1907. Sadly, Fuchs was to be one of many more, and his is one of the names on the Martyr's Memorial in Hamburg.

On 4 April 1936, a memorial to the radiology martyrs was placed in the garden of St. Georg's Hospital (Asklepios Klinik St Georg) in Hamburg (Figure 5.8). The monument had been suggested by Hans Meyer from Bremen, who was editor of the journal *Strahlentherapie*. The memorial stone recorded the names of 169 radiation workers who, whilst working with X-rays or radium, had died before 1936. All of the biographies of the deceased were printed in a book which was issued as a parallel supplement to *Strahlentherapie* and Meyer provided brief biographies. Fuch's name was one of 39 US names on the memorial, a surprisingly large number and only to be exceeded by France with 47 names. More names were added to the monument with side stones erected, and by 1959 the number of names recorded was 359. Percy Jones wrote biographies of 26 of the US martyrs in 1936[15] and reading them alongside Meyer's biographies of 1937[16] results in a sense of sadness at the suffering and yet at the same time a gratitude for their contributions to our knowledge of the new subject. This risk discouraged many from taking up X-rays as a speciality, and indeed the family of Kathleen Clara Clark (1898–1968) thought that she was mad when she took up radiography as a profession in the 1920s.[17] It is recorded that Clark was given an entire outfit of clothes made from silk by her parents, since it was believed that these were impenetrable to radiation!

THE LONDON HOSPITAL RADIOGRAPHERS

The first radiographers at the London Hospital suffered terribly from the effects of radiation. As already noted, the biological effects of radiation were entirely unexpected. The effects of radiation were confusing and difficult to explain, and indeed were seen as contradictory.[18] So as examples, warts could be both destroyed and produced, ulcers could be healed and produced (X-ray burns), and with malignancy it was known that rodent ulcers could be cured and yet epitheliomas of the hand were produced. This was very confusing and whilst the reasons may seem obvious to us now, it was not so at the time.

At the London Hospital, Ernest Harnack was the senior radiographer of the lay staff, and he had three assistants. These were Ernest Wilson, Reginald Blackall, and Harold Suggars, and all were badly damaged by overexposure to X-rays. Harnack had been appointed photographer to the hospital with Dr. William Hedley (1841–1930) in overall charge of the electrotherapeutic department. Harnack took the first radiograph that was of surgical value early in 1896, and this was of a needle in a nurse's foot which was successfully removed. The successful development of the work at the London Hospital was apparently due to the enthusiasm and keenness of Harnack, who for many years was responsible for all the X-ray work of the hospital. Harnack was a pioneer at a time when the dangers were unknown. Harnack developed radiation damage in his hands and continued his "good work" on the diagnostic side in spite of the fact that his hands gave him constant trouble and pain up to 1909, when he was pensioned off by the hospital authorities who awarded him a pension of £250.[19] It is remarkable that although his sufferings were terrible, he stuck so resolutely to his work. The hospital authorities fully appreciated how Harnack's labours and keenness had helped the London Hospital to become one of the leading X-ray departments of that time.

Ernest Wilson joined the radiographic staff of the department as a lay assistant in 1899. He was a first-class photographer and was vested with the responsibility of clinical photography as well as radiography. Harnack became responsible for the ward work and Wilson for the outpatient cases. Wilson was described as not being robust, and in 1898 had tubercular glands removed from his neck, and he also suffered from dental sepsis. Wilson's hands were eventually badly damaged from overexposure to the rays (Figure 5.9). Scott thought that this must have been due to gross carelessness, for in spite of the lack of protection around the X-ray tube in those days, Harnack's hands must have already shown signs of dermatitis at this time, and this should have been a warning to Wilson. It should be noted that X-ray protective gloves were not in general use until later. Scott thought that most of the damage to the hands of the lay staff was inflicted during the prolonged exposure to the rays necessary while removing foreign bodies, usually needles, under the fluorescent screen which was held for many hours each day. Wilson's hands showed signs of dermatitis within a few months of his starting work. By 1900, he had developed whitlows at the base of every nail, and in June 1904 the terminal phalanx of his right middle finger was removed. Further surgeries were performed in 1906 with no more surgery until 1910. There had been some improvement in his general health with an attempt at repair during several months of holiday. In 1910, there was evidence of a tumour,

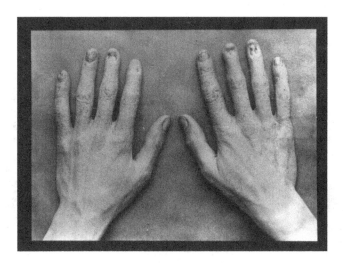

FIGURE 5.9   Wilson's hands were eventually badly damaged from overexposure to the rays. Author's collection.

an epithelioma. Wilson did not survive long after his retirement in 1910 and he died on 1 March 1911 at the age of 40 having developed secondary malignant deposits in his axillary lymph glands. At this time, the causes of the changes in the fingers were ill-understood. So, Scott speculated as to whether the changes were the result of necrosis in conjunction with nerve damage, were they the result of an alteration in the vascular and nervous nutrition of the bone, or were the changes malignant from the onset? There is a poignant set of radiographs that Wilson took of his fingers showing the progressive damage (Figure 5.10a–d). Even though these images were taken over 110 years ago, they are still disturbing and we can imagine his feelings as he recorded the progressive changes.

Scott described Reginald Blackall as another valuable member of the lay X-ray staff. He promoted X-ray therapy in the same way that Harnack had promoted X-ray diagnosis. He joined the treatment side of the department under Dr. Sequeira in 1900 and was the radiographer in charge when Scott was appointed to the London Hospital.

After 23 years of loyal service to the Hospital, Blackall also succumbed to complications secondary to X-ray dermatitis (Figure 5.11). Blackall had 15 operations and developed progressive injuries. By 1903, he was losing his fingernails, and then his fingers, followed by both hands because of the development of malignancy. Blackall remained cheerful and "not down and out by a long chalk".[20] Once again Scott commented that there must have been some degree of carelessness in exposing his hands to X-rays when the dangers were already well known, and noted that "Needless to say, the hospital authorities were in no way to blame for these tragedies; they were always only too willing to do anything for the welfare of their employees". It is difficult to imagine anyone taking up this position today. The Carnegie Trust awarded Blackall a Hero Testimonial and a £75 annuity in 1923. He had retired in

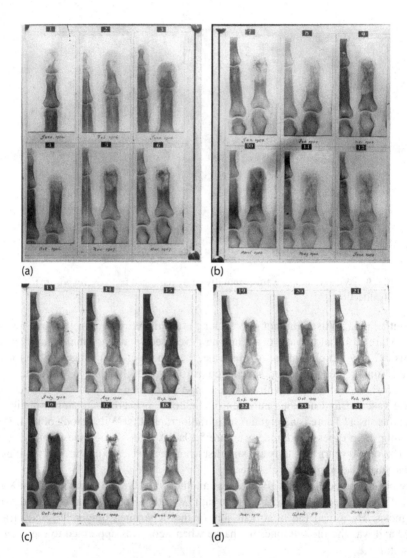

(a)   (b)

(c)   (d)

FIGURE 5.10 (a) Radiographs of Ernest Wilson's hands. Images 1–6, June 1904 to December 1907. (b) Radiographs of Ernest Wilson's hands. Images 7–12, January 1908 to June 1908. (c) Radiographs of Ernest Wilson's hands. Images 13–18, July 1908 to June 1909. (d) Radiographs of Ernest Wilson's hands. Images 19–24, September 1909 to June 1910. Author's collection.

1920 having worked for 23 years when the hospital authorities had told him that it would not be safe for him to continue his work.[21]

In his history of the London Hospital of 1910,[22] Morris noted that some of the operators in the X-ray department had terribly damaged hands. Morris stated that the dangers of radiation were known within three years with all of the original workers

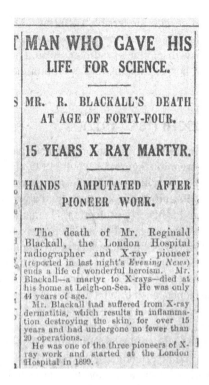

MAN WHO GAVE HIS
LIFE FOR SCIENCE.

MR. R. BLACKALL'S DEATH
AT AGE OF FORTY-FOUR.

15 YEARS X RAY MARTYR.

HANDS AMPUTATED AFTER
PIONEER WORK.

The death of Mr. Reginald Blackall, the London Hospital radiographer and X-ray pioneer (reported in last night's *Evening News*) ends a life of wonderful heroism. Mr. Blackall—a martyr to X-rays—died at his home at Leigh-on-Sea. He was only 44 years of age.

Mr. Blackall had suffered from X-ray dermatitis, which results in inflammation destroying the skin, for over 15 years and had undergone no fewer than 20 operations.

He was one of the three pioneers of X-ray work and started at the London Hospital in 1899.

FIGURE 5.11   Newspaper cutting of 1 December 1925 announcing the death of Reginald Blackall, from an album made by his colleague Sebastian Gilbert Scott. Author's collection.

having damaged hands which required multiple amputations, and also that nothing had been found to cure the malady. Somewhat surprisingly, Morris said that in 1910, that modern workers ran no risk. The X-ray tubes were surrounded by lead glass bowls from 1902 and from 1910 the operator could only make an exposure from within a lead-lined cabinet. The London Hospital obviously had better radiation safety practice by 1910; however, the claim that there was no longer any risk was sadly not the case.

While these men may be looked upon as pioneers of radiology, they were consequently up against unknown dangers. Nevertheless, those whom Scott knew personally resented being called X-ray martyrs, as the press was now calling them. A martyr would know what would happen to them, whereas these radiation victims did not know until it was too late. Scott commented that X-ray dermatitis of the X-ray operator was fortunately now a thing of the past.

And yet why did so many of the pioneers continue with their work even though they knew it was destroying them? The early radiographers were victims since their sufferings were unexpected, and yet were martyrs since even when they became aware of the dangers, they continued working. It was said of another American pioneer martyr, Mihran Kassabian (1870–1910), that his enthusiasm for his work never faded right up to his end despite his terrible sufferings. During the eight years of Kassabian's sufferings, he minimised any public awareness of his issues since he was concerned that the new discipline of radiology would be brought into disrepute. When Kassabian entered hospital for his first operation, he even used an assumed name.

DR. EMILIO TIRABOSCHI (1862–1912)

Emilio Tiraboschi was an Italian physician and pioneer radiologist working at the Ospidale Maggiore of Bergamo (Figure 5.12).[23] In 1895, the year that X-rays were discovered, his wife died leaving him with the care of their two children. He started working with X-rays in 1898 and was said to have used a very hard X-ray focus tube, and to take very little care for his own protection. He made only a small amount of money from his work in radiology, and in addition worked teaching physical and natural sciences at the College of St. Alexander in Bergamo, where he also worked as a doctor. Early radiology was not highly paid, and the college was Tiraboschi's best source of income. As early as 1907, after nine years of working in radiology, radiodermatitis had developed on the left half of his face and left hand. He also lamented his visual decrease in both eyes. His hair was white and he noted that his black hair was returning. Tiraboschi joked that "I'll be back young" (*Torno giovane*), and he had a stoic attitude to his sufferings. In spite of his illness, he is said to have had a cheerful character and he celebrated everyone. He was certainly aware of the consequences of his work with radiation and that he would suffer. He died at the young age of 49 having worked with X-rays for 14 years, and his elderly mother died shortly after. He had developed a loss of strength and three years before his

FIGURE 5.12   Dr. Emilio Tiraboschi (1862–1912). Public domain.

death had collapsed and been off work for a period of six months. His nutritional state remained good, but he showed increasing anaemia and pallor. He also developed bleeding from his gums. He kept working until the evening before his death and somewhat poignantly by the side of his bed was found a box of Blaud's pills. Blaud's pills were a mixture ferrous sulphate and potassium carbonate, and would unfortunately be useless for the treatment of radiation-induced anaemia.[24] Two of his colleagues published a detailed account of his post-mortem with the permission of his family.[25] His death caused international concern and was generally deplored; however, his death did enable a detailed analysis of the cause of his death.[26]

The autopsy results of Tiraboschi were interesting. The lesions on his face and hand were described as being slight. The spleen was small and hard and the glandular tissue was mostly destroyed with replacement by connective tissue. The ribs showed loss of the normal medullary cells. The testes were small and internally soft and yellow. Microscopically, the cells showed loss and replacement with fibrous tissue. The different cellular elements were difficult to distinguish. Tiraboschi died of radiation-induced aplastic anaemia with changes in the bone marrow, spleen, and testes. It is interesting that he did not develop more severe changes in his hands. Perhaps this was because he did not use his own hands to test the quality of the X-rays as was so common in the early pioneers, and the second generation of pioneers more commonly suffered from aplastic anaemia. Alternatively, it might be that his use of a hard X-ray tube limited his exposure to the biologically more harmful soft X-rays. The results of the autopsy were discussed at the First Congress of Radiology and presented to a very emotional audience. A memorial tablet to Emilio Tiraboschi is in the Radiological Pavilion of the old Ospedale Principessa di Piemonte in Bergamo. The tablet was unveiled in 1933 by the hospital management and a eulogy was given by his colleague Achille Viterbi.[27] Tiraboschi's name was one of nine Italians that were recorded on the Martyr's Memorial in Hamburg in 1936. Fortunately, by the 1930s, radiation protection in Italy was well developed.

THE FRENCH RADIATION MARTYRS

When the radiation Martyrs Memorial was erected in Hamburg in 1936, there were 47 French names out of the total of 169 names. This is 28% of the total which seems very high, and in contrast there were only 14 British names. It is sad reading through the lives and sufferings of these French pioneers. The Abbé Tauleigne (1870–1926) suffered in part from his exposure during the Great War. He continued to work in spite of progressive injuries to his arms. The Carnegie Foundation made awards to the Abbé in 1923 and 1924 and he received their silver medal and a sum of 5,000 francs. Jean Alban Bergonié (1857–1925) was a French oncologist, and the regional cancer research centre the Bergonié Institute of Bordeaux was founded by him.[28] His medal of 3 August 1923 shows his missing arm, Bergonié was to have amputations of both arms.

A prominent victim was the radiologist and double amputee Charles Vaillant (1872–1939) (Figure 5.13). His sufferings were much reported in the international media; the concern was such that in late 1922, Queen Elizabeth of Belgium visited

FIGURE 5.13 Charles Vaillant (1872–1939) following the amputation of his right arm. Newspaper cutting in an album made by Sebastian Gilbert Scott. Author's collection.

him at the Lariboisière Hospital where he had recently had his left forearm amputated.[29] Vaillant suffered progressive disease with a series of amputations and much suffering; however, he refused to be downcast and would cheerfully tell visitors "What is my life compared with the health of millions of my fellow beings for whom I gladly give it!".[30] Vaillant had started X-ray work when he was 23 years old in 1896 at the Lariboisière laboratory. He was described as tremendously enthusiastic with the hope that X-rays would assist in the abolition of disease. However, as he progressed as a radiologist, he became aware that there were problems when the nails of his hands started to fall off. Four years later, the index finger of his right hand was amputated. His skin became dry and burned and ulcerated, and in 1912 his left middle finger was amputated. Vaillant said that the one time that he lost hope was after this second amputation. Amputations continued at intervals of between four months to a year. Vaillant continued work following his amputations, and there is a poignant image of Vaillant signing the Golden Book of the City of Paris with a prosthetic arm on 22 February 1923. Even without a left hand, he could still use the stump to push down a lever. His amputations did not interfere with his teaching and he was a popular and sought-after lecturer. Following the amputation of his left hand, he was visited by the French President Raymond Poincaré (186–1934) who wanted to cheer him up. Vaillant was awarded the Légion d'honneur, which is the highest French order of merit, and the Prix Auddeford given to those injured in the performance of scientific duties, with its associated 15,000 francs. His disease was

FIGURE 5.14 Charles Vaillant (1872–1939) following the amputation of both arms and receiving the Carnegie Medal. Newspaper cutting in an album made by Sebastian Gilbert Scott. Author's collection.

relentless and his left arm was amputated and a poignant photograph of the patient in bed was circulated. His recovery was slow and he was unable to return to work. Unfortunately, the disease in his right arm progressed and he had a partial amputation of his right arm (Figure 5.14). It was noted by his colleagues at the Lariboisière Hospital that there were insufficient resources in France at that time to fund research into radiation protection and that further victims, such as Bergoiné and Vaillant who had a "blind sacrificial devotion to their fellows", would occur. What is remarkable about Vaillant are the number of accounts of his serenity in the face of his sufferings.

There were numerous fatalities in the 1920s, and in 1924 an X-ray fund was announced for all French X-ray victims and their relatives.[31] The aim was to obtain official recognition of their services to science and humanity. There were many beneficiaries, and the campaigners wanted victims to be assured of a comfortable income since many were unable to work. This was important since the salaries of French radiologists were not high, and as an example Vaillant only earned the equivalent of US$800 per year. There was also a need for those who have died to receive an official recognition, and a marble slab at the doors of the most important hospitals inscribed with the names of all victims was suggested.

THE BRITISH 1915 PROTECTION REGULATIONS

In November 1915, in response to progressive concerns, the Röntgen Society issued "Recommendations for the Protection of X-ray Operators" (Figure 5.15), and the seven recommendations are still relevant today.

The recommendations noted that the harmful effects of X-rays were cumulative and would take some weeks or months to develop. Previously, it had been believed that simple resting following exposure was all that was necessary before returning to work. All X-rays were to be seen as harmful and not just the soft or lower energy radiation. The use of the hand for testing the quality of an X-ray tube was condemned. This latter practice had been responsible for many injuries.

The guidelines recommended that all treatments should be performed by a qualified medical practitioner. All tubes should be covered by a protective cover with an adjustable opening for the rays to pass out. The use of a moveable screen with lead-containing gloves and aprons was recommended. In addition, the operators were warned that commercially obtainable radiation shields were often ineffective and

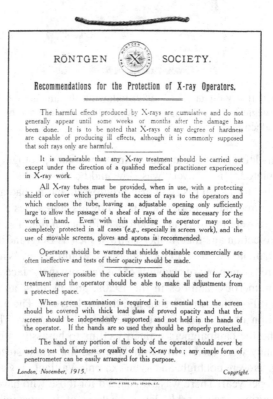

FIGURE 5.15 The Röntgen Society's "Recommendations for the Protection of X-ray Operators" of 1915, courtesy of the British Institute of Radiology. Author's collection.

would require testing. For fluoroscopy, the fluoroscopic screen should be covered with lead glass and independently supported and not held in the hands.

The cubicle system was recommended for X-ray treatments with the operator working from a protected space.

These recommendations of 1915 were in part a summary of current good practice and had been observed by the Anglo-Irish radiologist Florence Stoney (1870–1932)[32] during a visit she had made to the United States just prior to the outbreak of the Great War, specifically to observe up-to-date American practices in radiology, which were widely reported in the British radiological literature.[33] In America, she observed a major concern for radiation protection. Although knowledge of radiobiology at this time was poor, there were some basic principles that had been established. For example, Florence observed that nowhere was the operator left exposed to radiation apart from when performing fluoroscopic examinations. The X-ray control unit was always placed behind a metal screen, which was usually lined with lead, and the X-ray tube could be activated with the operator safely behind the switchboard as is the practice today. However, the apparatus was also not shockproof, and although the wires were simply insulated, touching the tube or cable would result in fatal electrocution. The cables supplying the tubes were run across the ceilings and were attached to the ends of the tube using a rheopore (a connector), and Florence described this as being both elaborate and excellent. She felt that the wires were quite well insulated and would not be constantly leaking current and sparking.

In most hospitals, there was only plain film radiography performed with no fluoroscopy at all. This was because of the dangers of fluoroscopy, and a justified caution about exposing the operators to radiation. In one office Florence observed that the fluorescent screen was viewed via a mirror and the operator stayed behind a lead screen, even for setting the photographic place in position. She thought this was elaborate but very efficient, and the aperture could be narrowed down to include only the actual region that was being observed and thereby obtaining an improvement in detail and in dose reduction.

IRONSIDE BRUCE AND THE DEVELOPMENT OF STANDARDS

It was following the shock at the death in 1921 at the young age of 42 of the radiologist William Ironside Bruce (1876–1921) from Charing Cross Hospital in London that the British X-ray and Radium Protection Committee came into being (Figure 5.16). In a similar manner to Tiraboschi, Ironside Bruce died of aplastic anaemia as was more common with the radiation deaths of that period. He was a well-known figure in British radiology and had made many contributions. With a passionate voice, the writer of his obituary exclaimed:

> Surely some steps, and those not feeble and indefinite, ought to be taken by those of us who are left, in an endeavour to check this appalling loss of life, to say nothing of the loss of health and maiming and disfiguring of X-ray workers.[34]

However, there were many reports in the 1920s of victims of radiation exposure, and the general public became very concerned. The British X-ray and Radium Protection

FIGURE 5.16 Dr. Ironside Bruce: "Noted radiologist falls victim in the prime of life". Newspaper cutting from 23 March 1921, in an album made by Sebastian Gilbert Scott. Author's collection.

Committee was active from 1921 to 1952. The committee recognised three sources of danger to X-ray workers: exposure to radiation, electrical risks from exposed cables, and toxic gasses from corona discharges. In spite of a limited knowledge of radiobiology, the committee defined a series of pragmatically reasonable precautions.[35] Developments were also taking place in other countries and as an example, the Safety Committee of the American Roentgen Ray Society made recommendations in 1922. Significant progress was to be made in radiation protection in the subsequent two decades.

An International Commission on Radiological Protection was conceived at the First International Congress of Radiology (ICR) that was held in London in 1925.[36] Discussions took place on X-ray and radium protection and appliances and the current recommendations were summarised.[37]

In 1928, the Second ICR was held in Stockholm, with the Section for Radiophysics and Medical Electrology being chaired by Rolf Sievert (1896–1966). Sievert was a Swedish medical physicist who worked on the biological effects of ionising radiation. The Radiophysics Section made proposals for protection which were adopted on 27 July 1928. These 41 paragraphs were essentially the British recommendations with minor amendments. Sievert said that the leading figure was George William Clarkson Kaye (1880–1941) who was Superintendent of the Physics Department of the National Physical Laboratory in England. Kaye had devised practical methods for radiation protection which were translated into the standards of the British X-ray and Radium Protection Committee. The ICR recommendations stated that the dangers of overexposure to radiation could be avoided by providing adequate protection and suitable working conditions. The proposals included recommendations regarding working hours, general X-ray recommendations, recommendations for X-ray protection, electrical precautions in X-ray rooms and radium-protective recommendations.[38] Workers in radiology were not to work more than seven hours a day for no more than five days a week, and were to have not less than one month's holiday a year. There was an emphasis on shielding from radiation, although no dose limits were recommended. At that time, there was more emphasis given to the development of units for the measurement of radiation than in radiation protection. The International X-Ray Unit Committee (later known as the International Committee for Radiological Units – ICRU) was proposed at the First ICR in London in 1925. The ICRU officially came into being at the Second ICR in Stockholm in 1928, and the roentgen unit (r and later R) to measure radiation exposure was adopted. In the pre-war period, the main emphasis of ICRP was in the application of the standards that had been set in Stockholm. It was at the meeting in Zurich at the Fourth ICR that the ICRP defined dose limits for workers. The committee said that the current evidence indicated that under satisfactory working conditions, a person in normal health could tolerate an exposure to X-rays of about 0.2 international roentgens per day.[39] During this period, the dose limit was seen, and referred to, as a permissible level of dose. This dose limit was chosen to prevent what became known as non-stochastic effects, which are the effects of radiation that increase in severity with the dose and for which a threshold dose was believed to exist. This approach lasted until 1977, when the former emphasis on dose limits was less important.[40] In 1977, the ICRP recommended a system of dose limitation, with the main purposes being to ensure that no source of exposure is unjustified in relation to its potential benefits or those of any available alternative, and that any necessary exposures are kept As Low As is Reasonably Achievable (ALARA), and that the dose equivalents received do not exceed certain specified limits and that allowance is made for future development. Whilst this principle of ALARA may be seen as a new development, this is far from the case and, as Allan Oestreich has noted, it was Paul Reyher (1876–1934) from Berlin who wrote in 1912 in relation to diagnostic and therapeutic radiology in children *"nur in allerkleinster Dosierung anzuwenden"* (use only the smallest possible dose).[41] This was in the first German textbook of paediatric radiology in which he described early experimental work on the effects of radiation in young animals which led him to conclude that radiologists should be reluctant to expose young children to radiation, and if it was necessary, then only the smallest dose should be used, and this is essentially ALARA and similar to the "Image Gently" campaign. The mission of the Image Gently Alliance is to improve the safe and effective imaging care of children. This also relates to the ethical principle of *"Primum non nocere"* (first, do no harm) as promoted by the Parisian pathologist and clinician Auguste François Chomel (1788–1858).[42] Essentially, radiation should harm neither practitioner nor patient and this was the aim of successive guidelines.

In the 1920s, it became apparent that radiation could cause chromosomal damage and the possibility that there might be a genetic effect became a known reality. Hermann Joseph Muller (1890–1967) worked on the physiological and genetic effects of radiation (or mutagenesis). In 1926, and using the fruit fly (*Drosophila*), Muller found a clear relationship between radiation dose and lethal mutations.[43] This created a sensation at the time. Muller was awarded the Nobel Prize for Physiology or Medicine in 1946 for the discovery that mutations can be induced by X-rays. His Nobel Prize was awarded in the time following both the dropping of the atomic bombs and also the promotion of the peaceful use of radiation or "atoms for peace". In his lecture for his Nobel Prize acceptance Muller stated that no threshold dose of radiation existed that did not produce mutations, and this led to the adoption of the linear no-threshold model of radiation for cancer risks.

Kathren and Ziemer have divided the first 50 years into three phases.[44] These were the pioneer era from 1895 to 1905, when gross changes were recognised with relatively simple means for coping. The dormant era was from 1905 to 1925, when the main concern related to the application of radiation with gains made in technical and biological knowledge. Finally, from 1925 there was the era of progress when radiation protection developed as a science and medical and hospital physics as a speciality was defined. In the United Kingdom, this resulted in the formation of the Hospital Physicists' Association in 1943. William Valentine 'Val' Mayneord (1902–1988) observed that the 1920s and 1930s were particularly favourable for the basic physics and the relevant engineering that was needed for medical applications and many significant advances were made.[45]

HAZARDS TO PATIENTS AND THE ADRIAN COMMITTEE

Whilst there were concerns about radiation injuries to practitioners that were expressed in the 1920s, there was also concern about injuries to patients. It only requires a superficial reading of newspapers of the period to find many legal cases following radiation injury. It should not be imagined that the cases related solely to radiotherapy, and many cases involved diagnostic exposures.

One particularly tragic case from Johannesburg in 1924 involved a claim of £5,000 against a Dr. Hamilton by Herbert Dale, a shaft timberman in a mine.[46] Dale alleged that the injury had been caused by an X-ray examination during which he had sustained burns, and he was now continually suffering. Dale claimed that the burn was the result of a fault in the apparatus. Dale complained of severe pains in his chest and threatened suicide on more than one occasion. During the court discussions, a Dr. Gilevall speaking for the defence stated that the apparatus had been issued with a certificate, and that a man of average intelligence after two hours' instruction or experience could use it appropriately. He also thought that patient idiosyncrasy might result in burns, and that he had seen a case with burns following a normal exposure. Another possible reason for the burns he said were voltage fluctuations. Hamilton believed that the burn was the result of faulty apparatus, and that the injuries were aggravated and their healing retarded by Dale's negligence. Dale said that the X-ray tube in its box was left by the expert who fitted in too near to the couch. A Mrs. Gatewood also claimed to have been injured by Dr. Hamilton. Miss Midler who had been matron of the hospital told the court how extremely painful X-ray burns were for the patient. The judge awarded £2,150 with costs to Dale and said that the fact that the burns were very severe and were caused by diagnostic work was enough to establish a case of negligence and shifted to the defendant the onus of proving that there was none.[47] The judge also said that Hamilton was responsible for having the tube in the correct position and should not try to shift the blame. Hamilton had not taken sufficient care, and that his lack of training as a radiographer did not enable him to use the apparatus successfully.

The lack of training of the operators of X-ray equipment was a concern in the 1920s and in the hands of unqualified practitioners, the X-rays, which were intended to cure diseases, could create new diseases. In 1922 Health Commissioner Copeland in New York announced that the operators of X-ray machines would need to be licensed by the Health Board. The concern was that the use of X-ray equipment by the unqualified was injuring practitioner and patient alike.[48] Copeland announced that forthwith the X-ray laboratory was to be under the charge and care of a licensed doctor or practitioner. There had been a similar concern in Great Britain about unqualified practitioners, and in 1920 the Society of Radiographers had been founded with a qualifying examination for membership.

Whilst it was obvious to the judge in the Hamilton case described above that a burn following a diagnostic exposure was the result of poor practice, what were the effects to the population when radiography was widely used and there was no apparent injury to the individual? This question became particularly important following the work of Muller on the chromosomal effects of radiation. The response in the United Kingdom was to set up the Committee on Radiological Hazards to Patients in 1956 under Edgar Douglas Adrian (1889–1977), Lord Adrian of Cambridge (Figure 5.17).[49] The committee considered both the biological action of radiation on the individual (or somatic effects) and also the possibility of genetic effects, that is the passing on of damage to future generations. It was noted that the effects on the foetus were somatic effects and could be caused by exposure to lower doses. Extensive surveys were carried out in 1957 and 1958 and reports were issued. Of interest is the use of an electronic computer for the analysis of results. Radiography was common, and at that time it was estimated that the average member of the public had a diagnostic X-ray examination every three to four years. The committee made various conclusions, and noted that the bone marrow was one of the most susceptible tissues and that the induction of leukaemia was one of the most important radiation effects. This correlated with the work of Alice Stewart on childhood leukaemia following foetal irradiation.[50] The recommendations were related to the indications for examinations and to radiographic technique. The general recommendations include having a clear indication for the examination, the need to avoid repeats, and an emphasis that the responsibility for radiography was as much the responsibility of the referring doctor as the radiologist. Approximately two-thirds of the recommendations related to good radiographic practice. Finally, the report saw no need for major restrictions in radiological work, and noted that the number and type of examinations must be dictated by the needs of the patient. The adoption of the recommendations would however materially reduce the dose to both bone marrow and gonads.

FIGURE 5.17   The Rt. Hon. Lord Adrian, OM, MD, FRS giving the opening address to the 22nd Annual Congress of the British Institute of Radiology in 1961. Courtesy of the British Institute of Radiology.

The recommendations of the Adrian report were sensible and were approved. Many articles appeared in the literature in the subsequent decades on the biological effects of radiation and many thousands of cancers and cancer deaths were predicted. As Hendee and O'Connor noted, the predictions were based on data arising from the Hiroshima and Nagasaki bomb survivors, and that this population is very different from those exposed to medical radiology.[51] They note that estimates of cancer deaths were obtained by multiplying very small hypothetical risks by large populations resulting in thousands of "cancer victims". The risks as assessed from the Japanese studies use a linear no-threshold model for radiation injury and this model may not be applicable to medical radiology. Hendee and O'Connor also note that various publications that estimated cancer deaths from radiology were often presented in the media in a sensational manner resulting in public anxiety and fear. This fear then results in a reluctance to undertake imaging procedures which itself may be more of a risk than the radiation risk, if any such risk existed. The recommendations of the Adrian report remain valid regarding appropriate indications for the examinations and good technique. This is reflected in the recent concerns regarding over-investigation which echo Adrian's conclusions.

THE RADIOLOGIST AS GUINEA PIG

Epidemiological studies may be made of radiologists as well as their patients. The results were reviewed by Roger Berry who described the changing patterns of disease and mortality in radiologists from the pioneers to modern times.[52] Berry noted the common sense recommendations that had been made for radiation protection in the 1920s, and also that the use of the nuclear bombs in Japan had resulted in early evidence of an increase in leukaemia in the bomb survivors. This had resulted in an increased interest in the health of radiologists. The epidemiologist Sir Richard Doll (1912–2005) and colleagues carried out the first proper epidemiological study of British radiologists using the membership details of the Röntgen Society which had been founded in 1897, and is the oldest radiological society in the world. The society continues as the British Institute of Radiology (BIR). Doll first published in 1958 looking at members of the BIR from 1897 until 1955, and continued acquiring data until 1997 giving a 100 years of observation of mortality from cancer and other causes.[53] The study found that the observed number of cancer deaths in radiologists who joined the BIR after 1920 was similar to that expected from death rates for all medical practitioners combined. Interestingly, there was evidence of an increasing trend in the risk of cancer mortality with time since first joining the BIR, such that there was a 41% excess risk of cancer mortality in those who were members for more than 40 years. Doll and colleagues thought that this was probably a long-term effect of radiation exposure in those who first joined during 1921–1935 and 1936–1954. They found no evidence for an increase in cancer mortality in radiologists who joined after 1954, and in this group the radiation exposures were likely to have been lower. The group also examined causes of death other than cancer in more detail than previously reported. They found no evidence that radiation caused diseases other than cancer even amongst the early pioneers, and this was despite the fact that the doses that they received were associated with more than a doubling in the death rate among the survivors of Hiroshima and Nagasaki.

X-RAY PROCESSING AND DARKROOM DISEASE

The chemicals used in radiography had potentially harmful effects, both within the department and also to the environment. The silver halide in the emulsion on the film was released into the fixing solution and therefore needed to be recovered. This was essential because of the high value of silver as a precious metal, the need to maintain the quality of the bath of fixative, and also to avoid the harm of environmental pollution. The methods either used metal exchange or were electrolytic.[54] An industry grew up to support silver recovery in X-ray departments, and in 1995 the company Lokas, which specialised in silver recovery from film and chemicals, presented a paperweight made of 99% pure recovered silver and 24% lead crystal (Figure 5.18) which was distributed at the Röntgen Centenary Congress in the UK.

In the initial period following the discovery of X-rays, the possibility of X-ray dermatitis, a skin inflammation, was unknown and the sore fingers experienced were attributed to the chemicals in the metol-developer. Moreover, corona discharges around exposed high tension were responsible for the formation of noxious nitrogen oxides and ozone.

Cases of poisoning and irritation in photographic workers had been known since at least 1911,[55] and it is therefore interesting that the US Army X-ray Manual of 1919 commented only that good ventilation was essential in the darkroom "not only

FIGURE 5.18 A paperweight made of 99% pure recovered silver and 24% lead crystal presented at the Röntgen Centenary Congress in the UK in 1995 by the company Lokas, which specialised in silver recovery from films and chemicals. Author's collection.

to increase the efficiency of the operator, but because a close, musty atmosphere is bad for the sensitive emulsion",[56] there being no mention of the possible chemical irritations. The possibility of disease in darkroom technicians was a concern until all wet processing was finished in radiology departments. Sarafis and others have reviewed what became known as darkroom disease.[57] Darkroom disease was said to be caused by the exposure of the radiology technicians (RT) to chemicals that were used in film processing, and had an adverse effect not only on the respiratory system but also in "every system of the human organism". The symptoms experienced would be that the voice would become more gruff than usual and day on day the sufferer would feel weaker and more exhausted. The victim would feel that the "heart sometimes pounds so fast and hard that you feel it will pop out of your chest". The sufferer would "have a pile of symptoms that pester you and destroy your well-being". It was also emphasised that the condition would be made worse because the doctors would not know the cause of the symptoms and that some of the sufferer's colleagues would even believe that the symptoms were all in the mind with no real physical existence.

In the UK, the first recorded evidence of what became known as darkroom disease was recorded in 1967 when the Secretary of the Society of Radiographers referred to correspondence from members complaining about the working conditions in the automatic processing darkroom at a hospital in Buckinghamshire.[58] Council agreed that the Secretary should write to the North West Metropolitan Regional Hospital Board suggesting that it take urgent steps to ensure that the working conditions in this hospital were substantially improved.

The disease became more common in the 1980s and Sarafis saw it as an example of "multiple chemical sensitivity" (MCS). MCS syndrome had first been described in Chicago, USA, in 1952, and the first case of "darkroom disease" was described in an RT in New Zealand in 1980 who presented with a number of symptoms that were not found in any recognised disease. She was diagnosed with an "acute sensitivity in the toxic effect of chemical substances of the dark room" and was ultimately awarded compensation. In the 1980s, a number of hazardous agents were present in the developing and fixing solutions of X-ray film processors, especially glutaraldehyde, diethylene glycol, and sulphur dioxide. The authors estimated that approximately 5% of the population of the United States had MCS symptoms. Nallon and others sounded a cautious note,[59] noting that the studies performed in New Zealand and the UK that reported a high incidence of symptoms within radiographers were non-controlled. Nallon's results clearly demonstrated that radiographers were no more symptomatic than a group of hospital staff who were not exposed to processing chemicals, and offered no support for the "darkroom disease" hypothesis. Regarding the cause of darkroom disease and MCS, some researchers saw the cause as primarily psychological rather than representing an organic syndrome. This was partially based on the fact that many symptoms were similar to the "sick-building disease", in which the psychological factor was the most prevailing aspect.

At an individual level, the symptoms could be quite distressing. One radiographer became very ill in the 1990s. She experienced tiredness when working in an unventilated darkroom. Following a move to work at a breast-screening unit, she started to feel quite ill, with tachycardia being a major symptom, and her balance was also affected. The film processor was located within the viewing room. She was admitted to hospital for tests but no cause could be determined. The Occupational Health Department was also unable to find an explanation for her symptoms. The Society of Radiographers became involved and referred her to a heart specialist but again no underlying cause was found for the tachycardia. Whilst off work she recovered, but when she returned to work she became ill again. She changed jobs and went to work in a bone densitometry unit where there was no film processing and remained well with no recurrence of the tachycardia. She was left with a legacy of multi-chemical sensitivity and had to be very careful of her environment. A number of radiographers where asthma was the predominant symptom received compensation; however, unfortunately, this did not apply to tachycardia and balance problems.

Once digital technology came into widespread use with the replacement of traditional film and processing, darkroom disease gradually disappeared, and we can only speculate on its exact cause. In general, most diseases have both physical and psychological causes; however, the chemicals used in automatic film processing seem to have been a particular problem.

NOTES

1. Thomas, A.M.K., Banerjee, A.K. 2013. *The History of Radiology*. Oxford: Oxford University Press.
2. Chick, H., Hume, M., MacFarlane, M. 1971. *War on Disease: A History of the Lister Institute*. London: Andre Deutsch.
3. Scott, S.G. About 1935. *X-Rays and Reminiscences*. Unpublished manuscript.
4. Abbott, F.C. 1899. Surgery in the Graeco–Turkish War. *Lancet*, January 14, 1, 80–83.
5. Leggett, B.J. 1928. *The Theory and Practice of Radiology, Volume Three: X-Ray Apparatus and Technology*. London: Chapman & Hall.
6. Kemerink, G.J., Kütterer, G., Wright, A., Jones, F., Behary, J., Hofman, J.A.M., Wildberger, J.E. 2013. Forgotten electrical accidents and the birth of shockproof X-ray systems. *Insights Imaging*, 4, 513–523.
7. Gunther, M.L. 1919. Précautions à prendre dans les installations radiologiques intensives. *Journal de Radiologie Electrologie*, 3, 544–545.
8. Winz, H. 1931. *Protective Measures against Dangers Resulting from the Use of Radium, Roentgen and Ultra-Violet Rays*. Geneva: League of Nations (Health Organisation).
9. Schall, W.E. 1932. *X-Rays: Their Origin, Dosage and Practical Application*. 4th ed. Bristol: John Wright & Sons Ltd.
10. Weinberger, B., Laskin, D.L., Heck, D.E., Laskin, J.D. 2001. The toxicology of inhaled nitric oxide. *Toxicological Sciences*, 59, 5–16.
11. Schall, W.E. 1940. *X-Rays: Their Origin, Dosage and Practical Application*. 5th ed. Bristol: John Wright & Sons Ltd.
12. Editorial Report. 1928. Electrocution in the X-ray room. *British Medical Journal*, ii, 1186–1187.
13. Lindell, B. 1996. *Pandora's Box: The History of Radiation, Radioactivity, and Radiological Protection Part I. The Time before World War II*. Trans. H. Johnson. Nordic Society for Radiation Protection (Privately Printed).
14. Grigg, E.R.N. 1965. *The Trail of the Invisible Light: From X-Strahlen to Radio (Bio) logy*. Illinois: Charles C. Thomas.
15. Brown, P. 1936. *American Martyrs to Science through Roentgen Rays*. Illinois: Charles C. Thomas.
16. Meyer, H. 1937. *Ehrenbuch der Röntgenologen und Radiologen aller Nationen*. (Band XXII) Sonderbände zur Strahlentherapie. Berlin und Wien: Urban & Schwarzenberg.
17. Thomas, A.M.K., 2020. "Our Katie": Kathleen Clara Clark MBE, MSR, Hon. FSR, Hon. MNZSR, FRPS (1898–1968). *The Invisible Light*, 47, 26–38.
18. Scott, S.G. 1911. Notes on a case of X-ray dermatitis with a fatal termination. *Archives of the Roentgen Ray*, 15, 423–424.
19. Unrewarded Hero. *Daily Mail*, 14 March 1924.
20. X-Ray Hero's £75 Pension (London) *Evening Standard*, 28 December 1923.
21. X-Ray Hero Award. *Daily Sketch*, 28 December 1923.
22. Morris, E.W. 1910. *A History of the London Hospital*. London: Edward Arnold.
23. Papagni, L., Uslenghi, C.M. 1995. Lombardia. In: *Immagini E Segni Dell'Uomo, Storia della Radiologia Italiana*. Ed. A.E. Cardinale. Napoli: Idelson-Gnocci, 731–748.
24. Editorial. 2004. Parenteral iron therapy: beyond anaphylaxis. *Kidney International*, 66, 457–458.
25. Gavazzeni, S., Minelli, S. 1914. L'Autopsia di UN Radiologo. *La Radiologia Medica*, 1, 66–71.

26. Editorial. 1914. The autopsy of a radiologist. *The Archives of the Roentgen Ray*, 18, 393–394.
27. Viterbi, A. 1933. Commemorazione Del Dottor Emilio Tiraboschi, Martire Della Radiologia. http://digitallibrary.usc.edu/cdm/ref/collection/p15799coll117/id/3646.
28. Hœrni, B. 2007. *Jean Bergonié, 1857–1925 - un grand médecin en son temp.* Paris: Éditions Glyphe.
29. *Queen Visits X-Ray Martyr.* Liverpool Post & Mercury, 15 December 1922.
30. *Dr. Vaillant, Martyr to Science, Loses Arms in Thirteen Amputations for X-Ray Burns.* New York Herald, 17 December 1922.
31. *X-Ray Victims' Fund Started in France.* North China Star, 26 February 1924.
32. Thomas, A.M.K., Duck, F.A. 2019. *Edith and Florence Stoney, Sisters in Radiology* (Springer Biographies) Switzerland: Springer Nature.
33. Stoney, F.A. 1914. X-ray notes from the United States. *Archives of the Roentgen Ray*, 19, 181–184.
34. Obituary. 1921. Ironside Bruce, M.D. (an appreciation). *Archives of Radiology and Electrology*, 23, 338–339.
35. The X-Ray and Radium Protection Committee. 1921. X-ray and radium protection. *The Journal of the Röntgen Society*, 17, 100–103.
36. Lindell, B. 1978. ICRP 1928–1978. *Radiological Protection Bulletin*, 24, 5–9.
37. Kaye, G.W.C. 1927. X-ray protective measures. *The British Journal of Radiology* (Röntgen Society Section), 23, 155–163.
38. Radio-Physics Section of 2nd International Congress of Radiology. 1929. *International Recommendations for X-Ray and Radium Protection.* A report of the Second International Congress of Radiology held in Stockholm 23–27 July 1928. *Acta Radiologica*, Supplementum III, Pars I.
39. IV Internationaler Radiologencongress Zürich 1934, 1935. *The Committee on Protection.* Leipzig: Kommissionsverlag Georg Thieme.
40. ICRP, 1977. *Recommendations of the ICRP.* ICRP Publication 26. Ann. ICRP 1 (3).
41. Oestreich, A.E. 2014. ALARA 1912: "As Low a Dose as Possible" a century ago. *Radiographics*, 33, 1457–1460.
42. Herranz, G. 2002. Rapid response: The origin of "Primum non nocere". *British Medical Journal*, 324, 1463.
43. Meggitt, G. 2016. *Genes, Flies, Bomb and a Better Life: In the Footsteps of Hermann Muller.* Cheshire: Pitchpole Books.
44. Kathren, R.L., Ziemer, P.J. 1980. Introduction: the first fifty years of radiation protection – a brief sketch. In: *Health Physics: A Backward Glance.* Eds. R.L. Kathren, P.J. Ziemer. New York: Pergamon Press, 1–9.
45. Mayneord, W.V. 1983. Introduction. In: *History of the Hospital Physicists' Association 1943–1983.* Newcastle upon Tyne: The Hospital Physicists' Association.
46. *When X-Rays Burn, Interesting Sidelights in Rand Case.* Cape Angus, 3 August 1924.
47. Burned by X-Rays. £2,150 Damages for Mine Worker. *The Star*, Johannesburg, 12 June 1924.
48. X-Ray Perils. State Action against Use by Unqualified Persons. *Pall Mall Gazette*, 11 April 1922.
49. Committee on Radiological Hazards to Patients. 1966. *Radiological Hazards to Patients: Final Report of the Committee.* London: HMSO.
50. Doll, R. 1981. Radiation hazards: 25 years of collaborative research. *British Journal of Radiology*, 54, 179–186.
51. Hendel, W.R., O'Connor, M.K., 2012. Radiation risks of medical imaging: separating fact from fantasy. *Radiology*, 264, 312–321.

52. Berry, R.J. 1986. The radiologist as guinea pig: radiation hazards to man as demonstrated in early radiologists, and their patients. *Journal of the Royal Society of Medicine*, 79, 506–509.
53. Berrington, A., Darby, S.C., Weiss, H.A., Doll, R. 2001. 100 years of observation on British radiologists: mortality from cancer and other causes 1897–1997. *The British Journal of Radiology*, 74, 507–519.
54. Chesney, D.N., Chesney, M.O. 1965. *Radiographic Photography*. Oxford: Blackwell Scientific Publications.
55. Hunter, D. 1975. *The Diseases of Occupations*. 5th ed. London: The English Universities Press Ltd.
56. Authorised by the Surgeon-General of the Army. *United States Army X-Ray Manual*. 1918 (reprint of 1925). New York: Paul B Roeber.
57. Sarafis, P., Dallas, D., Sotiradou, K., Stavrakakis, P., Chalaris, M. 2010. Dark room disease. *Journal of Environmental Protection and Ecology*, 11, 506–514.
58. Society of Radiographers, Council minutes of 26.07.1967.
59. Nallon, A.M., Herity, B., Brennan, P.C. 2000. Do symptomatic radiographers provide evidence for "darkroom disease"? *Occupational Medicine*, 50, 39–42.

6 Tubes, Plates, and Screens

Whilst having an appropriate tube to produce the X-rays was important, equally as important were the photographic plate and the fluorescent screen needed for detection. Röntgen was an accomplished photographer and took an immense number of photographs to record his holidays with his wife Bertha, as well as those needed professionally.[1]

THE X-RAY TUBE

In the 19th century, there was increasing interest in passing electrical discharges across evacuated glass bulbs, and various effects were observed. Sir William Crookes (1832–1919) investigated these phenomena extensively with a series of experiments on radiant matter. His tube (Figure 6.1) of 11 March 1879 used a concave electrode to focus the cathode rays on to a platinum target. These early vacuum tubes contained a small quantity of gas (and were called gas or ion tubes). The passage of the cathode rays (which were shown to be electrons) from anode to cathode depended on the ionisation of the gas. Some of these tubes were of the simplest possible form, and consisted of a length of glass tubing, with two metal electrodes to carry the high-tension current. However, an appreciable output of X-rays could be obtained from such a tube, although this fact was unknown at the time. As the tubes were used, the vacuum increased, the tube became harder, and it became increasingly difficult to pass a current through the tube. Current might then pass around the outside of the tube. Crookes developed a tube for regulating the vacuum which was shown at the Royal Institution in London on 4 April 1879 (Figure 6.2). The tube has two smaller tubes attached at either end containing solid caustic potash, that is potassium hydroxide, which could be heated as necessary. This was the earliest device for regulating the vacuum, and the principle was extensively used in the later gas X-ray tubes. And so, if the vacuum was too hard the current would not pass through the tube, and if the tube was soft with too much gas inside, there would be fluorescence; in both cases, the tube would be useless for producing X-rays.

In these early Crookes–Hittorf tubes, the anode and cathode were simple electrodes projecting into the bulb, and it was when using one of the large, pear-shaped tubes that Wilhelm Röntgen made his discovery (Figure 6.3). The cathode rays pass from the flat cathode plate and strike against the large end of the tube, resulting in vivid phosphorescence, the production of heat, and emission of X-rays. The focal spot for the production is large and so the image quality is impaired. The problems faced by the early manufacturers of X-ray tubes were producing a small focal spot, dissipating excess heat, electrical safety, and radiation safety.

DOI: 10.1201/9780429325748-6

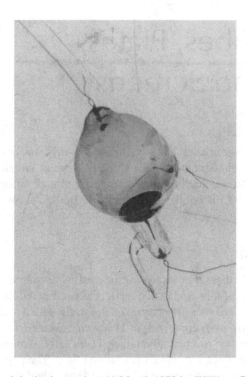

FIGURE 6.1   The original tube used on 11 March 1879 by William Crookes to investigate radiant matter. His tube used a concave electrode to focus the cathode rays on to a platinum target. Originally part of the BIR collection and now in the Science Museum. Original photograph in author's collection.

The life of these early tubes was very brief since, unfortunately, after a few exposures, the tube was either pierced by a spark, or cracked because of the heat generated by the bombardment of electrons. It was also generally thought that the production of X-rays was dependent on the phosphorescence of the glass, and it was this idea, although it proved to be a false one, that led to the discovery of radium and radioactivity, and to further discoveries in this field. It had been suggested by the French physicist Jules Poincaré (1854–1912) that if X-rays originated in the phosphorescing walls of the vacuum tube, it might be possible that X-rays would be produced by other phosphorescing bodies. This experiment was undertaken by Antoine Henri Becquerel (1852–1908), and resulted in the discovery of the radioactivity in the compounds of uranium, which was followed by the discovery of radium by Marie Salomea Skłodowska Curie (1867–1934) and the recognition of the other radio-elements.

Herbert Jackson (1863–1936) had been appointed as lecturer in chemistry at Kings College in London in 1879. Jackson had a lifelong interest in photography, and made many experiments in preparing and sensitising emulsions. In 1890, Jackson started his work on fluorescent and phosphorescent materials, and how they responded when exposed to ultraviolet light and discharges from tubes with both a low and high

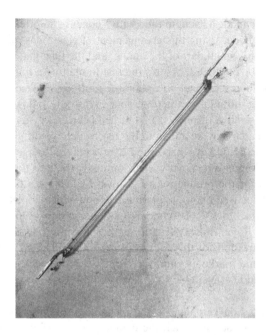

FIGURE 6.2   The tube for regulating the vacuum developed by William Crookes and shown at the Royal Institution in London on 4 April 1879. Originally part of the BIR collection and now in the Science Museum. Original photograph in author's collection.

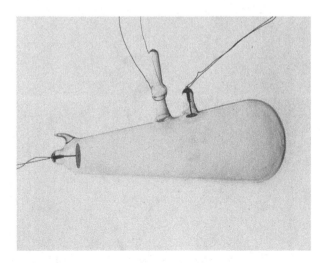

FIGURE 6.3   An early Crookes–Hittorf tubes in which the anode and cathode are simple electrodes projecting into the bulb. The tube was made in Germany and it was when using one of the large, pear-shaped tubes that Wilhelm Röntgen made his discovery. Originally part of the BIR collection and now in the Science Museum. Original photograph in author's collection.

vacuum. In 1894, he made a vacuum tube with a concave aluminium cathode and an inclined platinum anode. Using this arrangement of electrodes, the electron stream was concentrated on a small part of the anode, and the tube may be seen as a variation of Crookes' tube of 1879. Jackson's interest in making a fine-focus tube was to aid his studies of fluorescent and phosphorescent effects. Jackson believed that the tubes were emitting ultraviolet radiation, and it was only following Röntgen's discovery that he realised that the effects that he had been observing were related to the X-rays. Jackson's name may be added to the list, which would include Crookes and others, of those who could have discovered X-rays before Röntgen.

Shortly after Röntgen's discovery, Jackson made the first tube in England specifically designed for the production of X-rays. In Jackson's first focus tube, the cathode stream is focused upon the platinum anode which is placed at an angle of 45° and placed at the "focus" of a concave cathode, and in 1896 Jackson exhibited his tube before the Royal Society (Figure 6.4). The whole class of gas or ion tubes are all basically variations of this Jackson tube.[2] The tube had a small focal spot, and this resulted in a markedly improved image quality. Jackson believe that his improvements should be available to everyone and refused to take out a patent on his focus tube.

There was a rapid development in the technology of X-ray tubes. In February 1896, a tube was designed in which the cathode rays are focused on the platinum anticathode following Jackson's design, but with an auxiliary anode in the tube. The cathode is not withdrawn into the side tube, but is brought right out into the bulb.

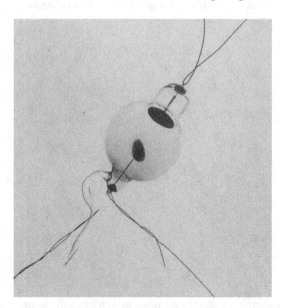

FIGURE 6.4 Herbert Jackson's first focus tube shown at the Royal Society in London in 1896. The cathode stream is focused on the anode which is at an angle of 45°. Originally part of the BIR collection and now in the Science Museum. Original photograph in author's collection.

On 17 March 1896, Alan Archibald Campbell-Swinton (1863–1930) designed a focus tube which had two spherical electrodes for use with alternating currents, with each electrode serving as cathode and anode in turn. This tube was the first of its kind. It was described by Swinton in the *Electrician* of 8 May 1896. Tubes with three electrodes became commonplace, with the usual arrangement consisting of a cathode and an anticathode attached to an anode which was in a side branch; this design helped to stabilise the tube performance.

The development of the ion tube was complex, and a very large number of designs were produced,[3] and it can now be difficult to determine why specific features were added.

Slight modifications in the shape and form of the electrodes and tube shapes were made by the various manufacturers, with each one claiming some special advantage for the particular form of tube. It soon became a matter of much perplexity for the busy practitioner to decide upon the particular make of tube to purchase. The prices ranged from 10 shillings (50 p) to about 5 pounds, and since multiple tubes and their life was short, this was a somewhat serious matter. Therefore, in January 1901, the President of the Röntgen Society, the pioneer Glasgow radiologist John MacIntyre (1857–1928), offered a gold medal for the best tube that can be produced for radiographic purposes by makers of any nationality. Although the tubes were to be tested for use in fluoroscopy, radiography, and penetrating power of the radiation, it was also emphasised that price would be a factor. A committee of experts was formed to act as judges, and 28 tubes were submitted by manufacturers, with 5 from Britain, 8 from America, and 15 from Germany. The results were announced in the Archives for August 1901.[4] The judges commented on the high quality of the submitted tubes, with some having features to take heavy discharges, and others having ingenious methods of regulating the vacuum. It was noted that under identical conditions, the output of X-rays from a selection of apparently faultless tubes could vary in both quantity and quality. After the preliminary tests, the number of tubes was reduced to six, and the final test was made photographically. The two best tubes were selected, and following further tests the cheaper of the two was awarded the prize. There was recorded considerable disappointment, and a good deal of grumbling, when the award was made to C.H. F. Müller of Hamburg. The tube itself possessed no distinctive features, and was typical of the tubes that were then used at the beginning of the 20th century. The award-winning tube was submitted by Harry W. Cox (c. 1870–1937) as a Cox "Record" tube, and was made by Müller (Figure 6.5). Cox imported German radiological apparatus, and sadly became a radiation martyr. It afterwards transpired that both of these final tubes were by the same maker. Müller of Hamburg was founded by Carl Heinrich Florenz Müller (1845–1912) and, like Philips, had made glass lamp bulbs.[5] In 1895, Müller made X-ray tubes for Röntgen at factory in Bremer Reihe, Hamburg, Röntgenröhrenfabrik. The Müller company merged with Philips in 1987.

When the dangers of radiation were realised, the tubes were enclosed in lead glass bowls or lead-lined wooden boxes. The supporting stand became more robust, and at the same time supported and held the fluorescent screen. The screen no longer had to be held and the operator's hands were protected. The apparatus became larger and

FIGURE 6.5   The prize-winning Cox "Record" tube, this one is numbered 54512. Author's collection.

heavier. The anticathode was increased in length, and taken outside the walls of the tube, and finishing in a metal radiator designed to disperse the heat. Other models sued a water jacket and reservoir for the same purpose, and a thermo-syphon system of circulating the water was also employed. A removable form of radiator, which could be cooled and replaced, or a hollow anticathode into which air could be blown with a small motor could be used. All of these systems were designed to remove the heat generated by the impact of the cathode stream upon the target, or anticathode. From 1921, a technique depending upon a large mass of metal for heat absorption was perhaps the most popular type of all, with the exception of the water-cooled tube.

Most X-ray departments possessed, among their "family" of tubes, one or possibly two tubes that were worth more their weight in gold owing to their remarkably good behaviour. In other words, for some unexplained reason, as long as they were not overrun, certain tubes remained perfectly steady for long periods without any regulation of the vacuum. Needless to say, those precious tubes were very carefully nursed and guarded, and their demise was the cause of mourning or cursing.

These Jackson-type tubes worked well, but their function was unpredictable and the radiographer needed to know the individual tube, and how a particular tube functioned in use. The function of the tube varied with its use and before the first use, the tube needed to be seasoned. It was not uncommon for one tube to be reserved for extremity work and the other for chest radiography. Alfred Barclay (1876–1949) said that these X-ray tubes, housed in lead glass bowls, were fickle and sensitive in the extreme, resenting any suggestion of overload (at, say, 2 or 3 mA!), and played odd pranks. The control of gas tubes was an art and a constant trial to the temper: "a globe of glass surrounded by a zone of profanity".[6] Until the introduction of the Coolidge tube in 1913, which eventually supplanted the old gas tube, radiologists and radiographers rivalled each other in the art of running in or "seasoning" these erratic and moody gas tubes. Undoubtedly, it was an art well worth developing. No experienced operator would think of putting a new tube, still green from the makers, into full work. This would certainly mean instability in working for the rest of its life. Barclay later wrote that the joy of tube training was lost to the modern operator, and the "running in" period would sometimes, with a difficult tube, extend for over a month. A small current gradually increased would be run through the tube by day for five minutes or so, the process being carefully watched and the rubbing conditions regulated, and each tube wanted different handling.

The only reason Barclay was given for working in artificial light, with the blinds down all day long, was that the operator was unable to tell whether the X-ray tube was functioning correctly in daylight. It should be remembered that the gas tubes used in those days were much more attractive than modern tubes; the inside of the bulb was lit up when activated by a charming apple-green light or fluorescence. The tube flickered unsteadily if the output of X-rays was poor, that is if the vacuum was too high. When functioning satisfactorily, the bulb was divided sharply into two distinct hemispheres, one half lit up brightly with the apple-green light already mentioned, the other half quite dark. The operator relied a good deal on the appearance of the bulb when activated to tell him the mood of his tube. This tube fluorescence was, of course, easier to see in the dark.

WILLIAM COOLIDGE AND HIS TUBE

William Coolidge (1873–1975)[7,8] produced a major improvement. He replaced the cathode with a heated spiral tungsten filament and molybdenum-focusing bowl. The filament could be heated and a current would pass through the tube even with a very low vacuum. The anode of the standard Coolidge tube was set at 45°. All modern X-ray tubes are variants of the Coolidge tube. With the Coolidge tube, the X-ray output of the tube was exactly predictable. In 1906, Coolidge made a major contribution when he discovered how to make molybdenum and tungsten ductile. Prior to this, these metals were thought of as being unworkable because they were far too brittle. The ductile tungsten was useful since it could be made into incandescent lamp filaments, replacing the earlier carbon filaments.

It was when working with X-ray tubes that Coolidge found a particular tube that worked well when the cathode became heated.[9] Coolidge then collaborated with Irving Langmuir, who was studying electron emissions from hot tungsten filaments. It was found that even in the highest vacuum, the electron emission was stable and reproducible. It occurred to Coolidge that this could be adapted for use in an X-ray tube. On 12 December 1913, Coolidge wrote:

> I L (Langmuir) tells me that in his study of the Edison Effect, current from the hot cathode is greater with vacuum of .01 or .02 micron than at higher pressure (except in case of argon). I will try this at once in an x-ray tube in which I can heat the cathode.

The Edison effect (also known as thermionic emission) is the emission of electrons from a hot cathode within a vacuum tube. This effect had initially been reported in

1873 by Frederick Guthrie in Great Britain. Guthrie was working with electrically charged objects. He found that a positively charged red-hot iron sphere would discharge, but this did not happen if the sphere had a negative charge. This effect was rediscovered by the American Thomas Edison on 13 February 1880, when he was trying to find why the filaments in his incandescent lamps were always breaking.

The team at GE then developed a high-vacuum tube with a heated tungsten filament acting as the cathode and a tungsten disc as the anode. The tubes were evacuated and initially the green fluorescence of the glass, which always took place when X-ray tubes were operated, was observed. As the vacuum increased, the fluorescence disappeared and the tube became stable and controllable. The ions in the tube that were previously needed became unnecessary. The limitations of the older gas tubes were largely secondary to the presence of these ions.

The first of these new tubes with a heated cathode was used by the well-known US radiologist Lewis Gregory Cole, from New York. The new tube was first demonstrated at a dinner in a New York hotel on 27 December 1913. It was enclosed in an open-topped lead glass bowl for radiation protection. Up to this point, the X-ray generators had an output considerably higher than the older X-ray tubes could endure and this was changed at a stroke. With his characteristic modesty, William Coolidge wanted to call the new design of tube the "GE Tube"; however, Lewis Cole proposed the name of "Coolidge Tube" and this is the name that stuck (Figure 6.6).

Coolidge wanted to test his new tube designs on human subjects and decided rather unwisely to use himself as a test subject, as so many had done before him with disastrous effects. Whilst this went well initially, Coolidge became concerned when the hair on his back started to fall out. Coolidge therefore obtained an embalmed leg from a local physician to use as a phantom. When Coolidge had finished his experiments, he took the leg to the company's incinerator for disposal and threw the leg into the incinerator without informing anyone. When the operator opened the incinerator as the covering came apart, he was horrified to see a human leg. He was

FIGURE 6.6 Typical Coolidge tube of classic design. Made in England by BTH, Number A2267. Author's collection.

convinced that he had come across a dreadful crime and so he called the police. A detective then visited Coolidge, and he had to give a very lengthy explanation!

Further experiments by Coolidge and his team on the new tube resulted in the observations that the radiographic contrast depended on the tube voltage and that the resolution depended on the size and position of the focal spot.

During the First World War, Coolidge became involved in producing a dependable and portable radiographic unit for military use. The individual elements were of necessity simple, light, and easily transportable. There was a petrol-driven generator attached to provide the current. The new tube had many benefits for military use and could be operated for long periods of time without overheating.

It might be imagined that the new Coolidge tube would rapidly sweep away the old ion/gas tubes. In fact, the gas tubes survived for a considerable period of time. The older designs were readily available and also quite cheap. For example, in GC Aimer's catalogue, a standard gas tube retailed at 7 pounds and 15 shillings (£7.75), whereas the Coolidge tube sold for £40, which was more than four times the cost. Aimer was forced to admit that "the Coolidge tube has certain advantages over the gas tube". Gas tubes were widely used until 1926, and on a number of the older installations until 1932.

Coolidge continued to work with X-rays and in the 1930s developed a 900,000-volt tube. He was also deeply involved in non-destructive testing and industrial radiography. Coolidge received many honours, but remained very modest. He said, "such honours as this I accept only if I can somehow share them with many others, since the entire staff of our research laboratory contributed to the success of this work".

Bouwers of Philips designed the Metalix tube in 1924. This self-protecting tube was a major improvement and enabled truly shockproof and portable apparatus to be produced. The Metalix tube incorporated the principle of line focus, the anode face being set at an angle of 19° to the cathode. This reduced the apparent size of the focal spot and increased image resolution.

Finally, the rotating anode tube was developed. It was designed by Bouwers and was marketed in 1929 by Philips as the "Rotalix". Siemens introduced a rotating anode X-ray tube with a tungsten anode disc in 1934. In the classic rotating anode tube, the anode target is a heavy tungsten disc that spins so that the focal spot of the cathode rays is changing and the heat is dissipated.

Contemporary tubes are variations on the theme of these earlier tubes. Modern tubes have to bear very high tube loading and this is particularly the case for tubes used for CT scanning and angiography.

X-RAY FILM

Photography was well developed by 1895 with the use of photographic glass plates, celluloid film, and paper. The first medical photograph was probably taken in May 1856 by Hugh Welch Diamond (1809–1886) who was interested in documenting facial appearances in psychiatric conditions and in patient identification.[10] Initially, radiographs were obtained using ordinary glass photographic plates. Unfortunately, the emulsion originally designed for exposure to light was sensitive to only a very small proportion of the X-ray energy that fell upon it, and the X-ray photons would pass through the emulsion. The first photographic material made specifically for radiographic use was X-ray paper produced by George Eastman's Kodak Company in 1896. By 1897, special X-ray film that was coated with emulsion on both sides was being used, with fluorescent intensifying screens on both sides. In 1901, Isenthal and Snowden Ward reviewed the photographic media,[11] and discussed their relative merits. The easiest to obtain was the photographic dry plate, on which a sensitive emulsion of gelatin and silver was coated. The old glass photographic plates used for X-ray work at this time had a disconcerting habit. On taking them out of the washing tank, the emulsion would be found for no apparent reason floating free in the water, having become separated from the glass base. This might occur with one batch of plates. The junior assistant had the pleasant job of putting candle grease round the edge of the glass before development, in order to prevent this peeling off of the film. The film was similar, with the emulsion coated onto a base of celluloid (that is cellulose nitrate). The celluloid film was flexible, but they were less sensitive than glass plates, were more expensive, and did not keep as long. There were double-coated films which were more sensitive with radiation exposure reduced by a factor of a half, but were more difficult to develop and required special handling. They were therefore abandoned and were only reintroduced in 1918. Paper film was cheaper than celluloid film, and was almost as durable as glass. One advantage of both types of film was that since they were practically transparent to radiation, six or more could be exposed at the same time resulting in a number of practically identical negatives. Edward Shenton (1872–1955), who was the pioneer in radiology at Guy's Hospital, created the hospital's service as a medical student. His day-book illustrates his practice. The radiograph as a negative was printed onto paper and one copy was pasted into the day-book, with a second sent to the referring clinician. The details of the request and report were also recorded.

The care taken by the individual operator could often minimise any difficulties quite considerably. In these early days, the skill of the X-ray operator was of real value, for on it depended not only the production of good X-ray photographs but also the smooth economic running of the apparatus. The pioneer radiologist was a photographer undertaking all the developing of the plates and necessary printing, his work very often kept them busy till late at night. The radiologist was, however, badly handicapped when in competition with the professional photographer. The various "tricks of the trade" were often used by photographers to improve an otherwise indifferent examination. As the technical quality of the X-ray photographs was supposed to indicate the value of the examination, these trade means of improving photographic results made the amateur work of the medical X-ray specialist by comparison significantly less impressive. The subsequent improvement in the apparatus eventually did away with the necessity of "touching up" the radiograph. No doubt, this technical advantage, which the professional photographer had over the medical radiologist, accounted for the lack of cooperation and support of the medical profession experienced by the latter, so noticeable at this period.

The darkroom at this early period, housed almost as badly as the X-ray room – some odd cupboards with no ventilation whatever – was usually considered efficient accommodation, and to make things worse, it was deemed essential that the walls should be painted black. The photographic plates made of glass were developed in porcelain dishes and only a few radiographs could be developed at a time. Metol developer was then in general use. The floor was continuously wet and not infrequently at the end of the day, and the radiologist's shoes were always soaked through. Altogether, the life of the early radiologist was not a happy one, as can be seen in reading the first-hand accounts. It should be realised that there were no trained radiographers or technicians in the early days to help, and all the work of the X-ray department had to be carried out by the medical personnel who had in a rash moment decided to take up X-ray work. The emulsion was still sensitive to light rays and consequently the film was still to be placed in light-tight envelopes or cassettes. In the early days, this loading-up of plates into these black light-tight envelopes, before they could be used, was a laborious business and later on, as the work increased, some of the larger hospitals employed one man, or rather boy, to do nothing else but "load up" plates.

The 1914–1918 war hastened many changes that would no doubt have occurred at some point in time. The "Wratten" X-ray plate had been produced from 1912; however, following the start of hostilities, the supply of photographic glass from Belgium ceased and, as an example, the Kodak factory in Great Britain completely switched production to film, introducing a single-coated X-ray film.[12] The plates were packaged singly and were placed in a light-tight envelope before use. Film was far more appropriate for wartime use since it saved weight and avoided breakages. Post-war the use of X-ray plates diminished. This change to celluloid film was stimulated in 1918 when the "Dupli-Tized" X-ray film was introduced, but there were a number of technical difficulties to be overcome, such as buckling of the celluloid, pressure marks on the emulsion, instability, etc. This film had a sensitive coating on both sides of the base and therefore gave two coincident images, resulting in a much higher contrast and sensitivity. This double-coated film became the international standard. They were also difficult to handle, as there were no film holders; in addition, they were highly inflammable. The use of double-coated film had been abandoned preciously, and even as late as 1914, there were no special facilities for the handling of developing film, with only small developing tanks being available. These films could be viewed when still wet and clipped into a holder. A doctor who wanted to see the radiograph urgently would ask to view the wet plates, and curiously it was common to see "WPP" (wet plates please) on request forms long after automatic processing was utilised.

The Patterson Screen Company introduced fluorescent intensifying screens with an improved performance, and in the 1920s added a protective coating to assist in cleaning.

Photographic glass plates were in general use up to about 1923. Even when these technical difficulties had been overcome, the introduction of the film technique to the medical profession was no easy task for the maker. There were years and years of prejudice to overcome. X-ray workers had got used to glass plates, and the makers did not realise what a hard job they were going to have before they would be able to convince them of the advantages of the film. Considering the disadvantages of the glass plate, especially in the larger sizes – 15 by 12 inches, their great weight, their fragility, their difficulty of storage and postage, etc., it must surprise modern radiologists to learn that it took some six years before the use of the film became general, and then it was only because the makers in 1923 were able to put a very fast film, far faster than any plate, on the market. In all diagnostic work, rapid exposure and a full detailed radiograph is essential for accurate work. Against their will, but in self-defence therefore, radiologists, in order to shorten their exposure and to reduce the wear and tear on their machines, eventually replaced their cumbersome glass plates for the easily handled and faster films. There was, however, still one serious drawback to these films, the celluloid base. This was made of cellulose nitrate, a highly inflammable compound. The fire risk was considerable when they came to be stored in large numbers. Many requests were made to the makers for a non-inflammable celluloid X-ray film, and from 1907 there were attempts to produce a less inflammable film. It required the death of some 130 individuals who were burnt alive in a huge fire involving the Cleveland Clinic at Ohio, USA, in 1929, caused by the spontaneous ignition of a huge X-ray film store in the basement of the Clinic, to force the change. This tragedy forced the makers' hand and in 1929–1930, in spite of manufacturing difficulties, a safety film, the base of which was cellulose acetate, was put on the market. Cellulose acetate film, made from cellulose diacetate and later cellulose triacetate, was introduced in 1934 as a replacement for the cellulose nitrate film stock that had previously been standard. When exposed to heat or moisture, acids in the film base begin to deteriorate to an unusable state, releasing acetic acid with a characteristic vinegary smell, causing the process to be known as "vinegar syndrome". The X-ray film was best kept upright in envelopes at an even temperature, and required large store rooms (Figure 6.7). The film needed to be kept dry and ideally the room should be ventilated. Problems could arise if the film was incompletely fixed, which results in deterioration and discolouration, and also a typical smell. The smell produced by the old films is related to the developer being left in the film and it comes out as fumes. The early film was brittle and needed be handled carefully to avoid cracking. Another problem was shrinkage of the film base which results in a crinkly crazed appearance. This can be difficult to stop and essential films needed to be copied for long-term storage. From the 1980s, polyester film stock was used which was thinner and had better overall properties. Film and screen technology continued to develop with a base material that absorbed ultraviolet light and the use of rare earth screens. However, just as conventional technology was perfected, all was to be swept away by the digital revolution.

AUTOMATIC PROCESSING

Film processing progressively developed from the early days with the use of dishes and tanks in a darkroom to automatic daylight processing.[13] The early processing was labour-intensive, and it was possible to optimise the image during processing. The films were labelled by hand using indelible white ink. The initial attempts at automatic processing were introduced by Pako in 1942, and duplicated the dunking of film in tanks and required a large unit. In the 1950s, Kodak introduced a

FIGURE 6.7 Typical X-ray film store. There were many rows of racks in this film store. Author's photograph.

unit that used rollers with continuous movement of the films and this was a major development.[14] By 1965, Kodak had a system that could process a completed film in 90 seconds. A darkroom was still required with the film cassette being unloaded of the exposed film and reloaded with new film all under safelight conditions, with the exposed film being fed into the processor for developing, fixing, washing, and drying. Whilst 90 seconds may not seem long, the processing of an angiographic series introduced a considerable delay into an examination, and made the performance of an interventional procedure time-consuming. Finally, by 1990, "Daylight" automatic processors were in use with the film removed from the cassette, then fed into the film processor, and the cassette was then reloaded with a new film to be used again, although traditional processors were needed for angiographic series.

PLATE AND FILM STORAGE

The storage of radiographs has always been difficult, whether on glass or film. In 1910, when Sebastian Gilbert Scott was appointed as director of radiology at the London Hospital, he became aware of the problem of film storage, which became acute about the year 1926. Scott was escorted into a fair-sized room. He was astonished by what he encountered. The greater part of the floor space was literally 12–18 inches (46 cm) deep in broken glass, and here and there Scott could dimly make out stacks, or pillars, of radiographic plates nearly 6 ft. high (1.8 m). Obviously, many of these pillars of glass had crashed to the floor over a period of years, probably due to the enormous weight, and so crushing the lower glass plates to powder. The few remaining pillars represented the more recent radiographs, and Scott was told that, if a radiograph that had been taken say one year ago was wanted for comparison, a member of the junior lay staff was given the task of finding it, and it was frequently two days before it was found. More often than not, the plate was either missing or broken. Scott admitted that he was "unable to resist the temptation of crunching my way over glass-littered floors, and deliberately pushing over the remaining stacks of radiographs. I then gave an order for the glass to be shoveled out".[15] Scott said that he had reason to remember this "crime" for some years after, whenever requests for previous radiographs of long-standing cases came from the medical staff. The answer he gave that these plates could not be found and that probably they had never been returned to the department was usually sufficient, "as the staff well knew that many a good radiograph found their way into the ward sister's private desk for future reference". Since it was the return of the film to the department that prompted radiologists reporting of the results, it was only after the introduction of computerised management systems that the scale of the number of films that were not being reported was recognised.

The X-ray films as hard copy required a complex process to optimise use; however, they were easier to store than glass plates. Many clerks were needed in X-ray departments to file and collate films, packets, and reports. Films needed to be collected from wards and clinics and filed in film bags. The reports were typed directly with a secretary taking direct dictation, or the reports could be handwritten. Three copies of the report were needed, one for the referring doctor, one to be kept in the film bag, and a third stored in a filing cabinet. A significant area of the X-ray department was needed to store film packets and reports. Part of the problem was that even in one institution, the patient would have multiple packets. For example, in the old Bromley Area Health Authority, there were four hospitals, and each hospital had its own film store and own packet, with additional packets for the ultrasound department and CT scanner, and as might be imagined, keeping track of all of these reports and film bags was a major task. Standards were defined as to how long the packets were to be retained, and the packets for children were kept for a longer time. The disposal of old radiographs had a complex process and an industry developed related to silver recovery from the unwanted film. The radiographs, and particularly the older films, had a significant quantity of silver halide in the emulsion and this was worth retrieving.

There were a number of solutions to the problems of film storage, including microfiling on film or tape; however, this required additional skilled staff and expense, and such solutions never had a wide uptake. By the early 1970s, in Guy's Hospital in London, the film storage room measured 717 sq. ft. (66.6 m²) with 200 ft. (60.9 m) of storage racks, each 15-storeys high. These racks were filled to overflowing and two assistants worked in the room retrieving the packets. The director of the department

Tom Hills devised an interesting solution, and developed a system that stored 4 × 4 inch (10.1 × 10.1 cm) miniaturized films. The large films were kept for a year and then miniaturized, and the important studies were retained in full size.[16] Whilst this system was ingenious, it fairly rapidly fell out of use. In reality, it was only the introduction of Picture Archiving and Communication Systems (PACS) that resulted in the demise of the many rooms full of racks of packets of X-ray film.

SOVIET ERA X-RAY AUDIO

Whilst the X-ray film was primarily intended for recording radiographic images, as a physical item other uses were possible. One such was for the bootleg recording of music in Soviet era Russia. These were called "bones" or "ribs",[17] and were produced from about 1946 to about 1964 (Figure 6.8). The music recorded was forbidden by the Soviet censor, and it seems poignantly symbolic that music that was to be hidden from the authorities was recorded on X-ray film that was intended to record the previously hidden interior of the body. The discs are a cross between the old-fashioned gramophone record and a radiograph. The discs were individually produced and each is unique, and as a dissident activity are termed *Roentgenizdat*, or private X-ray publications. There had been control of cultural activities under Joseph Stalin (1878–1953), and many styles of music were to appear on the "bones", and not just decadent Western music. The majority of the music on the discs was Russian in origin. The "bones" are palimpsests, that is, they are an object made for one use and modified for another. They are culturally important, and to have been caught in possession of one would result in imprisonment.

FIGURE 6.8 Forbidden music recorded onto X-ray film, in this case from a chest radiograph. The ribs can be seen clearly. Author's collection.

FLUOROSCOPY

When Wilhelm Conrad Röntgen discovered the X-rays in 1895, he observed their effect on fluorescent salts and then on photographic glass. Certain crystalline salts such as barium platinocyanide or calcium tungstate show fluorescence when exposed to X-rays. These fluorescent effects could be observed by holding a coated screen in front of an object. The first screens for clinical use were handheld, and by 6 February 1896 Enrico Salvioni the Professor of Experimental Physics from Perugia in Italy developed the cryptoscope, or "criptioscopio".[18] Salvioni described his simple apparatus which consisted of a small cardboard tube 8 cm high. On one end of the tube is a sheet of black paper on which was spread a layer of fish glue and calcium sulphide, which substance he found to be "very phosphorent under the action of Röntgen rays". On looking through the open end, even in a light room, he was able to see objects in a cardboard box. William Magie (1858–1943) was working independently in Princeton in the United States, and on 15 February 1896, he described the use of a sheet of black paper coated on one face with platinum baricyanine, located at the end of a tube or box.[19] He gave the advantage of the apparatus as avoiding the inconvenience of working in a darkened room, with the benefit of avoiding the long delay involved in the photographic process (Figure 6.9). The cryptoscope was often used to test the quality of the X-ray tube and the hand was a convenient test object, and this is one

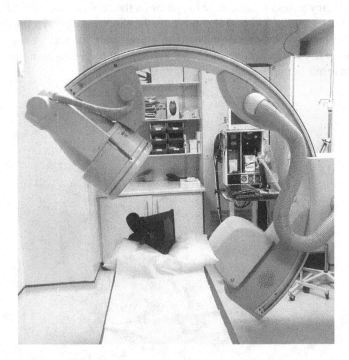

FIGURE 6.9 The black cryptoscope is seen on the table of the modern C-arm image intensifier, which could be used for diagnosis and interventions. They both serve a similar function with more developed technology in the modern apparatus. Author's photograph.

reason for the high incidence of injuries to the hand with early X-ray workers. With a strongly active X-ray tube, the cryptoscope would enable the operator to see the bones in the hands clearly when standing even 10–12 ft. (3.9 m) away from the X-ray tube. By 1897, the cryptoscope was in use in Paris railway station, and was used to examine luggage for contraband and harmful material. The cryptoscope developed in design and was in use until the 1950s. The cryptoscope could be held in the hand; however, an operating cryptoscope made for use by surgeons in the operating theatre. The cryptoscope was strapped to the head and could be tilted to allow either inspection of the fracture or fluoroscopy of the fracture to assist in reduction. In spite of the dangers, the cryptoscope remained in use for many decades into at least the 1950s. Presumably, this was partly related to its ease of use. There was a complete lack of electrical and radiation protection of the X-ray tube and absence of protection to the operators and viewers.

It was obvious that the transient image on the fluorescent screen might be captured, and several pioneers proposed this.[20] Edward Thompson[21] gave an account of the experiment of the New York pioneer Julius Mount Bleyer (1859–1915) who, on 7 April 1896, attached a camera to a cryptoscope and recorded the image. Bleyer called his device a photofluoroscope and took many images. This technique was an anticipation of the technique of photofluorography that was to be used in Mass Miniature Radiography that was used for population screening for pulmonary tuberculosis. Further work was done by the Italians A. Batelli and A. Garbasso. In Glasgow, John MacIntyre (1857–1928) used an ordinary camera to photograph a potassium platinocyanide screen, and he also allowed a sensitive film to pass underneath the aperture in a case of thick lead covering the cinematograph. The opening corresponded to the size of the picture, and was covered with a piece of black paper upon which the limb of an animal, such as a frog, could be photographed. The frog was under anaesthesia and the movements were slow; however, MacIntyre was able to cinematograph the movements of the frog leg, which he showed at the Glasgow Philosophical Society, and resulted in one of the most iconic of early radiographic images.[22] MacIntyre was a pioneer of instantaneous radiography and cineradiography.

The cryptoscope could only examine a small field. A larger fluorescent screen was held in the hands and the patient was examined on a special couch. The initial couches resembled a four-post bed with a canvas top (Figure 6.10). These X-ray tables gradually developed with initially separate units for upright or supine examinations. These could be combined in a pivoted table which could be moved into upright or horizontal positions as required. It was then logical to motorise the movement as the table and screen-holder became heavier. From being held in the hands, the screen was attached to the table. Fluoroscopy was used to observe an opaque meal or enema, and when needed the screen could be moved to one side and a series of X-ray plates could be exposed. Finally, the X-ray film in a cassette could be incorporated into the holder of the fluorescent screen and be easily moved across for exposure. The fluoroscopic image was of necessity quite dim and the examination had to be performed in a darkened room (Figure 6.11). A period of dark adaptation was needed to have a full ability to see the dim light, and the failures of fluoroscopy were mostly due to insufficient dark adaption.[23] In order to assist in dark adaptation,

FIGURE 6.10   The original Harnack–Dean X-ray couch used by Sebastian Gilbert Scott at the London Hospital in 1909. Note the absence of protection, either electrical or for radiation. There are over-couch and under-couch tubes. A Ruhmkorff induction coil can be seen on the table to the right of the couch. Author's collection.

the radiologist would wear red goggles and there are stories of these being worn when driving between hospitals (Figure 6.12). Detailed proformas were produced to record the results of the examinations. Visibility in the darkened room was difficult, and fluorescent cups for holding the opaque meal were developed by Siemens (Figure 6.13a and b).

X-RAY TELEVISION

The development of fluoroscopy both prior to image intensification and with image intensification is complex.[24] Electronic apparatus was developing rapidly following the Second World War, including image intensifiers and television technology. The first commercial image intensifier was available in 1953, and using it cineradiography became a practicable routine procedure, and had significant research uses with notable contributions by Gordon Ardran (1917–1994) at the Nuffield Institute in Oxford working with Frederick Kemp (1912–1976).[25] Ardran and Kemp studied the comparative anatomy and physiology of the throat and upper airways, and resulted in 80 joint publications on technique, normal anatomy and movements, and pathological states.[26] They were the first to point out that lingual function is essential to effective swallowing. These first image intensifiers used in radiology

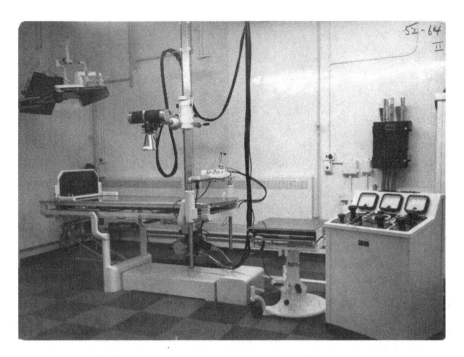

FIGURE 6.11 Fluoroscopic room, prior to image intensification, Room 2 at Farnborough Hospital, equipment installed in 1952, and replaced in 1964. Courtesy of the late David Rickard. Author's collection.

FIGURE 6.12 The red goggles belonging to Dr. Glendinning of Bromley and District Hospital. The goggles were supplied by Motteshed & Co., pharmaceutical chemists of Manchester, who also undertook radiographic services in the early period. Author's collection.

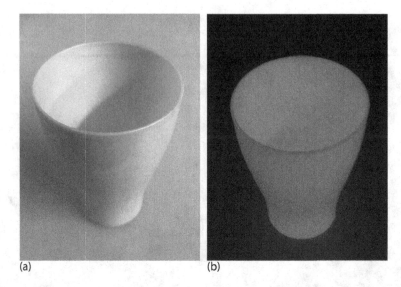

(a)   (b)

FIGURE 6.13 (a) Cup for holding the barium sulphate suspension in daylight, made by Siemens. (b) In the dark, the cup for holding the barium sulphate suspension would show a brilliant green, fluorescent colour. Author's collection.

had their output viewed using a mirror. This was somewhat awkward; however, the brighter image was ideal for cineradiography, and by 1954 Ardran was predicting that all cineradiography would be carried out by image intensification, although in 1954 the small field size of 5 inches made general use impractical.[27] It was a straightforward advance to view the fluoroscopic image with a television camera and display on a television, and by the early 1960s these were becoming available. Hanley and others at Hillingdon Hospital used and reviewed such a system in 1961.[28] They used the Marconi–Orthicon system using the Orthicon television camera, first developed in 1943. This system did not use an image intensifier, and the light gain from the screen was due to the sensitive camera. This system was more expensive; however, a 12-inch screen could be examined. The alternative system used a 9-inch image intensifier tube with a curved screen, with the output image scanned with a vidicon camera. Both systems presented the Image on a television screen. Hanley noted the versality of X-ray television, with numerous advantages. The image could be doubled in size, presented as positive or negative, and inverted if needed. Perhaps most significantly, no dark adaptation was needed, and as the examination could be performed in the light the performance of procedures was greatly facilitated. However, another significant difference is that in traditional fluoroscopy the patient is looked at directly, albeit through a fluorescent screen. With X-ray television, the attention is now focused on the television and not directly on the patient, and whilst this may not seem important, it is a separation from looking at the patient to an image of the patient; however, the room was no longer darkened (Figure 6.14).

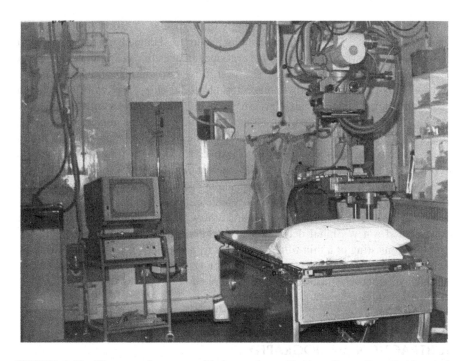

FIGURE 6.14   Fluoroscopic room, with image intensification, Room 2 at Farnborough Hospital, equipment initially installed in 1964. Courtesy of the late David Rickard. Author's collection.

DIGITAL RADIOGRAPHY

The conversion of the X-ray energy pattern into digital signals utilising scanning laser-stimulated luminescence (SLSL) paved the way for a new technique in digital radiography.[29] The basic principles were developed in 1983 by Kato and his group at the Technology Development Center, Miyanodai, Fuji Film in Japan. Kato explained that the inspiration for digital radiography was the success of computed tomography. It is interesting that computed tomography came first since it might have been imagined that digital radiography might have been introduced first. The experience gained by the use of digital data in computed tomography must have stimulated development in other areas. A prototype unit for digital chest radiography was developed by Barnes and his group in Birmingham Alabama with the Advanced Development Group, Picker International, Cleveland.[30] In their clinical study, 400 patients were examined during 1982–1983. They found the advantages to be improved exposure latitude and low-contrast detectability. Whilst there was prompt image reconstruction, there was lower image resolution. The Philips' ThoraVision was the first digital chest radiography machine using a selenium-based digital detector system and was announced at the European Congress of Radiology in 1993. The images produced by the ThoraVision were assessed by Floyd and others who found that the selenium-based system provided an image appearance that was significantly preferred by all of the radiologists, and perhaps surprisingly more strongly by those specialising in chest radiography.[31] The study from as long ago as 1995 demonstrated that a digital thoracic imaging system could routinely produce images that are perceived as equal or superior to conventional images. However, this good opinion of digital imaging was not held by all, particularly in musculoskeletal radiology. Again in 1995, Wilson asked why digital images as an alternative to plain radiographs had not been readily accepted.[32] Wilson concluded that although spatial resolution, image storage, and image transmission all presented potential problems, the greatest problem lay in image display. The commercially available workstations of that time did not compete well with the speed and simplicity of the conventional film and viewing box.

Accompanying the advances being made at the start of the digital information age, ideas were developed to build up a complete digital diagnostic system for radiography and fluoroscopy. In 1995, Lee and others from E.I. DuPont de Nemours & Co. patented a new digital detector for projection radiography using a multilayer structure consisting of a thin-film detector array, a selenium X-ray semiconductor, a dielectric layer, and a top electrode.[33] This prototype digital radiography technology electronically converts information from X-ray photons into digital image data.

DIGITAL FLUOROSCOPY AND DIGITAL SUBTRACTION ANGIOGRAPHY

Pioneer work in digital subtraction angiography (DSA) was done towards the end of the 1970s. In Kiel, Germany, in 1976, Heintzen and his group were able to enhance angiocardiographic images by digital preprocessing, including digitised image subtraction.[34] In 1978, Kruger, Mistretta, and others from Wisconsin noted that digital image processing has made possible a new class of images of high-contrast specificity and moderate spatial resolution.[35] They applied digital image processing techniques to standard fluoroscopic techniques, using image subtraction processing at the rate of 60 images per second. They obtained a mask as an image before contrast was administered, and this was subtracted from the images with contrast leaving only the contrast. In 1980, Christenson, Ovitt, and others from Arizona developed an experimental imaging system, using digital subtraction techniques, and demonstrated that it was possible to visualise the major arteries after intravenous contrast injection.[36] The impetus for developing such a system is to provide a means for detecting and quantifying atherosclerotic lesions in asymptomatic patients. At that time, the only means of showing such abnormalities was by using invasive angiography which had a morbidity. The group went on to publish human results displaying the low levels of intra-vascular contrast agent and the importance of high-speed digital computer processing, and display of the data was emphasised.[37] They recognised that there were still technical difficulties to be overcome.

Digital imaging techniques were further developed in the 1980s when analogue to digital (A/D) converters and computers were also adapted to conventional fluoroscopic image intensifier/TV systems. In 1980, Philips introduced the first digital subtraction angiography system, DVI (digital vascular imaging), which ushered in the digital age of conventional radiology.[38] It made use of the image intensifier and television technique and converted the analogue signal into digital format. In the early 1980s, a Philips DVI unit was placed at Lewisham Hospital in South London under Duncan Irving who with his wife, also a doctor, developed an interventional radiology service. After assessment, Irving concluded that because of the savings in time, radiation, and the volume of contrast medium used, which all lead to greater safety, as well as for the facility for the accurate monitoring of such procedures as angioplasty, that there would be a strong case for the siting of digital angiography units in theatres designed for interventional radiology.[39] A number of DSA systems were presented at the European Congress of Radiology in 1983, and by then several tens of units had been installed in many countries. Following a visit to Lewisham Hospital, such a system was installed by David Allison in the radiological vascular suite at Hammersmith Hospital. Whilst the system was used for intra-arterial examinations, considerable time was spent on intravenous DSA. Intravenous arterial imaging was never easy to interpret with confidence and by 1985, Hillman and Jewel were referring to disillusionment with intravenous studies and narrowing of its clinical applications.[40] Intravenous DSA, and indeed intra-arterial DSA, has now been replaced by other forms of non-invasive imaging, and intra-vascular studies are reserved for imaging guided interventions.

In 1997, John Rowlands and his group from the Imaging Research Program at Sunnybrook Health Science Centre, University of Toronto, Ontario, Canada, described the construction and evaluation of a projection real-time detector.[41] For dynamic X-ray imaging, that is, fluoroscopy, conventional X-ray image intensifiers were replaced with an active matrix flat panel device which was modified to include one of three innovative approaches. These were avalanche multiplication in selenium, intelligent pixels, or the novel photoconductor mercuric iodide.

As Hillman and Jewell have noted, both technical and clinical innovations have expanded the potential for digital imaging far beyond what was originally envisioned.

A BRIEF HISTORY OF PACS

The idea of Picture Archiving and Communication Systems (PACS) in the UK started in the early 1980s when there was the first attempt to create a fully filmless radiology department at St Mary's Hospital in Paddington, London. Oscar Craig (1927–2020) had met with Harald Glass, the Regional Scientific Officer, on 11 March 1982. Funding was requested from the Department of Health and Social Services, and in June 1985 the Minister of Health promised funding and in November of that year a paper entitled "Diagnostic radiology without films" was published in *The Practitioner* describing this new concept. However, it became apparent that the technology in 1985 was not adequate for the enormous task proposed. There was no filmless X-ray cassette, no clear means to rapidly transmit the vast amount of data, and little knowledge about the role of image compression. In 1987, Oscar Craig gave a talk at Hammersmith Hospital and David Alison became interested in developing an entirely filmless hospital. This was proposed in the late 1980s and coincided with technological developments. The British Government gave a grant in 1990 and computed radiography was introduced in 1993, with Hammersmith Hospital becoming filmless in 1996.

Developments were taking place in other hospitals in Europe.[42] In the Netherlands, Dr. Bakker and others worked on "The Dutch PACS Project". This was sponsored by the Ministry of Health Care and based at Utrecht University Hospital and was carried out from 1986 to 1989 with the aim of evaluating a Philips PACS prototype and to research the relationship between PACS and the Hospital Information Systems (HIS), the quality of diagnostic images, and assessment of technology. The first "digital reading room" was built at Utrecht University Hospital with the first coupling of PACS–RIS (Radiology Information System). They accomplished the complete digitisation of a small medical intensive care unit and the images and reports could be accessed at all times. The referring clinicians were very enthusiastic about the project since the radiological images were easily accessed without delay. The HIS and PACS were linked and the reports could be viewed on the PACS workstation. It was concluded that earlier availability of radiological images would increase the speed of diagnosis and treatment, which would reduce the average length of inpatient stay. At that time, it was concluded that PACS at Utrecht University Hospital would be approximately four times as expensive as conventional radiography; however, they estimated that by the year 2000 the cost of PACS would be the same as that of a conventional film-based system.

One of the few multivendor PACS installations began in 1986 at the University Hospital of Brussels working with the Multidisciplinary Research Institute for Medical Imaging Science. The University of Brussels was concerned about communication in hospitals and particularly communication between systems and also between users and systems. Working with others, they were concerned with modelling the PACS–HIS coupling, evaluating image transfer using high-speed networks, developing network structures, developing a multimedia software database which would enable intelligent information retrieval, and designing an adaptive user interface to increase diagnostic efficiency.

The first PACS project in Austria started in 1985 as a project between Siemens and the Department of Radiology in the University Hospital of Graz. The Department of Radiology at Graz had developed an in-house RIS and they contacted Siemens in 1985 to initiate a PACS pilot project. The aim of this PACS–RIS was for a sequential implementation which would meet the needs of both radiologists and clinicians. There would be digital acquisition, storage, and communication of radiological images from all imaging modalities and there was a gradual introduction of soft copy reporting and filmless working. The group at Graz were probably the first in Europe to produce an operational PACS–RIS coupling. This culminated at the end of the 1980s in the new Danube Hospital in Vienna, which was filmless and fully digital. It was led by Walter Hruby who was the head of the Radiology Department and who worked in cooperation with Siemens. The installation began in 1991, and it was proven that the costs of digital radiography were no greater than a conventional department.

The changes in computing technology and its influence on radiology have been astonishing with an evolutionary revolution or, as Water Hruby described it, a digital (r)evolution in radiology.[43] It has developed from a fruitful cooperation between radiology departments, industry, and government ministries, and with such a complex endeavour, no one group could do it by themselves. PACS is no longer confined to academic departments, but it is spreading to many hospitals and is profoundly changing all aspects of how we work. PACS gives us fast and reliable access to radiology and speeds up reporting, saves time, produces a better diagnosis, and results in an improvement in the quality of healthcare. Our film stores and racks of reports are now only a receding memory.

TELERADIOLOGY

The electronic transfer of images is as old as radiology itself, and indeed since the development of the electric telegraph, the transmission of images had been a goal. In 1895, Noah Steiner Amstutz (1864–1957), an American research engineer, invented what he called an electro-artograph, a machine which permitted the transmitting of photographs by telegraph.[44] Amstutz observed "that a picture, perfect in every detail, may consist of absolutely nothing but parallel lines". This is similar to the technique of engraving, and the process was automatic. Whilst the image was not digital, the artograph showed that images could be transmitted electronically (Figure 6.15a and b), and it was essentially a stage in the development of the fax machine.

The first use of teleradiology was probably in 1959 when Dr. Alberta Jutra linked two hospitals in Montréal. The hospitals were 5 miles apart and Jutra used a coaxial

Before Sending.   After Sending.
(a)   (b)

FIGURE 6.15 (a) The electro-artograph before sending. (b) The electro-artograph as received. From Story, 1899, see Note 44. Public domain.

cable to share videotaped fluoroscopic studies. Jutra was very perceptive and suggested a network connecting hospitals to doctor's offices to facilitate the exchange of radiological information, and he emphasised the need "to determine the efficiency, usefulness and economy of roentgenologic telecommunication". There were further developments in the 1960s and following an aeroplane crash at Logan Airport in 1962, a new medical department was set up at the terminal by Dr, Kenneth Bird. Bird connected Logan with the Massachusetts General Hospital, which was 2.7 miles away, using a telemedicine link. This microwave link was used by R. Murphy and co-workers in 1970 to transmit chest radiographs for remote interpretation. Murphy selected 100 chest radiographs of patients from a TB hospital, and a panel of three reached agreement in 92 of the cases and when compared to the radiology report, there was 77% agreement. Further studies were performed in the 1970s using a slow-scan television and images sent on ordinary telephone lines, and whilst images were acceptable for low-resolution nuclear medicine images, they were not acceptable for radiographs. By 1976, Lewis Carey in Ontario was using the Hermes satellite to investigate interactive telecommunication within a regional healthcare system. Over five months, Lewis Carey and his team provided 297 radiological consultations, and remarkably gave live supervision to 14 television fluoroscopic examinations which were being performed by a radiologic technician. Carey concluded that the consultation service was 90% effective when compared to direct film viewing.

There were more teleradiology studies in the 1980s in the United States[45] (Gitlin, 1986), with a large field trial in 1982 located at the Malcolm Grow Hospital in Andrews Air Force Base which served as a central location providing radiographic interpretation to one civilian and three military clinics. The radiographs were scanned using a video camera with a freeze frame constructing a $512 \times 512 \times 8$-bit matrix. The digitised X-ray images from the remote clinic were sent via telephone lines to the central medical centre and the radiological report was returned again by telephone lines. The results of the trial were successful and in 1984 the Uniformed Services University in Bethesda, MD, replaced Malcolm Grow Hospital as the hub linking the four sites. In the 1984 study, a matrix density of $1,024 \times 1,024$ pixels was used. The radiographs were scanned and converted to digitised images. The results of the 1982 and 1984 trials were very encouraging with only a few discrepancies between the video and film reports.

In 1985, Lewis Carey[46] reviewed teleradiology and predicted that improved access to radiologists, for 24 hours a day, will provide a level of service heretofore not possible. This proved to be the case and studies are now routinely reported by radiologists at home or transmitted between hospital sites for reporting. Images may also be reported in another country; however, there are clinical and legal concerns. There is a danger that teleradiology will adversely alter the relationship between the radiologist and clinicians and such a service needs to be used with caution. We now offer a 24/7 service with overnight scans being reported via a teleradiology link. However, technology continues to advance apace, and as an example it is now possible for radiological images to be reviewed on a tablet computer.

ARTIFICIAL INTELLIGENCE

Artificial intelligence (AI) is increasingly important in radiology. Oleg S. Pianykh has recently noted how significant it was that in 1955, John McCarthy said: "Probably a truly intelligent machine will carry out activities which may best be described as self-improvement".[47] This sounds very much like continuously learning AI.

The transformation of our world was initiated by two remarkable individuals, John von Neumann (1903–1957) and Alan Turing (1912–1954). They were working on integrating pure and applied sciences, and this led to the development of electronic computing. They had met in Cambridge in 1935 and Turing had recently written a paper improving on von Neumann's work. Turing may be considered the father of computing in the UK, and von Neumann in the United States. Turing defined intelligence as the state when an observer interacting with both a computer and a human could not tell the difference between them, and this is the "Turing model".

In 1986, Laurens Ackerman reviewed automated image analysis.[48] He saw artificial intelligence as the application of computer techniques that allowed them to take on the characteristics of human intelligence. Before this period, artificial intelligence and computerised image analysis were entirely separate. Computerised image analysis can be divided into signal processing, pattern recognition, and image understanding. It would seem likely that the development of computerised image analysis was stimulated by the US space programme with its need for the analysis of highly noisy images sent from space. In the 1970s and 1980s, Ackerman and others were using a TV camera to allow a computer to examine xeromammograms. The computer was able to extract parameters from mammographic lesions which could be classified into either benign or malignant. The machine had no intuitive idea as to the characteristics of the breast lesions. The essence of pattern recognition and signal processing is that there is no real understanding by the computer of the picture in terms of natural language, and so no knowledge base is created. It was the advent of true artificial intelligence that was to transform everything.

It is the advent of more developed applications that allows complex medical technology to be made universally available. As an example, at the end of the 19th century, Willem Einthoven (1860–1927) was investigating electrical activities in the body and in 1901 using his string galvanometer was producing electrocardiograms (ECGs) of remarkable quality. The first commercial ECG machine was made in 1912 by Sir Thomas Lewis and was about the size of a Smart car. It was a complex piece of apparatus to be interpreted by an expert and only used in a centre of excellence. However, by the 1950s, the first attempts were being made to automate ECG analysis. It was soon expected that digital computers would have an important role in ECG processing and interpretation, and now many models incorporate embedded software for analysis and interpretation and are the size of a laptop or smartphone.

Since IBM's Watson computer was able to beat the winning human competitors in the US game show Jeopardy! in 2011, what might now be the limits for the digital replacement of doctors by AI? We can foresee a CXR being performed by a radiography assistant under supervision and the digital image being accompanied by a provisional computer-generated report. Technology never stands still.

CONCLUSION

For the purpose of analysis, PACS, teleradiology, and artificial intelligence may be seen as separate subjects. However, such divisions have gradually become artificial with technical developments continuing to blur the distinctions. This is shown in the Pennine Acute Hospitals NHS Trust region-wide approach to the analysis of X-rays, MRI scans, CT scans, mammography, and other diagnostic images. In October 2020, a total of eight hospital trusts across Greater Manchester announced a project for a region-wide cooperation.[49] This is interesting for several reasons. The access of patient information by specialist centres will be greatly facilitated as will reviews by multidisciplinary medical teams, and the possibilities for research will be greatly enhanced. However, information governance will be critical, and patient confidentiality and the individuals control of their own data will need to be considered. Technological innovations raise ethical and philosophical issues and these cannot be ignored.

NOTES

1. Busch, U., Müller, C. 2017. *Wilhelm Conrad Röntgen – Photographien. Photographs.* Remscheid: Bergischer Verlag.
2. Gardiner, J.J. 1909. The origin, history & development of the X-ray tube. *The Journal of the Röntgen Society*, 5, 66–80.
3. Rønne, P., Nielsen, A. 1986. *Development of the Ion X-Ray Tube.* Copenhagen: CA Reitzel Publishers.
4. *Report of Tube Competition*, 1901. Archives of the Roentgen Ray, 6, 10–11.
5. Fehr, W. 1981. *C.H.F. Müller ... Mit Röntgen began die Zukunft (... with Röntgen began the future).* Hamburg: Überliefertes und Erlebtes.
6. Barclay, A.E. 1942. The passing of the Cambridge diploma. *The British Journal of Radiology*, 15, 351–354.
7. Liebhafsky, H.A. 1974. *William David Coolidge: A Centenarian and His Work.* New York: John Wiley & Sons, Inc.
8. Anderson, J.M. 1963. *Yankee Scientist, William David Coolidge.* Schenectady: Mohawk Development Service.
9. *X-Ray Studies.* 1919. General Electric Company, Schenectady.
10. Fitz-Simon, C.T. 1989. A historical review of the camera's use in medicine. *Medical Historian, Bulletin of the Liverpool Medical History Society*, 2 (July 1989), 19–26.
11. Isenthal, A.W., Ward, H.S. 1901. *Practical Radiography: A Handbook for Physicians, Surgeons, and Other Users of X-Rays.* 3rd ed. London: Dawbarn and Ward, Ltd.
12. Bentley, T.L.J. 1946. The evolution of radiography. *Radiography*, 12, 1–9.
13. Sprawls, P. 2018. Film-screen radiography receptor development: a historical perspective. *Medical Physics International Journal*, Special Issue, *History of Medical Physics*, 1, 56–81.
14. Taylor, N. 2021. X-ray processing: a brief "Personal" history and experiences as a dark room technician. *The Invisible Light, The Journal of the British Society for the History of Radiology*, 48 (May 2021), 5–11.
15. Scott, S.G. *X-Rays and Reminiscences.* Unpublished manuscript.
16. Anon. 1971. Miniaturisation of X-ray records at guy's hospital. *British Hospital Journal and Social Service Review*, December 4.

17. Coates, S. 2015. Roentgenizdat – an introduction. In: *X-Ray Audio*. Ed: S. Coates. London: Strange Attractor Press.
18. Salvioni, E. 1896. Investigation on Röntgen rays. *Nature*, 53, 424–425.
19. Oestreich, A.E. 1995. Professor William F. Magie and the American discovery of the fluoroscope, 1896. *American Journal of Roentgenology*, 165, 1060–1063.
20. Glasser, O. 1933. *Wilhelm Conrad Röntgen and the Early History of the Roentgen Rays*. London: John Bale, Sons & Danielsson, Ltd.
21. Thompson, E.P. 1896. *Roentgen Rays and Phenomena of the Anode and Cathode*. New York: D. van Nostrand Company.
22. MacIntyre, J. 1896. X-ray records for the cinematograph. *Archives of Clinical Skiagraphy*, 1, 37.
23. Zimmer, E.A. 1954. *Technique and Results of Fluoroscopy of the Chest*. Springfield: Charles C. Thomas.
24. Balter, S. 2019. Fluoroscopic technology from 1895 to 2019 drivers: physics and physiology. *Medical Physics International Journal*, (Special Issue, History of Medical Physics) 2, 111–140.
25. Guy, J.M., Golding, S.J. 2005. Head and neck radiology in Oxford: a brief and incomplete history. *The Invisible Light, The Journal of the British Society for the History of Radiology*, 23 (November 2005), 15–21.
26. Ardran, G.M. 1964. The value of cineradiography in medicine. *British Journal of Radiology*, 37, 819–825.
27. Ardran, G.M. 1954. The clinical value of X-ray image intensification. *Occupational Medicine*, 3, 324.
28. Hanley, H.G., Greenwood, F.G., Scott, M.G. 1961. Recent advances in X-ray television with special reference to urology. *British Medical Journal*, i, 1310–1313.
29. Sonoda, M., Takano, M., Miyahara, J., Kato, H. 1983. Computed radiography utilizing scanning laser stimulated luminescence. *Radiology*, 148, 833–838.
30. Fraser, R.G., Breatnach, E., Barnes, G.T. 1983. Digital radiography of the chest: clinical experience with a prototype unit. *Radiology*, 148, 1–5.
31. Floyd, C.E., Baker, J.A., Chotas, H.G., Delong, D.B., Ravin. 1995. Selenium-based digital radiography of the chest: radiologists' preference compared with film-screen radiographs. *American Journal of Roentgenology*, 165, 1353–1358.
32. Wilson, A.J. 1995. Filmless musculoskeletal radiology: why is it taking so long? *American Journal of Roentgenology*, 165, 105–107.
33. Lee, D.L., Cheung, L.K., Jeromin, L.S. 1995. A new digital detector for projection radiography. *Proceedings SPIE*, 2432, 237–247.
34. Brennecke, R., Brown, T.K., Bürsch, J.H., Heintzen, P.H. 1976. Digital processing of videoangiocardiographic image series using a minicomputer. *Computers in Cardiology*, IEEE Computer Society, Long Beach, 255–260.
35. Kruger, R.A., Mistretta, C.A., Houk, T.L., Riederer, S.J., Shaw, C.G., Goodsitt, M.M., Crummy, A.B., Zweibel, W., Lancaster, J.C., Rowe, G.G., Flemming, D. 1979. Computerized fluoroscopy in real time for noninvasive visualization of the cardiovascular system: preliminary studies. *Radiology*, 130, 49–57.
36. Ovitt, T.W., Christenson, P.C., H.D. Fisher III, Frost, M.M., Nudelman, S., Roehrig, H. 1980. Intravenous angiography using digital video subtraction: X-ray imaging system. *American Journal of Roentgenology*, 135, 1356–1141.
37. Christenson, P.C., Ovitt, T.W., Fisher III, H.D., Frost, M.M. Nudelman, S., Roehrig, H. 1980. Intravenous angiography using digital video subtraction: intravenous cervicocerebrovascular angiography. *American Journal of Roentgenology*, 135, 1145–1152.
38. Ludwig, J.W. 1986. Introduction. In: *Digital Subtraction Angiography in Clinical Practice*. Philips Medical Systems, Best.

39. Irving, J.D. 1984. Intra-arterial digital imaging with special reference to its use in inter-ventional techniques. In: *Digital Imaging, Physical and Clinical Aspects*. Eds: R.M. Harrison, I. Isherwood. London: Hospital Physicists' Association.

40. Hillman, B.J., Newell II, J.D. 1985. Foreword. In: *Symposium on Digital Imaging*. Eds. B.J. Hillman, J.D. Newell II. *The Radiological Clinics of North America*, 23, 175.

41. Zhao, W., Rowlands, J.A.1995. X-ray imaging using amorphous selenium: feasibility of a flat panel self-scanned detector for digital radiology. *Medical Physics*, 22, 1595–1604.

42. Lempe, H.U., Romeny, B. ter H., et al. 2000. Early development of PACS in Europe. In: *PACS Design and Evaluation: Engineering and Clinical Issues*. Eds. C.J. Blaine, E.L. Siegel. *Proceedings of SPIE*, 3980.

43. Hruby, W. 2001. *Digital (R)evolution in Radiology*. New York: Springer Wein.

44. Story, A.E. 1899. *The Story of Photography*. London: George Newnes.

45. Gitlin, J.N. 1985. Teleradiology. *Radiologic Clinics of North America*, 24(1), 55–68.

46. Carey, L.S. 1985. Teleradiology: part of a comprehensive telehealth system. *Radiologic Clinics of North America*, 23(2), 357–362.

47. Pianykh, O.S., et al. 2020. Continuous learning AI in radiology: implementation prin-ciples and early applications. *Radiology*, 24, 1–9.

48. Ackernan, L. 1986. Towards automated image analysis: future possibilities in historical perspective. *Radiological Clinics of North America*, 79–85.

49. https://medical.sectra.com/news-press-releases/news-item/7B8609C8688FC4DC/ (accessed 13 June 2021).

7 Radiologically Guided Intervention

During episodes of sickness and accident, there has always been a need for a medical intervention, which involves the traditional techniques of history taking and physical examination followed by an action. The interventions may be performed by those formally trained as a physician, and also by lay healers such as traditional bone-setters. Traditional imaging was visual, and the art of inspection was, and is, taught in medical schools. As technology and knowledge advanced, they were applied to medical interventions. For example, as knowledge of electricity increased, it was applied to medical treatment. John Wesley (1703–1791), the founder of Methodism, had an electrical machine (Figure 7.1), which may be seen in his house next to his chapel in London,[1] and he was a firm believer in its therapeutic value. Wesley's well-known book *The Desideratum: Or, Electricity Made Plain and Useful, by a Lover of Mankind and Common Sense* was published anonymously in 1759.[2] Wesley gave an account of both the current knowledge of electricity and its therapeutic utilisation. Wesley did not make extravagant claims, and always strived to do good, or at least to do no harm, which is something that all practitioners should have as an aim.[3] Electrical departments were gradually set up in hospitals and doctors became very used to electrotherapy. The electrical departments included apparatus for both diagnosis and therapy, and the history of electrotherapy is completely intertwined with that of early radiology.[4] The medical founder of electrotherapeutics was Guillaume Benjamin Amand Duchenne (1806–1875), known as Duchenne de Boulogne. In 1847, Duchenne had shown that an electrode placed on certain points of the skin, which he called "points d'election", would cause muscular contraction.[5] These points were later shown to correspond to the position of motor nerves. In 1887, the Electrical Department in Glasgow Royal Infirmary was opened,[6] and probably for the first time in hospital practice electrical wires were carried from the department to all the hospital wards to avoid the need to carry heavy apparatus around the building. Electrical treatments were initiated, and a powerful electromagnet was built and installed for the detection and extraction of metallic foreign bodies in the eye. These electrotherapy departments were quick to acquire the new X-ray apparatus and using it for medical interventions was obvious. The new Electrical Pavilion opened in Glasgow in 1897 and had a "localizer of the most recent design for detecting the situation of foreign bodies". The electrical and Röntgen ray aspects of the departments were completely enmeshed, and many of the early textbooks were as much concerned with electrotherapy and what came to be called radiology. Mihran Krikor Kassabian (1870–1910) was an Armenian-American physician and director of the Roentgen Ray Laboratory at Philadelphia's General Hospital. Kassabian was vice president of both the American Roentgen Ray Society and the American Electro-Therapeutic

FIGURE 7.1   John Wesley's electrical machine at Wesley's house in London. Author's photo.

Association. His well-known 1907 textbook was equally concerned with electrotherapy and with the Röntgen rays.[7] Kassabian was an expert in both electrotherapy and radiotherapy. Sadly, Kassabian was to become a radiation martyr at the early age of 39.

As indicated, the idea of using Röntgen's new rays for treatment and intervention was obvious to the pioneers. Intervention may be classified into medical interventions that are informed by radiology and interventions that are guided by radiology. Guided intervention may be related to the intermittent taking of a radiograph to monitor the procedure, or performing in real time using fluoroscopy.

IMAGING INFORMED INTERVENTION

While fractures may be assessed and managed using physical examination alone, the assessment of more complex fractures is difficult. For example, it was only following the work of John Poland (1855–1937) that the complexity of epiphyseal fractures in childhood began to be understood.[8] Poland noted that epiphyseal separations were continually mistaken for ordinary fractures or dislocations, and that in many cases suitable treatment was not adopted. His was one of the first surgical books to be illustrated with radiographs, and in the first section there is a simple bone age atlas.

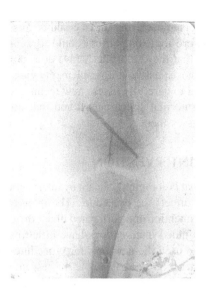

FIGURE 7.2 The outcome following the complex orthopaedic intervention of an epiphyseal fracture dislocation of the knee treated with open reduction and internal fixation at Beckenham Cottage Hospital in South London in 1913. Author's collection.

The radiograph (Figure 7.2) from 1913 shows the outcome following the complex orthopaedic intervention of an epiphyseal fracture dislocation of the knee treated with open reduction and internal fixation at Beckenham Hospital in South London that would have been unimaginable before radiography.

In early 1896, Howard Marsh saw a 22-year-old man who had injured his elbow when his horse had rolled over him.[9] The elbow was very much swollen and clinical assessment was very difficult. Marsh thought that it was probable that there was a fracture of the lower humerus involving the elbow joint. Marsh obtained a radiograph which showed an uncomplicated backwards dislocation of the elbow, which was confidently reduced under chloroform anaesthesia. Marsh noted that some of his colleagues regarded the Röntgen photography as a scientific toy, and interestingly warned that caution was needed to avoid either positive or negative conclusions which were erroneous. However, when it was a question as to the precise nature of the injury that had been sustained, Marsh stated that the method will often be valuable in the highest degree.

Poland himself continued with a cautious note in his Hunterian Oration of 1901[10] noting that unfortunately the discovery soon attracted much public attention, and that many imaginative minds drew uncertain conclusions and spread abroad erroneous impressions on the subject. Poland commented that the natural consequence of this has been great discredit of the method, which had been the fate of many other discoveries. Remarkably, he also said that no fracture of the joints, especially of the elbow, should be treated until an accurate diagnosis has been made and the X-rays employed. This was only six years after the discovery by Röntgen, and showed how widespread and rapid was the uptake of radiography. Orthopaedics underwent a profound change, with Poland declaring that it could be questioned whether treatises on fractures written before the Röntgen era could now be regarded as authentic. Poland also maintained that the use of the X-rays does not supersede the ordinary methods of clinical diagnosis, and that it supplements them in a most valuable way. So, we learn from Poland to carefully assess new techniques, to avoid an uncritical enthusiasm, to learn the normal appearances, and that imaging should always be combined with clinical assessment.

IMAGING GUIDED INTERVENTION

Guided intervention is what is commonly meant by interventional radiology (IR) when a real-time visualisation directs the procedure. The intervention may be both diagnostic and therapeutic in intention and performed under different imaging modalities. Whilst most reviews of guided interventions have little to say before Dotter's work in the 1960s, the pioneers used the new rays for procedures such as fracture reduction and foreign body extraction using a simple fluorescent screen or a cryptoscope. The cryptoscope, a hooded fluorescent screen, could be held in the hand to observe the object of interest, or the cryptoscope could be strapped to the surgeon's head as an operating cryptoscope. The screen could be hinged to allow an alternating direct vision or fluoroscopy and make real-time imaging easier. However, excessive use of the fluorescent screen was to result in significant morbidity to the early radiologists.

PNEUMOTHORAX ASPIRATION

In 1907, John Fawcett (1866–1944) who was a surgeon at Guy's Hospital in London aspirated a pneumothorax under X-ray control.[11] The patient was a 22-year-old man with a right-sided pneumothorax that had been confirmed by radiography. After 19 days, the patient was unchanged, and an aspiration was performed. The patient was placed on his back and lying on an X-ray couch. An intensifying screen was placed over the front of his chest, and a trocar and cannula were inserted into the pleural cavity. The lung was seen compressed towards the spine, and Fawcett could therefore determine the exact distance that the trocar should be passed to avoid contacting the lung surface. The trocar was withdrawn, and the cannula was connected to an under-water seal in a flask. Air was seen to bubble through into flask with each inspiration. A negative pressure was applied, and the lung was at once seen on the X-ray screen to expand. As the lung expanded, the cannula was gradually withdrawn until, when the lung appeared to be fully expanded, it was removed, and the puncture wound was sealed. A second radiograph was taken which showed the lung to be fully expanded. The patient had bed rest for two days and was discharged from the hospital 12 days later, when re-examination of the thorax during that time with the X-rays showed that the lung continued to act perfectly. Pneumothorax was managed conservatively in this period, and usually resolved by itself within three weeks; however, a proportion of patients became chronic invalids. This patient was symptomatic and was not improving, and so an intervention was reasonable. The use of a trocar for aspiration was recommended by Sir William Osler (1849–1919) in his standard textbook *The*

Principles and Practice of Medicine[12] in cases of tension pneumothorax, although not for spontaneous or traumatic pneumothorax.

FOREIGN BODY REMOVAL

Foreign bodies could be located either using a purely radiographic technique or else fluoroscopically, and this was performed from the earliest days. In the fluoroscopic technique, the patient was placed on a stretcher and an under-couch tube is used. By moving the fluorescent screen, and using a simple parallax technique, the depth of the foreign body may be determined. The fluoroscopic technique had the advantage of speed and the avoidance of processing of X-ray plates; however, there was a higher radiation dose to the operator. The localisation of bullets was to prove of special value in military surgery.[13] An example is the Tirah Campaign of 1897 in the north-west of India on the frontier with Afghanistan. Surgeon-Major Walter Beevor (1858–1927) was a regimental surgeon with the Coldstream guards. He used his own Röntgen Ray apparatus, as was typical during this period, and this enabled him to show the exact position of a bullet whose course in the body could not otherwise be followed.[14] In one instance, a soldier, who was wearing a sheepskin coat, was hit by a bullet which became covered by a complete envelope of sheepskin. When the wound was probed by the medical officer, a soft object was felt which was not thought to be a bullet. Radiography showed that this was a bullet, and it was easily extracted. The negative results of radiography were also seen as excellent, since the absence of any foreign substance was demonstrated, and unnecessary exploration was avoided. By the time of the Great War, the extraction of projectiles was commonplace, and the radiographic procedure was described as being of almost mathematical precision.[15] The use of fluoroscopy has been implicated in the early deaths of the Franco-Polish physicist Marie Curie and the Anglo-Irish doctor Florence Stoney. Both women were heavily involved in military radiology on the Western Front during the Great War where the rapid localisation of foreign material was essential, and both spent many hours with the fluoroscope.

Florence's work in the Great War was particularly concerned with the localisation of bullets, and her knowledge of anatomy greatly helped the surgeons. This was particularly illustrated in a case described by John Lee who was an Australian RAMC surgical colleague at the Fulham Military Hospital. Lee describes a soldier injured in France in 1915 who had developed an intracranial abscess with a fragment of retained shrapnel.[16] A piece of shell had entered the skull in the middle of the left temporal region. Lee had tried unsuccessfully to extract the shrapnel using an electromagnet, and then it occurred to him that it might be possible to see and remove it under X-ray fluoroscopic guidance. Lee discussed the patient with Stoney, and a combined surgical procedure was therefore performed. With Florence's assistance, it was possible to see both the surgical probe and the fragment and their relation to each other. The fragment had been pulled by the magnet and was lodged within the cerebral substance. Forceps were passed by Lee to a depth of 4 inches and the shrapnel was removed. This is an early example of radiologically assisted neurological intervention and Lee recorded his indebtedness to Florence.

EARLY NATURAL ORIFICE INTERVENTIONS

Apart from the manipulations of fractures or the removal of foreign bodies, early radiological interventions also included what we would now call natural orifice interventions. In other words, the natural openings of the body are used for access, and there was no need to create an opening artificially. This could be in the renal tract using an electric cystoscope and catheters, or by using the natural openings of the alimentary tract.

OESOPHAGEAL INTERVENTION

As can be seen from the experiences of Stoney and Fawcett, radiological guided interventions are older than is commonly presented. For example, in 1988, when Grundy and Belli described their series of patients who had received balloon dilatation of upper gastrointestinal tract strictures, they stated that such techniques were derived from angioplasty, that is, treating arterial strictures,[17] and indeed this was first proposed in 1981. None of the references quoted by Grundy and Belli were earlier than 1980. Of the 30 patients that they treated, 5 had achalasia of the cardia, and they were apparently unaware of the series presented by Alfred E. Jordan in 1923.[18] Jordan treated achalasia of the cardia, otherwise called cardiospasm, using a dilating bag or balloon. The lower oesophageal narrowing was diagnosed by swallowing a meal of bismuth carbonate emulsion. Many of the patients had advanced disease having had symptoms for decades and suffered from malnutrition. In one patient, Jordan could see the entire opaque meal still in the oesophagus six days after the initial bismuth meal. Clinically, such patients were thought to have a dilated stomach and were regularly washed out until the opaque meal revealed the true cause of the problem. Jordan noted that the narrowed area could be traversed with a surgical bougie with a metallic acorn located at its end, and that this could be followed on the fluorescent screen. A gum elastic catheter was used with a spindle-shaped dilating bag attached to the lower end above the acorn. The bag was passed into the patient when empty and using the acorn as a guide. When the acorn was in the stomach, the balloon was dilated. The balloon was then withdrawn through the lower oesophageal sphincter (Figure 7.3). With a more severe narrowing seen when the cardiospasm was long-standing, a reel of silk tread ligature attached to a small piece of lead shot was used. The shot with attached silk was swallowed and after an hour or so was seen in the lower oesophagus on the screen. After hours or days, the lead shot passed into the stomach and then into the intestines. Jordan said that the patient would carry the silk reel in his waistcoat pocket, and the silk projecting from the mouth is passed over one ear. When the lead shot had passed from the body naturally, the thread could be used as a guide. The slack in the thread was withdrawn and when taut, the perforated metallic acorn of the dilator was threaded onto it. Under screen control, the dilator is passed through the cardiac reason using the thread held taut as a guide. The process of dilatation was then carried out.

Jordan commented that relief was immediate, and the patient was now able to take solid food. The procedure needed to be repeated several times before the patient

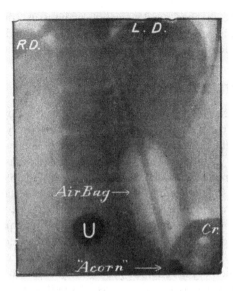

FIGURE 7.3   Dilating airbag in the stomach. U is the umbilicus, RD and LD are right and left hemi-diaphragm, Cr is the iliac crest. From Jordan, 1923, see Note 18. Public domain.

was cured, and the time between dilatations was gradually increased. Jordan was unaware of the cause of achalasia; however, he thought that the great improvement following the dilatation did not prove that a simple spasm was the primary cause.

LARGE BOWEL INTERVENTION

The treatment of large bowel intussusception avoiding operative surgery has a long history and has been reviewed by Frush and colleagues.[19] Intussusception is when the bowel folds in upon itself causing a blockage, and is usually seen in children. Treatments were varied, including oral medications, enemas, bloodletting, electrotherapy, and manual manipulation via the rectum. Enemas as treatment could either be hydrostatic using water or pneumatic using gas. Frush reported the case of Ulhoom in 1741 when rectal tobacco smoke was used to inflate the bowel. Apparently, better-quality tobacco was more efficacious, and tobacco leaves could also be added to the enema fluid. With the development of asepsis and anaesthesia in the nineteenth century, surgery became the most common treatment. Bedside diagnosis of intussusception was however difficult and following the discovery of X-rays, it became possible to make a diagnosis using an opaque enema. In 1913, Ladd was able to obtain a radiograph following an attempt to reduce an intussusception using a bismuth enema. There then followed an increased interest in non-operative therapy, and from the 1920s fluoroscopy during an opaque enema could be used to guide reduction. There were controversies about using either liquid contrast medium or air and of the exact role of transabdominal manual palpation as an adjunct to the enema. Bismuth as an enema was replaced by barium, and the barium enema remained the gold standard for the diagnosis and reduction of intussusception being gradually replaced by ultrasound in the 1980s. By 1996, Daneman and Alton were recommending ultrasound for diagnosis, and an air enema for reduction.[20]

RENAL TRACT INTERVENTIONS

Edwin Hurry Fenwick (1856–1944) was a urologist at the Royal London Hospital and a pioneer of electrical cystoscopy.[21] Following the discovery of the X-rays, Hurry Fenwick recognised the potential of Röntgen's discovery and became an enthusiastic supporter of the new technique. That the new rays could be used to investigate the urinary tract was appreciated soon after their discovery in 1895. Before the use of radiography, the investigation of upper renal tract disease was not easy. For example, the surgeon could perform a cystoscopy and a ureteric bougie with wax fixed on to its tip would be passed up the ureter. Any obstruction to the passage of the bougie could be felt and the distance inserted noted, and when removed if the wax was seen to be scratched, then this was evidence of a stone. The early apparatus was of low power and visualisation of the abdomen was not easy, although abdominal compression devices were of some assistance. It was only following the development of a more powerful X-ray apparatus from about 1905 with "instantaneous radiography" that the image quality improved. The X-ray shadow pictures, or skiagrams, that were obtained were often confusing and it was difficult to define and differentiate the nature of the calcifications. The calcifications had a number of origins, including calcified lymph nodes, calcified atheroma, ureteric calculi, or phleboliths. It should be noted that traditionally our radiological techniques were used to confirm a suspected clinical diagnosis, and since the technique was often quite invasive, it was only applied when there was a reasonable chance of the examination being positive.

Using the cystoscope, it became possible to introduce a ureteric bougie, which could be combined with abdominal radiography. This was performed by Schmidt and Kolischer independently in 1901, having been suggested by Tuffier in 1898.

In 1905, Hurry Fenwick had developed ureteric bougies with their walls impregnated with a metal for increased radiographic contrast (Figure 7.4). The opaque bougie labelled EHF stands out in good black shadow; however, the bougie labelled foreign shows only a faint shadow. Fenwick notes that it was important to secure a radiographic bougie which throws the darkest shadow. Following the positioning of the bougie at cystoscopy, radiography would demonstrate the course of the ureters and the potential urinary tract location of the calcifications. The position of an opacity in relation to the ureter could be determined with confidence, and a phlebolith could be confidently distinguished from a ureteric calculus (Figure 7.5). In 1897, he had also used a small fluorescent screen, the cryptoscope, to examine diseased kidneys at the time of operative surgery. In addition to the radiopaque bougies as a positive contrast, Hurry Fenwick also used a negative contrast with air inflation of the bladder to make the pelvic area less dense. The radiopaque impregnated bougies developed into the radiopaque catheters that are essential for angiography.

Hurry Fenwick commented on the distressing situation with the failure of operative surgery when a kidney was opened, and therefore was damaged to some extent,

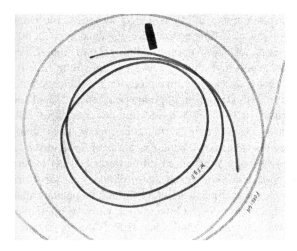

FIGURE 7.4 A comparison radiograph of two ureteric radiographic bougies. The opaque bougie labelled EHF "stands out in good black shadow", and is the Marshall bougie made by Bell and Croydon. The bougie labelled "foreign" shows only a "faint shadow". From Fenwick, 1908, see Note 22. Public domain.

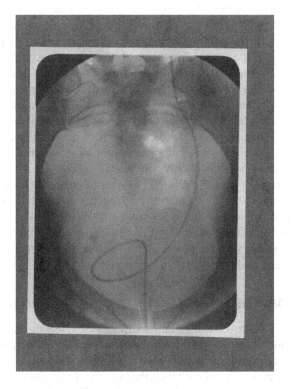

FIGURE 7.5 The use of a radio-dense ureteric catheter in 1914 by Fenwick. The relationship of any shadow to the position of the ureter may be determined easily. Author's collection.

to remove a stone when it was in fact no longer located in the kidney and was now within the ureter. Fenwick estimated that this happened in about 30% of cases when the X-ray expert was not called upon to help in the diagnosis. Fenwick's opinion was that the X-ray expert (or radiologist as they came to be called) could guide the urinary surgeon (that is urologist) with a precision unattainable before the introduction of the (X-ray) method, and this opinion was without cavil (or disagreement). Fenwick was writing in 1908 when the techniques used were still quite primitive and before the introduction of either retrograde or intravenous pyelography. In 1908, Fenwick produced his well-known book *The Value of Radiography in the Diagnosis and Treatment of Urinary Stone* in which he described his experience with radiography.[22] This book is a radiological and urological classic. Hurry Fenwick was one of the first to practice clinical-radiological-pathological correlation, and related the clinical findings with radiography and then with the operative findings. Quite remarkably, he was teaching operative electrical cystoscopy using a bladder phantom before 1900.

MODERN IMAGE GUIDED INTERVENTION

The topic of IR can be seen to be as old as radiology itself. David Allison, a pioneer of therapeutic embolisation, reviewed IR in 1983[23] and noted the development of a new subspeciality of IR. In the same way that diagnostic radiology and radiotherapy separated, IR is separating from diagnostic radiology with separate training and accreditation. In the United Kingdom, IR gained subspeciality status in 2010. The subspeciality training takes place over six years with a certificate of completion of training in clinical radiology with interventional radiology subspecialisation. In reality, image guided interventions have never been the sole provenance of radiologists, and as an example, IR in urology was developed by urologists as shown by the life and work of Hurry Fenwick. This is true for both interventional diagnostic and interventional therapeutic studies. Before the speciality of neuroradiology was developed, it was clinicians who commonly performed the myelogram or encephalogram with the radiologist reporting the study. As another example, the London neurologist Peter "PK" Thomas (1926–2008) used to perform his own carotid arteriograms by direct puncture in the neck in the 1950s and early 1960s. Even as late as 1980, an abdominal trans-lumbar aortogram by direct needle puncture might be performed by a surgical registrar, with the new registrar teaching the old registrar how to perform the procedure, and the resultant image reported by the radiologist.

THE BIRTH OF MODERN INTERVENTIONAL RADIOLOGY

The key date for modern IR was 16 January of 1964. On this day, Charles T. Dotter (1920–1985) with his trainee Melvin P. Judkins (1922–1985) performed the first-ever percutaneous transluminal angioplasty (PTA) at Oregon Health and Science University. In the early years of radiology, the purpose of vascular catheterisation was solely for diagnosis; however, following the work of Dotter, catheterisation increasingly became a prelude to intravascular intervention, and today vascular studies are performed non-invasively and diagnostic angiography has disappeared. At that time in 1964, the ability to diagnose vascular disease was far greater than the ability to treat the problem, and this concerned Charles Dotter. Arteriosclerotic obstructions were common, and Dotter was looking for improvements in non-surgical interventions. Dotter developed an idea for arterial dilatation, and tested it using a cadaver. Dotter was making guide wires out of piano wire, and a colleague was able to pass a guide wire past a stenosis in a cadaveric artery. Dotter next passed a catheter over the wire, and then a larger catheter over the first catheter.[24] In 1963, William "Bill" Cook (1931–2011), with his wife Gale, started what became Cook Group in a spare bedroom in their apartment.[25] In the November of 1962, Bill Cook and Dotter met at the Cook's rather low-budget booth at the convention of the Radiological Society of North America in Chicago. Cook's company was then only four months old, and on the stand Cook had wire guides, needles, a blowtorch, and was making catheters in front of his fascinated visitors. Dotter asked to borrow Cook's blowtorch for the night and returned the next day with ten perfectly made catheters. Cook recounted that he sold the catheters for $10 each and that this was enough to pay for the booth! Bill Cook and Charles Dotter developed a lifelong friendship, which was to prove mutually beneficial. Following a discussion about catheter and wire guide manufacture, Cook visited Dotter in Oregon. Cook could not afford the airfare and so Dotter paid his expenses. Dotter had his own laboratory where technicians made their own wire guides. Dotter was also producing his own catheters using Cook's Teflon tubing. This was long before the contemporary period with pre-packaged and preformed sterile catheters. The catheters were supplied unsterile as a long loop and could be cut and formed as desired (Figure 7.6). There were a variety of recommended shapes for angiography for selective studies of particular arteries (Figure 7.7). The fabrication technique for angiographic catheters was time-consuming and required skill and patience (Figure 7.8). The stages were as follows:

1. *Tip forming:* A forming wire was placed in the catheter, the catheter was warmed, and the catheter was pulled from both ends.
2. *Tip finishing:* The tip was tapered and cut.
3. *Shaping:* This was performed by inserting a preformed wire and placing the tube in warm water. The catheter was then quenched in cold water.
4. *Side hole forming:* This was achieved using a sharpened hold punch cannula.
5. *Flaring:* The catheter was cut to a preselected length. The tubing would automatically flare when placed near an alcohol lamp.
6. *Sterilising:* The tubing was filled with a cold sterilising solution and also fully immersed. Gas sterilisation could also be used. The tubing was not to be autoclaved.

It was to be the relationships between the new breed of radiologists interested in interventions and the various companies that made needles, wires, and catheters that was to prove so fertile and productive. Cook gathered a group of radiologists who were all interested in access to the body using the Seldinger technique, including the

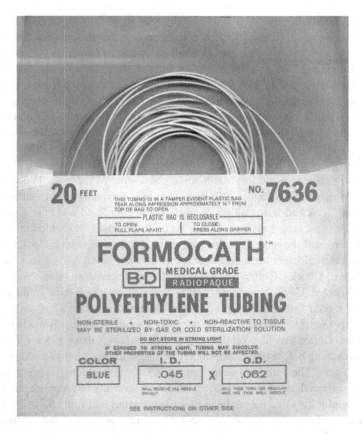

FIGURE 7.6 Formocath polyethylene tubing, made by Becton, Dickinson and Company and sold in 20 ft. (6.1 m) lengths. From the author's collection and reproduced courtesy of Becton, Dickinson and Company.

Italian Cesare Gianturco (1905–1995), the German Andreas Gruentzig (1939–1985), the Czechoslovakian Josef Rösch (1925–2016), and Charles Dotter.

Dotter's first patient was a bedridden 82-year-old woman with peripheral vascular disease and a painful, ischaemic, and ulcerated foot. The angiogram showed an atheromatous obstruction of the superficial femoral artery at the level of the adductor hiatus. The vessel lumen was much reduced, and the peripheral vessels filled only slowly. The surgeon had recommended amputation; however, the patient had refused surgery. Dotter performed a percutaneous trans-femoral catheter dilatation of the narrowed area, and he recorded that it took only a matter of minutes and was without difficulty. When the dilating catheter was removed, good pulses were palpable in the lower leg and foot. Angiography revealed that the stenosis was no longer present. There was an immediate diminution in pain, discoloration, and coldness of the foot. During the week after the procedure, there was a healing of the ischaemic skin changes, including the ulceration of the lower leg. A follow-up angiogram was performed three weeks after the procedure and the lumen remained patent. At eight

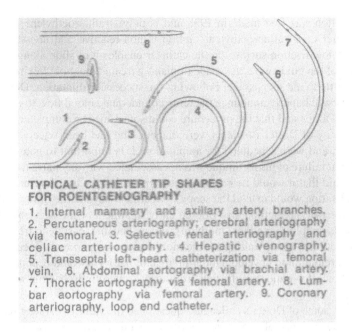

TYPICAL CATHETER TIP SHAPES
FOR ROENTGENOGRAPHY
1. Internal mammary and axillary artery branches.
2. Percutaneous arteriography; cerebral arteriography
via femoral. 3. Selective renal arteriography and
celiac arteriography. 4. Hepatic venography.
5. Transseptal left-heart catheterization via femoral
vein. 6. Abdominal aortography via brachial artery.
7. Thoracic aortography via femoral artery. 8. Lumbar aortography via femoral artery. 9. Coronary
arteriography, loop end catheter.

FIGURE 7.7 Typical catheter shapes for angiography. From the author's collection and reproduced courtesy of Becton, Dickinson and Company.

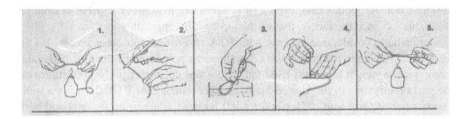

FIGURE 7.8 Catheter fabrication methods. From the author's collection and reproduced courtesy of Becton, Dickinson and Company.

months, the patient was ambulatory, the ulceration was gone, the gangrenous toes had separated, and the sites were healed. The patients' pain disappeared, and she was able to walk until she died three years later.

The paper by Dotter and Judkins on this transluminal treatment of arterial obstruction appeared in November 1964, and they described ten patients.[26] In the opening paragraph, they say: "Despite the frequency and importance of arteriosclerotic obstruction, current methods of therapy leave much to be desired. Nonsurgical measures, however helpful they may be, provide the patient little more than an opportunity to live with his disease". The paper gives a detailed account of the technique that they used. The therapeutic procedure involved an antegrade catheterisation and dilatation of the atheromatous region using a tapered and radiopaque Teflon dilating catheter. Teflon was first made in 1938 and is polytetrafluoroethylene, a synthetic fluoropolymer of tetrafluoroethylene. Teflon is non-reactive and non-stick and by providing a low-friction surface on the catheter enables it to slide along the artery. The centre of the guide wire was made of piano wire and the more flexible tip could be bent a little to aid its passage. Following the successful dilatation, Dotter noted: "the patient will happily announce the return of adequate blood flow to the troubled extremity". Dotter said that the procedure could sometimes be completed in 10 or 15 minutes. Many of Dotter's patients were diabetic and had been rejected for definitive surgery, and were scheduled for amputation. It is important to note that Dotter recorded that failure of his technique was not associated with any harm to the patient.

Dotter said that it would be reasonable to expect that the transluminal technique for recanalisation would extend the scope of treatment beyond the limits of the present-day surgery. The method also offered early treatment of the ischaemic leg, and in view of its simplicity and low morbidity, it now became feasible to treat intermittent claudication without waiting for the more serious symptoms to occur or for the collateral circulation to develop. Dotter predicted that PTA would become the treatment of choice for patients suffering from arteriosclerotic ischaemia of the lower extremities.

However, some of Dotter's colleagues questioned the value of catheter recanalisation since the presence of coexisting disease in distal arterial branches would defeat its purpose. However, Dotter knew that by treating a proximal, and pressure gradient producing stenosis, the result would be an increase in the distal blood pressure and therefore an increase in the flow of all patent run-off branches, whether they are narrowed or not. As Dotter said, even a rusty sprinkler may prove capable of doing a creditable job once the faucet, that is, the tap, is fully opened!

Dotter had considerable success with PTA in Europe and other places, but in the United States the technique was not to become popular until the mid-1970s and many were to view it as a short-lived fad. There was not surprisingly a reluctance on the part of surgeons in the United States to refer patients for PTA. There is a well-known request form sent to Dotter for a left femoral arteriogram in 1964 and both underlined and in capitals on the request form is written "VISUALIZE BUT DO NOT TRY TO FIX!"

In 1976, Geraldus van Andel published a comprehensive account of the new technique, which he termed the "Dotter procedure".[27] The book is a comprehensive account of PTA and is a classic text of interventional radiology. Van Andel starts his book by quoting Langenbeck who said that it is less important to find operative methods than to seek ways and means of avoiding them. There had been presentations on PTA in the Netherlands in the early 1970s, and in 1973 van Andel was fascinated to hear Dotter speak on PTA at the XIIIth International Congress of Radiology held in Madrid. He returned to the Netherlands and with the cooperation of the surgeons at the Diaconessenhuis in Eindhoven developed the technique. The technique of PTA, unlike vascular surgery, could be readily acquired by radiologists who were already skilled in arterial catheterisation. It was simple to perform, few instruments were needed, and there was no need for a large, trained staff. The hospital stay was short, the patient could be mobilised quickly, and it caused little discomfort. The procedure had few complications, the patient's symptoms were seldom made worse, and the procedure could be repeated if necessary. The procedure could be performed on the elderly and frail, and general anaesthesia was not needed. Van Andel noted that the treatment costs would be less for the individual patient and for the community. He also said that the vascular surgeon would also be able to concentrate on procedures that only the surgeon could perform. In reality, the "turf wars" between radiology and surgery have persisted, and indeed the boundaries between interventional radiologists and clinicians who use imaging for interventions have also become indistinct.

Many have followed Charles Dotter in developing vascular intervention, including the use of balloon catheters and stents. As Dotter predicted: "No doubt the interest and ingenuity of others will lead to refinements of technic as well as further clarification of the role of this attack on arteriosclerotic obstructions".

FURTHER DEVELOPMENT OF INTERVENTIONAL RADIOLOGY

Charles Dotter reviewed interventional radiology in Ben Felson's (1913–1988) iconic publication *Seminars in Roentgenology* in the January of 1981.[28] Both the January and April issues of Seminars were devoted to intervention and Dotter's was the first article. Dotter commented that the term "interventional radiology" as a term was imperfect but useful, and with Allison was recognising it as a recently emergent subspeciality of radiology. The term had been coined by Alexander Margulis (1921–2018) in 1967 and referred to various percutaneous techniques that were imaging guided and were alternatives to surgery. Dotter had a particular admiration for Porstmann's technique, which was the non-operative and permanent closure of the patent ductus arteriosus and called it a classic example of interventional radiology.[29] He noted that over 300 people had their ductus closed non-operatively, which was clearly a better way to treat that lesion. Porstmann reported few failures, no recurrences, low morbidity, and zero mortality. In 2003, Borges and colleagues presented the 30-year follow-up of a now 68-year-old woman who had undergone a transcatheter closure of a patent ductus arteriosus (PDA), which was performed percutaneously using an Ivalon closure plug with a metallic skeleton as the occluder device.[30] The publication was in memoriam of Werner Porstmann (1921–1982), who introduced the method and was the first to perform this procedure at the Charité University Hospital of Berlin in 1966. Dotter sees interventional radiology as starting in the early 1960s, and then undergoing near exponential growth in the 1970s. As a reason for the growth, Dotter identifies the impetus as coming from better tools, better techniques, and better image monitoring systems. The needle and catheter system would have two roles. It could be used either for "mechanically attacking" a lesion as in angioplasty, foreign body, or gallstone retrievals or for drainage of an abscess, obstructed renal tract, or biliary tree. Alternatively, the aim may be to deliver a drug, occlusive material, or a specific device.

If the 1970s could be viewed as a time of exponential growth, the momentum was continued with perhaps even more vigour into the 1980s, and innovation continues today. There were a number of factors that advanced interventional radiology in the 1980s. The fertile relationships between radiologists and the companies making needles and catheters meant that pre-packages and sterile devices were available for immediate use, and complicated and time-consuming preparation was not needed. Secondly, the newer non-ionic contrast agents were becoming available. The older ionic contract agents were painful to inject, and it was difficult for the unanaesthetised patient to remain still. The most significant factor inhibiting the use of the new non-ionic agents was their prohibitive cost; however, the costs gradually became acceptable in the 1980s, and by the 1990s the ionic agents were gradually passing out of use.

It has been the use of digital radiography that has also resulted in the major leap forward in imaging guided intervention. The traditional simple fluorescent screen was viewed in a darkened room and the operator needed to have dark adapted eyes to see the screen. Performing procedures in a darkened room was not easy. X-ray television with image intensification had been introduced in the 1960s and would have facilitated the growth of intervention in the 1970s. Whilst the television monitor was bright, it only persisted whilst the patient was being irradiated. Whilst the rapid film-changing devices needed for angiography were used, and indeed Charles Dotter had been involved in their creation, the film processing of the cut film still resulted in a significant delay before the angiographic run or series could be viewed. Abdominal or visceral angiography was a laborious process, and subtraction methods to show the blood vessels alone had to be performed following the procedure using the traditional photographic methods.

Digital fluoroscopic X-ray units with digital subtraction arteriography (DSA) were first developed for vascular applications and enabled such structures to be seen more clearly (Figure 7.9).[31] The patient was positioned (Figure 7.10) with the arm

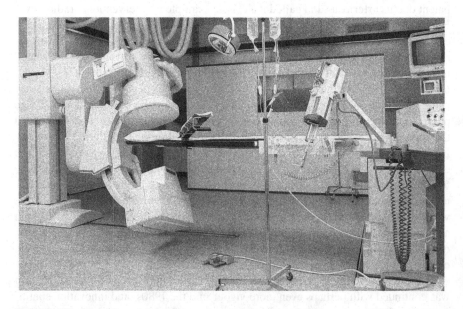

FIGURE 7.9 The Diagnost Arc A: a microprocessor controlled angiographic stand. From De Vries, 1986, see Note 31. With permission of Philips Medical Systems.

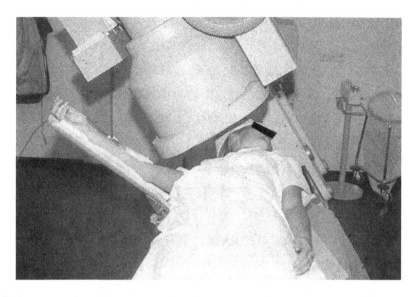

FIGURE 7.10 Patient positioned for examination of the right carotid bifurcation. From De Vries, 1986, see Note 31. With permission of Philips Medical Systems.

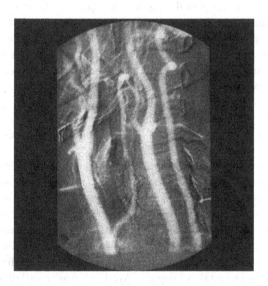

FIGURE 7.11 Left oblique position with apparently normal left and with carotid bifurcations. From De Vries, 1986, see Note 31. With permission of Philips Medical Systems.

extended for a peripheral venous injection, and both left oblique (Figure 7.11) and right oblique (Figure 7.12) projections of the carotid vessels could be obtained. In the early 1980s, a Philips fluoroscopic unit with DSA was installed at Lewisham Hospital in South London. In 1983, the radiologist Duncan Irving reviewed his

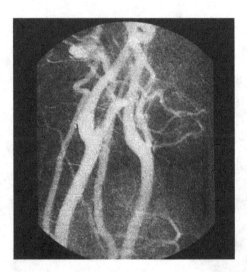

FIGURE 7.12   Right oblique position with the head turned to the left gives a better view of the right carotid bifurcation. From De Vries, 1986, see Note 31. With permission of Philips Medical Systems.

experience of intra-arterial digital imaging with special reference to its use in interventional techniques.[32] In DSA, a mask is made without a contrast medium, the contrast is injected, and the mask is subtracted from the contrast image, and the resulting image is of the vessel alone. Such units were made with two monitors, with one monitor showing the live image and the second displaying a selected still image. In describing his experience of angioplasty, Irving noted that having the static image of the arterial stenosis displayed on one monitor whilst performing the angioplasty was of great aid in positioning the dilating balloon. However, for Irving, the real value of DSA was in the application of "roadmapping" techniques. For a roadmap, a mask was made in a continuous mode whilst injecting a dilute contrast medium. This information is stored, and all subsequent images taken in conventional fluoroscopy are superimposed and continuously subtracted. This technique would make the passage of a difficult stricture safer and easier, would help with super-selective catheterisation, and in puncturing impalpable arteries. Irving's conclusion was that, whilst the main role at that time was for the performance of outpatient angiography using intravenous injections, there was a strong case for siting DSA units in theatres designed for interventional radiology, and this has proven to be the case. Following a visit to Lewisham Hospital, a Diagnost Arc U 14 was installed at Hammersmith Hospital with an immediate impact on the services offered (Figure 7.13).

THE DEVELOPING SCOPE OF INTERVENTIONAL RADIOLOGY

In 1929, the first documented human cardiac catheterisation was performed by the 25-year-old Werner Forssmann (1904–1979) in Eberswalde, Germany.[33] He directed a catheter through a vein in his left arm towards his heart and recorded it with a

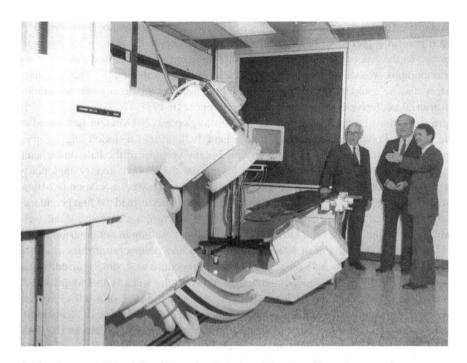

FIGURE 7.13   The opening of the Philips Diagnost Arc U 14 at Hammersmith Hospital on 19 December 1984. Robert Steiner, a pioneer of cardiac angiography, is seen on the left; a Philips representative centrally; and David Alison, a pioneer of intravascular interventional radiology, is seen on the right. Author's collection. Unknown photographer.

radiograph. Although his older colleagues were aghast and Forssmann was removed from his hospital post, his work became the basis for cardiac catheterisation. In 1941, André Cournand (1895–1988) and Dickinson Richards (1895–1973) developed Forssman's method and employed the cardiac catheter as a diagnostic device for the first time using cardiac catheterisation to measure the cardiac output. Forssmann, Cournand, and Richards were jointly awarded the Nobel Prize for Medicine and Physiology in 1956.[34]

In 1953, Sven Seldinger (1921–1998) described his new method of angiography using catheters rather than the direct use of the needle or the insertion of a catheter using a cut-down.[35] In what might seem obvious, a needle punctures the vessel, the wire is passed down the needle, the needle is withdrawn, and a catheter is inserted over the wire. What is obvious retrospectively is seldom so obvious prospectively. It should be noted that Seldinger commented that progress in catheter angiography was hampered by the lack of a suitable thin-walled catheter that could be used percutaneously.

The diagnostic coronary angiogram was pioneered by Mason Sones (1919–1985) in 1958 using a catheter injection into the right and left sinuses of Valsalva adjacent to the coronary arteries using a brachial artery approach.[36] As has been described, transluminal angioplasty was introduced in 1964 by Charles Dotter. Melvin Judkins used shaped catheters for the performance of selective coronary angiography in 1967.[37] Until that time, there had been a reluctance to expose patients with ischaemic heart disease to the risk of coronary arteriography; however, it was realised that coronary arteriography was of no more risk than selective angiocardiography. That coronary artery disease could be treated had increased the need for the detailed demonstration of arterial occlusive disease, or indeed its absence. In 1974, Dr. Andreas Grüntzig performed the first peripheral human balloon angioplasty.[38] Grüntzig had heard of the ideas of Porstmann who had placed a latex balloon into a slotted angiographic catheter. Grüntzig and his team had designed many versions of the balloon catheter. In 1976, Grüntzig presented the results of his animal studies on coronary angioplasty in a poster at the 49th scientific session of the American Heart Association in Miami Beach, which generated great interest. In 1977, Grüntzig performed the first percutaneous transluminal coronary angioplasty in Zurich.[39] Intravascular coronary stents were introduced in the 1980s, with Ulrich Sigwart (b. 1941) making major contributions.[40] Sigwart said that whilst most stenoses of coronary and peripheral arteries could be traversed and treated by balloon angioplasty, the procedure was compromised by late occlusion or restenosis. He recommended an intravascular stent as an endoprosthesis to hold the artery open, and in his 1987 paper described the use of a self-expanding woven stent made from a surgical-grade stainless steel alloy. The use of stents in the peripheral vasculature has been pioneered by the studies of Julio Palmaz (b. 1945) and Ernst-Peter Strecker. The first stents developed by Dotter were made of nitinol. Cesare Gianturco (1905–1995) introduced his self-expandable Z-stent, Strecker a knitted tantalum stent, and Palmaz a balloon expandable stent. The insertion of stents in the peripheral vascular system is now routinely performed after angioplasty and stents can also be used in many other areas, including the trachea (Figure 7.14).

In the biliary system, percutaneous transhepatic cholangiography was first described in 1937 with opacification of the biliary tree via a needle placed into the liver. If the dilated biliary tree could be demonstrated, it could also be drained radiologically, and in 1981 Peter Mueller (b. 1947), Eric van Sonnenberg, and Joseph Ferrucci (1907–1994) described their series of 200 consecutive percutaneous biliary drainages.[41] H. Joachim Burhenne (1925–1996) developed a technique for non-operative retained biliary tract stone extraction using a steerable catheter with a basket on the end. Burhenne had a success rate of 95% in 661 patients which he treated between 1972 and 1979.[42] Stents were used in the biliary system as early as 1978. In 1985, Palmaz and colleagues presented experimental work on expandable intrahepatic portacaval shunt stents.[43] Stents have been devised to use in the peripheral vascular system, in the tracheo-bronchial tree, the biliary tract, and in the urinary tract.

Interventional radiology in a relatively short time has developed in a remarkable manner, and as Werner Porstmann says, when he did his first transfemoral PDA closure in 1966, the method was not considered as an alternative to conventional surgery, and that the new method was only considered applicable to extraordinary situations, such as recurrence following a surgical ligation. Writing in 1981, Porstmann noted that the percutaneous transluminal method offered a superior alternative to surgical PDA closure. In reality, the radiological and the surgical approaches should

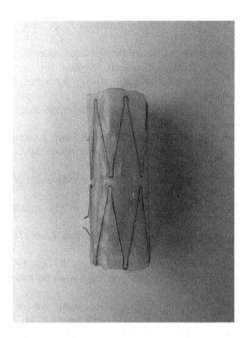

FIGURE 7.14   Covered tracheal Z-stent made by Cesare Gianturco himself. The thin material covering the stent is silicone, and as an alternative silicone and HYPAN could be used. Courtesy of Anna-Maria Belli. Author's collection.

not be seen as being deeply competitive but rather as synergistic, and the clinical wisdom is in deciding the most appropriate approach for the individual patient.

NOTES

1. https://www.wesleyschapel.org.uk (accessed 1 May 2021).
2. Wesley, J. 1759. *The Desideratum: Or, Electricity Made Plain and Useful, by a Lover of Mankind and Common Sense.* 2nd ed. Many editions including (1871). London: Baillière, Tindall, and Cox.
3. Wesley Hill, A. 1958. *John Wesley among the Physicians: A Study in Eighteenth-Century Medicine.* London: The Epworth Press.
4. Colwell, H.A. 1922. *An Essay on the History of Electrotherapy and Diagnosis.* London: William Heinemann.
5. Duchenne de Boulogne. 1855. *De l'Électrisation localisée et de son application à la physiologie, à la pathologie et à la thérapeutique.* Paris: Chez J-B Baillière.
6. Macintyre, J. 1903. The electrical pavilion, Glasgow royal infirmary. *Archives of the Roentgen Ray,* 7, 101–102.
7. Kassabian, M.K. 1907. *Röntgen Rays and Electro-Therapeutics with Chapters on Radium and Phototherapy.* Philadelphia: J.B. Lippincott Company.
8. Poland, J. 1898. *Traumatic Separation of the Epiphyses.* London: Smith, Elder & Co.
9. Marsh, H. 1896. A case of Roentgen photography. *British Medical Journal,* II, 1318–1319.

10. Poland, J. 1901. *A Retrospect of Surgery during the Past Century.* London: Smith, Elder & Co.

11. Fawcett, J. 1908. Pneumothorax treated by aspiration under the X-rays. *Proceedings of the Royal Society of Medicine,* 1 (Clin Sect), 38–40.

12. Osler, W. 1905. *The Principles and Practice of Medicine.* 6th ed. New York: D. Appleton and Company.

13. Thomas, A.M.K. 2007. The first 50 years of military radiology 1895–1945. *European Journal of Radiology,* 6, 214–219.

14. Shadwell, L.J. 1898. *Lockhart's Advance through Tirah.* London: W. Thacker & Co.

15. Ombrédanne, L., Ledoux-Lebard, R. 1918. *Localisation and Extraction of Projectiles.* London: University of London Press.

16. Lee, J.R. 1916. Removal of intracranial foreign body under X rays. *British Medical Journal,* I, 47.

17. Grundy, A., Belli, A. 1988. Balloon dilatation of upper gastrointestinal strictures. *Clinical Radiology,* 39, 299–235.

18. Jordan, A.E. 1923. *Chronic Intestinal Stasis (Arbuthnot Lane's Disease): A Radiological Study.* London: Henry Frowde.

19. Frush, D.P., Zheng, J.-Y., McDermott, V.G., Bisset III, G.S. 1995. Nonoperative treatment of intussusception: historical perspective. *American Journal of Radiology,* 165, 1066–1070.

20. Daneman, A., Alton, D.J. 1996. Intussusception, issues and controversies related diagnosis and reduction. *Radiologic Clinics of North America,* 34, 743–756.

21. Fenwick, E.H. 1893. *The Cardinal Symptoms of Urinary Disease, Their Diagnostic Significance and Treatment.* London: J & A Churchill.

22. Fenwick, E.H. 1908. *The Value of Radiography in the Diagnosis and Treatment of Urinary Stone: A Study in Clinical and Operative Surgery.* London: J&A Churchill.

23. Allison, D.J. 1983. Interventional radiology. In: *Recent Advances in Radiology and Medical Imaging.* Ed. Steiner, R.E. Edinburgh: Churchill Livingstone.

24. Geddes, L.A., Geddes, L.E. 1993. *The Catheter Introducers.* Chicago: Mobium Press.

25. Hammel, B. 2008. *The Bill Cook Story, Ready, Fire, Aim!* Bloomington: Indiana University Press.

26. Dotter, C.T., Judkins, M.P. 1964. Transluminal treatment of arteriosclerotic obstruction: description of a new technique and a preliminary report of its application. *Circulation,* 30, 654–670.

27. van Andel, G.J. 1976. *Percutaneous Transluminal Angioplasty: The Dotter Procedure: A Manual for the Radiologist.* Amsterdam: Excerpta Medica.

28. Dotter, C.T. 1981. Interventional radiology, review of an emerging field. *Seminars in Roentgenology,* 16, 7–12.

29. Porstmann, W., Wierny, L. 1981. Percutaneous transfemoral closure of the patent ductus arteriosus: an alternative to surgery. *Seminars in Roentgenology,* 16, 95–102.

30. Borges, A.C., Gliech, V., Baumann, G. 2003. Images in cardiology, thirty years after percutaneous closure of patent ductus arteriosus. *Heart,* 89, 650.

31. De Vries, A.R. 1986. Digital subtraction angiography and catheter angiography in examinations of the carotid artery. In: *Digital Subtraction Angiography in Clinical Practice.* Philips Medical Systems, Best.

32. Irving, J.D. 1984. Intra-arterial digital imaging with special reference to its use in interventional techniques. In: *Digital Radiology, Physical and Clinical Aspects.* Eds. R. M. Harrison, I. Isherwood. London: The Hospital Physicists' Association.

33. Forssmann, W. 1972. *Experiments on Myself: Memoirs of a Surgeon in Germany.* Trans: H. Davies. New York: Saint Martin's Press.

34. Schlessinger, B.S., Schlessinger, J.H. 1996. *The Who's Who of Nobel Prize Winners 1901–1995*. Phoenix: Oryx Press.
35. Seldinger, S.I. 1953. Catheter replacement of the needle in percutaneous arteriography: a new technique. *Acta Radiologica*, 39, 368–376.
36. Sones, F.M., Shirey, E.K., Proudfit, W.L., Westcott, R. 1959. Cine coronary arteriography. *Circulation,* 20, 773–774.
37. Judkins, M.P. 1967. Selective coronary arteriography. *Radiology*, 89, 815–824.
38. Grüntzig, A.R., Hopff, H. 1974. Perkutane Rekanalisation chronischer arterieller Verschlüsse mit einem neuen Dilatationskatheter. *Deutsch Medizinische Wochenshrift*, 99, 2502–2505.
39. Grüntzig, A.R. 1978. Transluminal dilatation of coronary-artery stenosis. *Lancet* ii, 263.
40. Sigwart, U., Puel, J., Mirkovitch, V., Joffre, F., Kappenberger, L. 1987. Intravascular stents to prevent occlusion and restenosis after transluminal angioplasty. *New England Journal of Medicine*, 316, 701–706.
41. Mueller, P.R., van Sonnenberg, E, Ferrucci, J.T. 1982. Percutaneous biliary drainage, technical and catheter related problems in 200 procedures. *American Journal of Roentgenology*, 138, 17–23.
42. Burhenne, H.J. 1973. Non-operative retained biliary tract stone extraction: a new roentgenologic technique. *American Journal of Roentgenology*, 117, 388–399.
43. Palmaz, J.C., Sibbitt, R.R., Reuter, S.R., Garcia, F., Tio, F.O., et al. 1985. Expandable intrahepatic portacaval shunt stents: early experience in the dog. *American Journal of Roentgenology*, 145, 821–825.

8 Contrast Media

INTRODUCTION

Contrast media have been used since the earliest days of radiology,[1] and subsequent developments in medical imaging have not removed the need for their use as might have been expected. Ideally, medical imaging should be performed without the need for introducing foreign material into the body, and such use is of the nature of a failure although an understandable one. The history of contrast media is complex and interesting, and has recently been reviewed by Christoph de Haën.[2] The need for contrast media was well expressed by the pioneer radiologist Alfred Barclay when he said in 1913 that the X-rays penetrate all substances to a lesser or greater extent and that the resistance that is offered to their passage is approximately in direct proportion to the specific gravity.[3] Barclay continued by noting that the walls of the alimentary tract do not differ from the rest of the abdominal contents in this respect, and consequently they give no distinctive shadow on the fluorescent screen or radiogram. Barclay clearly stated the essential problem that confronts radiologists. The density differences that are seen on the plain radiographs are those of soft tissue (which is basically water), bony and calcified structures, fatty tissues, and gas. The liver has the same density as the heart and therefore the two structures cannot be separated on plain films. It was only when the CT scanner was invented by Sir Godfrey Hounsfield[4] that density differences within soft tissues could be readily appreciated, and even with CT scanning the administration of contrast media are commonly needed. We identify structures radiographically when a border is present between tissues of differing radiodensities, and when the tissues are of the same density, that border is lost. This is the basis of the silhouette sign that was popularised by Benjamin ("Benny") Felson (1913–1988) from Cincinnati, whose textbook on chest roentgenology is one of the great books of radiology.[5] This sign was first described by H. Kennon Dunham (1872–1944), also from Cincinnati, in 1935. Dunham noted that if the left heart border is not visible, then this implies disease in the adjacent lung, the lingual segment of the left upper lobe. The basis of contrast media consists in the artificial manipulation of tissue density so that specific structures are revealed, and Barclay noted that the method depends on filling the alimentary cavities with some substance that differs as widely as possible in density from that of the tissue structures, i.e. by something very heavy such as a bismuth salt, or by inflating them with a gas.

TYPES OF CONTRAST MEDIA

Contrast media may be divided into positive contrast media that are of high atomic number and negative contrast media that are of low atomic number (Table 8.1).

DOI: 10.1201/9780429325748-8

TABLE 8.1
Positive and Negative Contrast Media

	Example	Use
Positive contrast	Bismuth subnitrate, Barium sulphate (both as suspensions)	Gastrointestinal tract (opaque meal, opaque enema)
	Iodinated contrast media (as a solution)	Intravascular studies, urography. Used to image the vessel that it contains, and to measure the perfusion of the organ that the vessel supplies
Negative contrast	Room air, carbon dioxide, oxygen	Encephalography, myelography, cystography, retroperitoneal air studies
Combined positive and negative	Iodinated contrast or barium combined with a gas	Double-contrast studies, including barium meals, barium enemas, CT colonography, and arthrograms

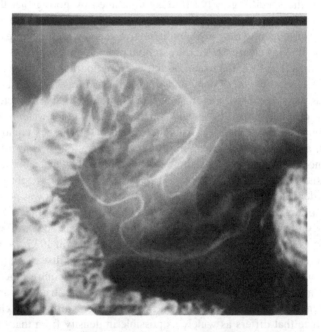

FIGURE 8.1   A traditional double-contrast barium meal showing the duodenum to diagnose a possible ulcer. Author's collection. Author's photograph.

Material used for radiographic contrast may be solid, liquid, or gaseous. Liquid contrast media may be found as solutions or suspensions. The double-contrast barium meal is an example of combining a positive contrast (a barium suspension), which is coating the bowel wall, and a negative contrast (carbon dioxide released from a swallowed effervescent agent) distending the duodenal cap (Figure 8.1).

THE GASTROINTESTINAL TRACT

Following the discovery of Röntgen's new rays in 1895, many considered how it would be possible to demonstrate the oesophagus and stomach. In March 1896, Wolf Becher was able to show the stomach and intestine of the guinea pig using a lead sub-acetate solution; and in April 1896, Carl Wegele proposed radiography after passing a stomach tube that contained a thin metal wire. In the February of 1897, Lindermann from Hamburg was able to demonstrate radiographically the greater curvature of the stomach using a gastric tube that was covered with a fine wire netting. Obviously, only limited information could be obtained using an opaque gastric tube. In the June and July of 1897, Jean-Charles Roux and Victor Balthazard in Paris were able to observe gastric peristalsis in the frog and dog, and then in a human. They used bismuth subnitrate at 200 mg/ml mixed in with food, and in all three cases could divide the stomach into an upper inactive reservoir and an active pre-pyloric region.

The most important of these early workers was Walter Bradford Cannon (1871–1945), who was a first-year medical student at Harvard Medical School. Cannon worked with his fellow student Albert Moser, and following the suggestion of the Harvard physiologist, Henry P. Bowditch (1840–1911), started researching the gas-trointestinal tract using the new rays. Cannon used bismuth mixed in with bread, meat, mush, and viscid fluids. This was then fed to a goose, and peristalsis was shown clearly with waves of contractions moving regularly down the oesophagus to the stomach. There was no evidence of any squirting action from the mouth. This contradicted the current belief that food was pushed into the stomach by the actions of the mouth and pharynx. As a medical student, Cannon worked with Francis Williams (1852–1936), who was the pioneer radiologist at Boston City Hospital. In 1898, Cannon assisted Williams in looking at the stomachs of two children.[6] They used a fluorescent screen covered with a sheet of celluloid, which could be marked directly using a pencil with the position of the stomach. Radiography was not possible in the very early days because of the low power of the apparatus. Cannon and Williams observed changes in the position of the stomach between the prone position and standing, the movements of the stomach on respiration, and the changes in the shape of the stomach during digestion. Whilst such observations may seem obvious to us today, at that time these basic anatomical and physiological changes had not previously been observed. These observations were innovative and important. Canon made further useful observations on the nature of peristalsis.[7]

The early radiologists used bismuth subnitrate to outline the human alimentary tract since attempts to use a gas alone were unsuccessful. In the initial period, drawings were made of the outlines, and as the power of the apparatus improved, it became possible to obtain radiographs. The subject was reviewed in 1917 by Russell Carman (1875–1926) and Albert Miller, both from the Mayo Clinic, Rochester.[8] Bismuth sub-nitrate was toxic and resulted in some fatalities, and so its use was abandoned. It was replaced by bismuth subcarbonate, which was subsequently used extensively. The oxychloride of bismuth was used occasionally since it was lighter and could be more readily held in suspension. Bismuth salts had been used therapeutically for indigestion since they are alkaline and would neutralise the gastric acid. It was therefore believed that bismuth would also suppress peristalsis and that this would be of aid in radiography. Other agents used for the opaque meal were oxides of zirconium (marketed as kontrastin) and thorium, and also the magnetic oxide of iron. The swallowed bismuth could be followed through the gastrointestinal tract to demonstrate the stomach, small bowel and large bowel in one examination (Figure 8.2a–d). The bismuth study illustrated was performed in January 1918 by the radiation martyr William Ironside Bruce (1876–1921) from Charing Cross Hospital.

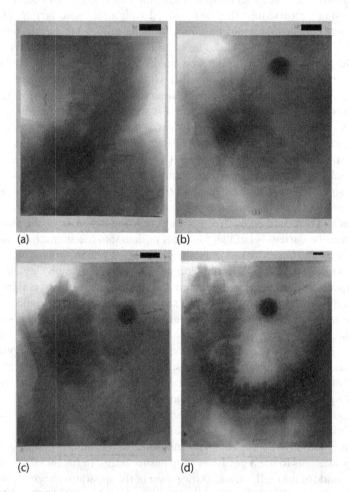

(a)   (b)

(c)   (d)

FIGURE 8.2   (a) Gastrointestinal bismuth study of January 1918 "with Dr. Ironside Bruce's compliments": The stomach is shown, and the umbilicus and duodenal "cap" have been labelled by Bruce. (b) The bismuth has passed down the bowel. The small bowel is shown, and the ileum and umbilicus have been labelled by Bruce. (c) The bismuth has continued to pass down the bowel. The ascending colon is shown, and the hepatic flex(ure), caecum, and umbilicus have been labelled by Bruce. (d) On this final image of the series the bismuth is now completely in the large bowel. The transverse colon and umbilicus have been labelled by Bruce. Author's collection.

Bismuth salts were gradually replaced by barium salts. Barium was found to be equally as satisfactory as bismuth and was less than a tenth of the cost. The barium was manufactured as a finely divided powder, and had to be free of the soluble salts which were toxic. Barium neither inhibited nor suppressed peristalsis, and by the time Carman and Miller were writing in 1917, it had largely superseded the use of bismuth.

As the use of barium progressed, there was the need to show fine mucosal detail and so a dense barium reconstituted with water was used. For a double-contrast study, pioneered in Japan where there was a high incidence of gastric cancer, an effervescent substance was swallowed to distend the stomach and duodenum with carbon dioxide (Figure 8.1). The double-contrast method that was first applied to the colon was promoted in Japan by Shirakabe, Ichikawa, and Kumakuru who, by looking at the mucosal pattern on the stomach, were able to diagnose early gastric cancer. The method was both accurate and reliable. The double-contrast barium meal was able to demonstrate the stomach and duodenum with remarkable clarity. A different barium formulation was commonly used for studies of the small bowel, and an agent was added to prevent flocculation.

The Japanese company Fushimi Pharmaceutical Co. of Kagawa manufactured barium as Barytgen de luxe, which was distributed by Eisai Co. of Tokyo. The formulation was successful, and promoted for its high stability against stomach acids, its optimal adhesiveness to stomach mucosa, its constancy in colloidal suspension, and its pleasant taste and odour. They recommended mixing the Barytgen de luxe to water and mixing thoroughly, and preparing the mixture on the night before the day of administration.

The company Schering marketed "X-opaque" barium as a powder in a 300 g sachet which was to be mixed with 70 ml of water. This resulted in a high-density suspension, approximately 216% w/v, with a low viscosity. The barium was a blend of precipitated and crushed barium sulphate of varying particle sizes that was said to be essential for good mucosal coating. The barium sulphate was to be powdered into rough particles with jagged edges, and having a size range of 0.5–30 μm. For the double-contrast technique, an effervescent gas-producing agent was used and this needed to be compatible with the barium sulphate preparation. The effervescent granules typically consisted of sodium bicarbonate 44.8%, citric acid 18%, potassium acid tartrate 26.9%, with the addition of a sweetener and flavouring. Anti-foaming and de-foaming agents were needed for both double-contrast barium meals and enemas since a bubble might simulate a polyp, and typically 12% w/v of simethicone was added.

For the opaque enema, barium was again mixed with a variety of compounds, including condensed milk, fermented milk, or starch. Carman preferred a mixture of mucilage of acacia, condensed milk, and barium. Mucilage of acacia is a viscid liquid used as a soothing agent in inflammatory conditions of the respiratory, digestive, and urinary tract. In his classic 1933 textbook on gastrointestinal radiology,[9] the Cambridge radiologist Alfred Barclay recommended adding tragacanth for both meals and enemas. Tragacanth is a plant and an extract is used to treat both diarrhoea and constipation.

A colonic activator was sometimes added to the barium mixture in the period before the adoption of the double-contrast technique. Agents used were oxyphenisatin[10] (which was marketed as Veripaque) or tannic acid. Veripaque in a dose of 3 g was added to 1–2 l of the barium mixture. The colon was completely filled with the mixture and images were obtained. The barium was then voided and the image of the contracted colon gave mucosal detail. Tannic acid and oxyphenisatin stimulated the contracture of the colon, and made the barium sulphate adhere to the bowel wall. The concentration of tannic acid recommended for use in barium enema examinations varies between 0.25% and 3.0%. However, neither oxyphenisatin nor tannic acid was without their complications.[11]

In the 1980s, May & Baker marketed EPI-C, for use as a barium enema, as a liquid dispersion for a 1:1 dilution with water and for combination with a foam control preparation. The preparation was a barium sulphate suspension 70% w/w (150% w/v) and formulated to provide optimal coating of the mucosa of the large bowel in double-contrast barium enemas. The data sheets give little information about what was added to the barium. As with many medications and pharmaceuticals, in the early period they were made locally by pharmacists or doctors and with development they were produced in factories with minimal local preparation needed, if any at all.

As is the case in so many instances, just as a technique is perfected, it becomes obsolete. Examples include intracranial air studies, oral and intravenous examinations of the biliary tree with iodinated contrast agents, and optimisation of barium for gastrointestinal examinations.

THE RENAL TRACT

RETROGRADE PYELOGRAPHY/PYELOURETEROGRAPHY

The impregnation of catheters and bougies with material of high atomic number as a contrast material is now used almost universally. The anatomical position of the catheter or medical device may be identified with confidence, and this is essential in angiography, both for diagnosis and to guide intervention. Either the entire catheter may be rendered opaque, or specific parts may be opaque, depending on the function of the device. So, for example, the gastric Ryle's tube had metal balls or a marker at its tip for location, and a tracheal intubating bougie has barium in its tip for improved X-ray guidance. A further use of contrast material to identify the presence of a device is the use of a radiopaque thread in a surgical swab.[12] If the swab is retained following surgery, its presence can be shown by abdominal radiography. It is therefore important for the radiologist to know the differing radiographic appearances of medical devices.

Once the use of radiopaque bougies was appreciated, the injection of a liquid radiopaque material via an opaque catheter was an obvious progression, as suggested by Klose in 1904. It was already appreciated that the alimentary tract could be outlined with a radiopaque material such as a bismuth salt, and so a similar technique in the renal tract was a logical progression. For retrograde pyelography or pyeloureterography, which is demonstration of the pelvis of the kidney and ureter from below, a suspension of bismuth subnitrate was initially used; however, this procedure was difficult and it was not easy to remove the bismuth from the renal tract. The technique of retrograde pyelography was refined by Voelcker and von Lichtenberg in 1906 and they produced the first complete outline of the ureter and renal pelvis.[13] They were trying to outline the bladder with colloidal silver and on a radiograph noted that the solution had entered the ureter and renal pelvis. They were encouraged by this and therefore injected a 2% solution of Collargol (colloidal silver) which was increased to a 5% solution (Figure 8.3). The technique was again not without its problems. These were related to the difficulties in inserting ureteric catheters and the toxicity of the contrast agents used; however, the technique gradually gained acceptance. The early history of pyelography was well recorded by William Braasch from the Mayo Clinic in 1915[14] and 1927.[15] Other workers used Argyrol, which was a 40% or 50% solution of silver nitrate. These silver-based compounds were toxic to the kidneys and if excessive pressure was used for injection, they sometimes resulted in renal necrosis and some fatalities occurred. In the United States, Braasch investigated these compounds extensively and showed areas of renal necrosis. These severe toxic effects demonstrated the need for safer contrast agents. Because of the problems with the silver compounds, in 1907 Burkhardt and Polano injected oxygen into the renal pelvis, but the radiographic shadow produced was difficult to distinguish from bowel gas. It was also difficult to maintain a full distension of the pelvis and ureter during the exposure.

By 1915, thorium salts were being used with good radiographic opacification, but an advance took place in March 1918 when Douglas Cameron from Minnesota

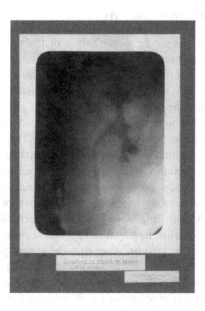

FIGURE 8.3 Retrograde pyelogram, performed by Sebastian Gilbert Scott at the London Hospital, showing a large normal kidney with collargol in its pelvis. Author's collection.

recommended the use of sodium or potassium iodide as a 25% solution for retro-grade pyelography.[16] Cameron was aware that halogen salts were of a sufficiently high molecular weight to give good contrast on radiographs. After some investigations, Cameron recommended sodium iodide as the medium of choice, but erroneously thought it was non-toxic even when introduced into the circulation. The recommended 25% solution was hypertonic to blood plasma, and since he thought that this might cause toxicity, he used a 13.5% solution. Further experiments showed the advantages of the iodide salt over the bromide for pyelography. A 14.56% sodium iodide solution was isotonic to a 10% sodium bromide solution and the higher molecular weight of iodide produced a greater radio-opacification. Braasch therefore recommended a 12% sodium iodide solution in his book of 1927. Braasch emphasised the importance of the sterility of the solution, which could be boiled and kept in individual containers for each patient. For making larger amounts of the solution, sterilisation could be performed in bulk by the addition of 1 g of mercuric iodide for each 3 l of the 12% sodium iodide solution.

Following the introduction of retrograde pyelography, Alexander von Lichtenberg (1880–1949), professor of urology at St. Hedwick's Hospital in Berlin, started extensive laboratory work with an aim of developing clinical intravenous pyelography (IVP), but with no success. The nearest successful approach was achieved by Hryntschalk of Vienna in 1929[17] who succeeded in producing good radiographic visualisation of the renal calyces and pelves in laboratory animals following an intravenous injection of iodinated pyridine compounds, probably synthesised by Binz and Räth; however, his work was not fully accepted by the medical establishment.

INTRAVENOUS PYELOGRAPHY (IVP)

Achieving reliable, safe, diagnostic visualisation of the urinary tract was a major objective. Retrograde pyeloureterography was an invasive surgical examination, and a procedure that could be performed as an outpatient was essential. In 1923, a multidisciplinary team at the Mayo Clinic described the use of intravenous and oral sodium iodide to visualise the urinary tract.[18] One of the groups had noticed that the urinary bladder was visible on radiographs when patients took large doses of oral or intravenous sodium iodide for the treatment of syphilis. The visualisation was poor, but they could correlate the dose of iodine against the urinary iodine concentration and bladder radiopacity. Sodium iodide was too toxic for use in the clinical environment. Other workers used sodium iodo-urea, but these compounds could not be given in large-enough doses to produce adequate visualisation.

In 1925 and 1926, Arthur Binz (1868–1943) and Curt Räth (b. 1893), from the Agricultural College in Berlin, had synthesised many organic iodine and arsenical preparations based on the pyridine ring with the aim of producing an improved drug for the treatment of syphilis and other infections. The pyridine ring is a six-pointed ring made up of five carbon atoms and one nitrogen atom. Linkage to this ring greatly detoxified the arsenic and iodine atoms, and Binz and Räth synthesised more than 700 of these compounds. One group of iodinated pyridine compounds was found to be selectively excreted by the liver and kidney and was therefore called the Selectans. Some of these synthesised pyridine drugs were therefore sent to several clinicians for evaluation for the treatment of biliary and renal infections.

In 1928, Moses Swick (1900–1985), a urology intern at Mount Sinai Hospital in New York, was awarded the Libman Scholarship to carry out medical research overseas.[19] He chose to work with Professor Leopold Lichtwitz (1876–1943) at the Altona Krankenhaus in Hamburg, Germany. Lichtwitz had some clinical success in the treatment of biliary coccal infections using Binz and Räth's iodinated Selectan drugs. These drugs contained iodine, and it occurred to Swick that they might possibly visualise the renal tract by radiography. Swick then made radiological, chemical, and toxicological studies in laboratory animals and patients. The initial studies were encouraging and Swick transferred his work to von Lichtenberg's urological department at St Hedwig's Hospital in Berlin where the first successful human intravenous urograms were obtained. The non-ionic N-methyl-5-iodo-2 pyridone (Selectan neutral) was used; however, Swick preferred the less toxic and more soluble salt 5-iodo-2-pyridone-N-acetate sodium (Uroselectan) that had been patented by Räth in May 1927. This new compound Uroselectan produced excellent quality intravenous urograms with relatively little toxicity.

Swick and von Lichtenberg presented the work to the Ninth Congress of the German Urological Society in September 1929. Swick presented the first paper based on animal work, but also showed several excellent quality human studies exhibiting various disease processes such as hydronephrosis and horseshoe kidney. Von Lichtenberg and Swick together presented the second paper on the human clinical applications with von Lichtenberg reading the paper. Both papers were published in November 1929 in *Klinische Wochenschrift*.[20,21]

Unfortunately, Swick and von Lichtenberg could not agree on who should be accorded priority of discovery of this new and remarkable technique.[22] Assigning priority to any discovery is always difficult, and in the case of the IVP, there were many steps involving many workers. The reality is that Arthur Binz and Curt Räth synthesised the agents used, Moses Swick performed essential clinical and laboratory research, and Alexander von Lichtenberg had a long-term goal of developing the IVP and also provided the facilities and resources that Swick needed. The IVP, or intravenous urogram (IVU) as it became known since the examination demonstrated more than the renal pelves, was the result of the work of many different groups over many years. It was a highly successful procedure and was in use for many decades, only passing out of use with CT urography. The CT urogram is essentially an IVU performed using a CT scanner rather than using plain films.

CONTRAST MEDIA DESIGN

EARLY CONTRAST MEDIA

Within two years following the introduction in 1929 of Uroselectan into clinical practice, Binz and Räth made modifications to the pyridine ring. These were marketed as diodrast (Diodone) and neo-ipax (Uroselectan B, Iodoxyl) (Figure 8.4). Each molecule contained two iodine atoms. Binz and Räth were fully supported by

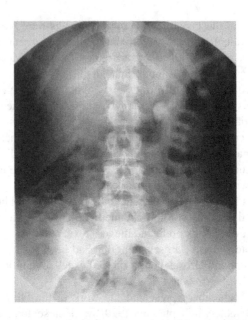

FIGURE 8.4   An intravenous pyelogram performed on a 29-year-old female on 11 November 1932 at the Red Cross Clinic for Rheumatism in London using Uroselectan B. The upper part of the left kidney is abnormal and a final diagnosis of tuberculosis of the kidney was made. Author's collection.

the Berlin-based company Schering Kahlbaum in the development of these pyridine agents, and as a result Schering became the world's leading manufacturer of intravascular contrast agents. These compounds, and their variants, were highly successful, and became the standard intravascular and urologic contrast media for the next 20 years.

Moses Swick continued his interest in contrast media on his return to the United States. He worked at Mount Sinai Hospital in New York with Vernon Wallingford (b. 1897), a research chemist from Mallinckrodt Chemical Works of St. Louis. In 1933, they introduced the six-carbon atom benzene ring as the iodine carrier instead of the heterocyclic pyridine ring used by Binz and Räth. The use of the benzene ring was an important innovation and in 1933, they were awarded the Billing's Gold Medal of the American Medical Association.

There are a number of changes that had to be made to the benzene ring before its iodinated derivatives were suitable for clinical use. About 20 years later, Wallingford demonstrated that if an amine group were to be introduced into the *meta*-position (C3) on the benzene ring, then it must allow three iodine atoms to be introduced at the C2, C4, and C6 locations. An amine in the *ortho* (C2) or *para* (C4) position would only allow two iodine atoms to be introduced. In 1953, Wallingford demonstrated that an amine (-NH_2) group at the C3 position allowed a side chain such as acetyl (-$COCH_3$) to replace one of its hydrogen atoms.[23] This acetyl-amino group significantly reduced the toxicity of the tri-iodo compound and sodium acetrizoate (marketed as Urokon and Diaginol) was introduced clinically in 1952 by Mallinckrodt. This was the first tri-iodinated contrast medium with three iodine atoms to one molecule.

In 1956, Hoppe and colleagues with others demonstrated that a second acetyl-amino group could be added to the benzene ring at the C5 position to produce a fully substituted tri-iodinated acid radical.[24] The toxicity was reduced even further, and this compound sodium diatrizoate was introduced in the mid-1950s and marketed as Urografin (Schering AG, Germany), Renografin (Squibb, USA), and Hypaque (Sterling Drug). Sodium diatrizoate and its derivatives became the standard intra-vascular contrast agents until the development of the lower osmolar and non-ionic agents in the early 1970s. Urografin remains in use as a 30% solution for bladder examinations or cystography and retrograde urography (pyelography).

In 1959, the small Norwegian pharmaceutical company Nyegaard & Co. of Oslo were accused by Schering of infringing their patent on diatrizoate, which they thought had not been patented in Norway. Following this, Nyegaard tried to synthesise diatri-zoate by another means, and developed a new fully substituted tri-iodinated benzene ring compound (metrizoic acid), which they then marketed as Isopaque (Triosil).

The conventional tri-iodinated contrast agents (diatrizoate, iothalamate, metri-zoate) designed for intravascular use are ionic monomeric salts of tri-iodinated fully substituted benzoic acids and are referred to as high-osmolar contrast media (HOCM) because of their very high osmolality. The only chemical difference found between them is in the nature of the substituted side chains.

It had become apparent by the 1960s that the cation used was also important. Each of these types of intravascular contrast media was a salt, comprising the tri-iodin-ated ring and one cation. The cation could be either sodium or N-methylglucamine (meglumine), or a mixture of the two, as the non-radio-opaque cation necessary to produce the salt molecule. Meglumine produced less pain and less vasodilatation when injected into arteries, but it produced more diuresis and was therefore not ideal for urography. Vosse in 1960 showed that the sodium salt produced more damaging effects on the blood–brain barrier and in 1964, Gensini and di Giorgi demonstrated an increased cardiac toxicity when pure sodium salt solution was used for coronary arteriography.[25]

For coronary angiography, a mixture of sodium and meglumine salts is essential to minimise cardiac arrhythmia. Sodium cations produced less viscous solutions than meglumine and therefore a mixture of the two was often preferred. The bal-ance of cations was further investigated by Nyegaard and in 1963 and 1965, they introduced several versions of Isopaque with a balance of sodium, meglumine, mag-nesium, and calcium salts, different formulations being recommended for cerebral, coronary, vascular and urinary tract visualisation. Contrast media were marketed in various formulations and concentrations, depending on the precise clinical need.

LOW-OSMOLAR CONTRAST MEDIA

Torsten Almén (1931–2016) was a Swedish radiologist who became interested in contrast media and was working at Malmö (Figure 8.5). He studied the pharmacol-ogy of contrast agents and believed that the very high osmolality of the contrast

FIGURE 8.5   Torsten Almén (1931–2016). Author's photograph.

media, that is up to eight times physiological osmolality, was responsible for much
of its toxicity. Almén performed angiography on a daily basis and noted how painful
the patients found the injections. Almén knew that an arterial injection of contrast
medium that was isotonic to serum, such as a suspension of thorium dioxide or an
emulsion of iodised oil, did not produce any pain. Almén grew up on the most southern coast of Sweden and recalls a family holiday taken as a boy in Bohuslän on the
west coast of Sweden. He found swimming in the water uncomfortable because as
soon as he opened his eyes they started to hurt. The salty water at Bohuslän made
his eyes sore, whereas the brackish water around Ystad did not cause any discomfort.
He reasoned that "a plasma-isotonic aqueous solution of contrast medium molecules
might not cause pain, and should therefore be created!" Almén reasoned that an isotonic contrast medium would both cause less pain and also be less toxic.

Almén learnt chemistry and suggested reducing the osmolality of contrast media
by substituting the non-radio-opaque cation with a non-ionising radical such as an
amide. His concept was theoretical and was unsupported by either chemical or clinical research and was made when he was a Research Fellow in Philadelphia from
1968 to 1969. His paper on this topic was rejected by leading radiological journals
but was eventually accepted and published by the *Journal of Theoretical Biology* in
1969. This was a journal of which most radiologists were unaware.[26] It seems more
than a little unfortunate that the most important paper on contrast media since Moses
Swick's 1929 paper was not published in a radiological journal.

Almén's ideas were rejected by several pharmaceutical manufacturers, but Hugo
Holtermann, the Research Director of Nyegaard, encouraged his team to attempt the
synthesis of some of Almén's theoretical molecules. The research team was not fully
convinced that Almén's proposal could be implemented in practice, and Holtermann,
who had developed Isopaque, was unconvinced about a likely success; however, they
were willing to try Almén's ideas. Almén also made known his ideas as to how these

compounds might be constructed to facilitate water solubility and hydrophilicity and to reduce their viscosity. It is remarkable that fewer than 6 months were to elapse between the first meeting of Almén and the Nyegaard research group in June 1968 and the production of the first compound.[27] The team produced 80 different compounds. A review of Almén's 1969 paper commented:

> The general principles of Dr. Almén's proposal are probably sound. The implementation of it is probably impractical. He seems to be unaware that the ionic nature of the iodinated compounds is an essential property for their solubility in water – so part of his proposal, namely using non-ionic hydrophilic compounds, may be invalid.

In November 1969, after biological and pharmacological testing, compound 16 (nicknamed "Sweet Sixteen") was shown to be the most promising and it was marketed as Amipaque, "A new principle in X-ray contrast media" enabling water-soluble myelography to replace oily Myodil (Pantopaque) myelography (Figure 8.6). Amipaque was the first of the low-osmolar contrast medium (LOCM). Amipaque was based on the glucose amide of Isopaque (metrizoate) leading to its generic name of metrizamide (Amipaque). As it contained the glucose radical, metrizamide could not be autoclaved. Because of the complex nature of its production, it was expensive and inconvenient to use, being presented as a freeze-dried powder with a diluent. It was, however, a major toxicological improvement on all pre-existing water-soluble myelographic and vascular agents and in the late 1970s, it became the internationally recognised agent for myelography, enabling water-soluble myelography to replace oily Myodil (Pantopaque) myelography. Although it had an advantageous intravascular profile, metrizamide was generally regarded as too expensive and too inconvenient

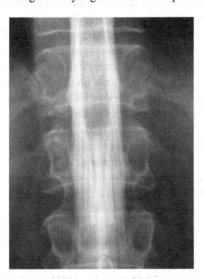

FIGURE 8.6 Lumbar radiculogram with metrizamide (Amipaque) showing beautiful visualisation of the nerves silhouetted by the surrounding iodinated contrast. Author's collection.

for vascular studies. In recognition of his achievement, Torsten Almén was presented with the Antoine Béclère Prize at the 1989 World Congress of Radiology.

Amipaque had major advantages over conventional ionic contrast agents. Being non-ionic, it did not dissociate in an aqueous solution and was therefore of considerably lower osmolality. It was better tolerated causing less subjective discomfort. During angiography, there was either no pain or markedly reduced pain. When used for coronary angiography, there were less changes in blood pressure, heart rate, and coronary circulation time. When used for cerebral angiography, there was less slowing of the heart rate and less changes in the electro-encephalogram. There was less pain in endoscopic-retrograde pancreatography, and it was unlikely to cause phlebitis or thrombosis following venous injection.

In the mid-1970s, metrizamide was replaced by the second generation of low-osmolar contrast media. These were Iohexol (marketed as Omnipaque by Nycomed, who had been previously called Nyegaard) and Iopamidol (marketed as Niopam by Bracco of Milan), and were easier to synthesise and were therefore much less expensive. They did not contain the glucose radical and could therefore be autoclaved and were stable in solution. These two second-generation LOCM, together with similar molecules, became the contrast media of choice for all intravascular procedures in the mid-1990s.[28] Omnipaque was almost completely excreted by the kidneys and was of very low toxicity.

MYELOGRAPHIC AGENTS

Prior to 1970, only iodinated oils such as iophendylate (Myodil, Pantopaque) were available for myelography. Ionic compounds such as meglumine iothalamate (Conray) and methiodal (Abrodil) were generally considered too irritant and toxic, although they were occasionally used for lumbosacral radiculography.

The French company Guerbet, following original research performed by Mallinckrodt, developed the ionic compound meglumine iocarmate (Dimer-X) combining two tri-iodinated benzene rings into one large dimeric molecule containing six atoms of iodine and so reducing the osmolality. Dimer-X could only be used in the lower portions of the spinal canal below the spinal cord for radiculography, but it produced excellent quality radiographs of the lumbosacral nerve roots. It was presented as a 60% w/v solution, and unfortunately its high osmolarity was responsible for some of the adverse reactions. It was promoted for use in lumbosacral radiculography, cerebral ventriculography, and double-contrast knee arthrography. Though much less toxic than the previous aqueous contrast media, it had to be used with great care and in a strictly limited dose. By contrast, metrizamide could be used throughout the spinal canal and was much less toxic than meglumine iocarmate, which it replaced for myelography and radiculography in the late 1970s.

THE SECOND-GENERATION LOW-OSMOLAR CONTRAST MEDIA

The introduction of metrizamide revolutionised the use of contrast agents and marked the boundary between the older conventional ionic high-osmolar media

(HOCM) and the modern low-osmolar compounds (LOCM). Iohexol and iopamidol were the first two second-generation non-ionic LOCM agents to be synthesised and in the 1990s were the intravascular and myelographic agents of choice.

In 1977, the French company Laboratoire Guerbet produced a new contrast agent of low osmolality, which was a derivative of meglumine iocarmate (Dimer-X). This new molecule consisted of two tri-iodinated benzene rings that were linked together, and it was therefore a dimer. The dimer had one carboxyl group replaced by a non-ionising radical. The second carboxyl group was attached to either a sodium or a meglumine cation. The resulting product (sodium and meglumine ioxaglate, marketed as Hexabrix 320)[29] proved a good arteriographic agent but often caused nausea and vomiting on intravenous injection. Being ionic, Hexabrix was not suitable for myelography or radiculography.

NON-IONIC DIMERS

In order to reduce the osmolality even further, two molecules of non-ionic monomers were linked to produce a large non-ionising molecule containing six atoms of iodine. Such products include visipaque (Iodixanol) and iotrolan (Isovist), which are of physiological osmolality at all concentrations. These non-ionic dimers were believed to have advantages for myelography and be beneficial for arteriography. These new agents have additional benefits and are significantly less nephrotoxic.[30]

Torsten Almén has reviewed the development of the non-ionic contrast media.[31] Development has resulted in agents isotonic with plasma and causing less pain and toxicity. The current agents in use for X-ray examinations are tri-iodinated, non-ionic contrast agents. It has been the development of these safe contrast agents that has greatly facilitated the development of modern radiology in general and of interventional radiology in particular.

THE GRAHAM TEST AND BILIARY CONTRAST AGENTS

In the early 1920s, the diagnosis of gallbladder disease was largely related to having a typical history and to physical examination. Everts Graham (1883–1957), who was Professor of Surgery at Washington University in St. Louis, knew that the opaque meal could outline the alimentary tract and was looking for an opaque substance that could be introduced into the gallbladder. In 1909, Abel and Rowntree had noted that 90% of orally administered phenoltetrachlorophthalein was excreted by the liver. Graham and Warren Cole thought that bromide and iodide compounds of phenopthalein could be tried experimentally. The first compound used was tetraiodophenolpthalein, and good results with opacification of the gallbladder were obtained in dogs. The technique was introduced into clinical practice as the Graham test after it was announced in 1924[32] (Figure 8.7). In 1933, Alfred Barclay described both the oral and intravenous administration of the agent.

The agents were slowly perfected, with the introduction in the 1970s of Endobil (the N-methylglucanine salt of iodoxamic acid) for cholecystography and cholangiography, and Biliscopin (meglumine iotroxinate) for intravenous cholangiography.

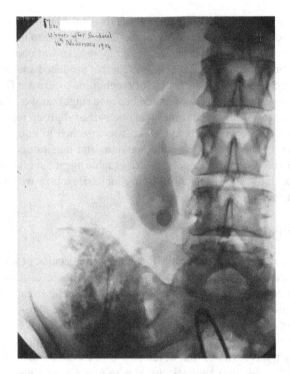

FIGURE 8.7 Graham's test (oral cholecystogram) from 16 November 1934. The pear-shaper gallbladder is seen centrally containing contrast and a stone. Author's collection.

The examination was frequently accompanied by plain tomography to show the duct, and CT was also used when the resolution improved. It must have been quite frustrating for the contrast media companies who had devoted time and money in the synthesis and testing of these newer agents, and that as they were optimised, they became obsolete being replaced by ultrasound and magnetic resonance imaging (MRI).

ANGIOGRAPHY

POST-MORTEM ANGIOGRAPHY IN VIENNA

It is remarkable how rapidly ideas about visualisation of the soft tissues of the body with contrast media developed. In the case of the alimentary tract, it was possible immediately to demonstrate anatomy and physiology *in vivo*. This was not to be the case in the vascular system until non-toxic agents were developed. However, that angiography could be performed post-mortem using non-physiological and toxic agents demonstrated that vessels could be visualised, and held out the possibility of clinical use when physiological agents were developed. There is considerable current interest in post-mortem angiography[33] and virtual autopsy or "virtopsy",[34] and it should be remembered that angiography developed from post-mortem studies.[35] The first angiographic procedure was performed in January 1896 by physicist Edward Haschek and his medical friend D. Th Lidenthal. They injected a calcium carbonate emulsion (Teichmann's mixture) into an arm that had been separated from a cadaver. The exposure time was 57 minutes, which is not unreasonable when one remembers the low power of the apparatus that was then available. This procedure was performed in Vienna, and the radiograph can be seen at the Museum in the Josephinum. The hand used for the experiment was provided by Dr. Julius Tandler who later became the professor of anatomy in Vienna.

The angiographic work in Vienna was soon followed by the work of the group in Sheffield, England. Prof. Hicks, who was the principal of Firth College in Sheffield, and Dr. Addison, achieved both a renal and a hand arteriogram.[36] The vascular injections had been performed on 6 February 1896 using the ordinary red lead mass, which was used in the dissecting rooms, and gave good radiographic images of the arteries in the hand and kidney. If the arteries could be demonstrated post-mortem, then at some point angiography of the living might be possible.

EGAS MONIZ AND THE PORTUGUESE SCHOOL OF ANGIOGRAPHY

From the earliest days of the application of X-rays, *in vitro* or post-mortem angiograms had been obtained,[37] and its early history has been reviewed by Dolby.[38] The main problems encountered related to the nature of what was injected and to its toxicity. Direct puncture and/or cut-down arteriography was achieved in the early 1920s by Brooks using sodium iodide and bromide solutions,[39] and by Berberich and Hirsch who injected strontium bromide into the femoral artery of a living subject and obtained useful images.[40] The major breakthrough in angiography was achieved in the Santa Marta Hospital in Lisbon, Portugal, on 28 June 1928 when the first successful human carotid arteriogram was performed.

Thanks to the work of Portuguese radiologists, the goal of practical angiography in the living was finally realised. The two key names are those of António Egas Moniz (1874–1955) and Reynaldo dos Santos (1880–1970). Veiga-Pires and Grainger have reviewed the outstanding contribution of the Portuguese School in the development of arteriography.[41] The charismatic leader of the Portuguese team was Egas Moniz, who was professor of neurology in Lisbon. Moniz was a brilliant polymath, author, politician (he was the Portuguese Foreign Secretary), researcher, and clinician. Moniz went on to develop the now discredited technique of pre-frontal leucotomy and for this Moniz received the Nobel Prize in 1949. Moniz was severely handicapped by gout which affected his fingers, and although he was unable to make any injections himself, he meticulously planned his research project on the diagnosis and localisation of cerebral tumours.

Moniz was dissatisfied with the recently developed technique of ventriculography, which he found could make a correct diagnosis in less than a third of patients. At this time, there was little known about the use of intravascular injection of radio-opaque substances. Moniz had been aware of the pioneer work of the Frenchmen Jean Sicard and Jacques Forestier in early angiography. They had tried with an intravascular injection of Lipiodol, an oily contrast medium, but their experiments were unsuccessful. Moniz's desire to directly show the brain prefigures modern cross-sectional imaging with intravenous contrast enhancement. Moniz and his team made experiments on animals and cadavers and showed satisfactory radiographic appearances with arterial injections. Moniz produced a classification of the cerebral arteries in 1928 based on his cadaveric studies, and this was to prove useful in interpreting angiograms in the living subject.

Moniz had reasoned that if he could concentrate radiopaque material within the brain, then the brain would become visible on radiographs. Moniz made experiments in 1926–1927 with dozens of substances in various concentrations placed in small rubber tubes and radiographed inside a dry skull. The tubes were of a bore similar to that of the larger cerebral arteries. He knew that bromides were used as sedatives, so since they accumulated in the brain, they might show up on radiographs. Moniz gave large amounts of bromides orally but showed nothing. He then tried injecting bromine directly into a carotid artery but apart from giving the patient a headache, he again showed nothing. Moniz then attempted opacifying the brain itself by intravenous or parenteral administration of a variety of agents, giving large doses of lithium bromide and strontium bromide. The patients chosen suffered either from severe epilepsy or Parkinson's disease, since it was thought that at the very least such patients would benefit from the injected bromine. When 10 ml of a 30% solution of strontium bromide was administered, the patient had no symptoms. When higher concentrations were used, there was a feeling of warmth, and at over 40% the patient developed more generalised symptoms, but they were not prolonged. Moniz finally determined that 10 ml of a 70% solution of strontium bromide produced satisfactory opacification of the cerebral arteries.

Moniz then made direct injections into the carotid artery in four patients using a percutaneous injection, but showed little of note: partly because of patient movement related to pain. He then tried exposing the carotid artery, and 4 ml of 70% strontium bromide was injected. It was felt that the agent was being diluted, and so an injection was made following a temporary ligation of the carotid artery below the point of injection. There was some visualisation of the arterial tree, and the first cerebral arteriogram in the living was obtained. However, the patient developed severe post-procedure symptoms and unfortunately died, partly thought to be due to the carotid ligation and strength of the contrast. The death was a great shock to the team and Moniz wrote:

> This accident, the only one we had in the course of our early investigations before arriving at the desired conclusion, was a great shock to us. We thought much about it, but considering the films obtained, we gave heed to the opinion of some competent colleagues to the effect that we should continue, though more cautiously, the experiments we had begun. They gave us their valuable support.

Moniz further wrote:

> The main idea in our work to obtain cerebral arteriography as the following. With a precise picture of the normal arrangement of the cerebral arteries made opaque to

X-rays, we thought it would be possible to make the diagnosis of the localization of the majority of tumours through the alteration in the normal arterial pattern in the cerebrum. Many tumours, or at least the very vascular ones, should also show their own circulation.

There was about a month of indecision and it was decided to use a new group of substances opaque to X-rays, the iodides. After these techniques failed, he tried using intra-arterial injections using an iodide salt. Moniz chose iodine because of its higher atomic weight compared to bromine. The team again made preliminary experiments using iodides of ammonium, sodium, potassium, and rubidium. For patient studies, a 25% solution of sodium iodide was used and the effect of dilution and arterial capacity were determined. It was felt that 5 ml would be enough. In two patients, injections of 3 ml of 25% sodium iodide were made with limited success. In the third patient, an intracarotid injection with a temporary ligature was made using the rapid injection of 5 ml of 25% sodium iodide. The injection was successful with arterial filling and their positions were altered due to the presence of an intracranial tumour. His successful patient, on 28th June 1927, was the ninth in his series, a young man with a pituitary tumour. Moniz wrote and published describing the new technique in detail. Moniz developed the understanding of the anatomy as shown by angiography and published many labelled angiographic studies, typically using Thorotrast as a contrast medium.

Following the successful examination, there was a period of development of the technique, including the use of stereoscopic angiography, and a deepening of the understanding of the appearances.[42] Perhaps not unsurprisingly, the cerebral arteriograms and subsequent venograms revealed anatomy that differed from classical descriptions and therefore supported Morton's prediction.

In 1929, Moniz's surgical colleague Reynaldo dos Santos, professor of surgery in Lisbon, introduced percutaneous trans-lumbar aortography (TLA) by direct abdominal aortic puncture with injection of a sodium iodide solution. Dos Santos described the technique.[43] Punctures were made into the aorta with a long needle in a variety of positions, including above the coeliac trunk, above the kidneys, above the inferior mesenteric artery, and above the origin of the common iliac artery. In his early studies, Dos Santos used a 100% solution of pure sodium iodide as a contrast medium, which was quite toxic. The injection was painful even with the later classical ionic contrast agents, and therefore required general anaesthesia. The TLA became the standard vascular examination for decades and was being routinely performed over 50 years after the procedure was first described. There were surprisingly few complications from the procedure which persisted into the 1980s. The TLA has been replaced by non-invasive studies, often using ultrasound or MRI, and performed as an outpatient with no sedation or anaesthesia.

Other members of Moniz's team were equally innovative and successfully introduced angiopneumography (that is pulmonary angiography) by de Carvalho,[44] lymphography by Monteiro, phlebography (that is venography) by João Cid des Santos who was the son of Reynaldo, and portal venography by Pereira. In 1929, Werner Forssmann had introduced a well-oiled ureteral catheter via an antecubital vein into his own right atrium, and it was in 1931 that Moniz, de Carvalho, and Almeida Lima using the Forssmann method demonstrated the pulmonary vasculature with an injection of sodium iodide. The Portuguese School therefore introduced many aspects of clinical angiography during 1930–1950, but the international adoption of their techniques was severely delayed by the Second World War.

THOROTRAST IN ANGIOGRAPHY AND ITS CONSEQUENCES

In 1931, Thorotrast was introduced as a contrast medium and in October it was first used in cerebral angiography. The new contrast medium was seen as a great improvement since it was not irritant and gave an excellent opacity. In 1950, Almedia Lima described Thorotrast and compared its use favourably to the organic iodine derivatives used at the time.[45] Thorotrast was a colloidal suspension of thorium dioxide, being a stable aqueous colloidal solution containing 25% thorium dioxide by volume in a tapioca-dextrin medium. A preservative of 0.15% methyl p-hydroxybenxoate was added to the solution. Lima wished for a better contrast medium; however, he felt that at that time none was available. Lima knew of no other substance that gave such satisfactory results as Thorotrast. He had not seen any serious disturbances following its use and had personally performed 2,000 angiograms. He saw the problems as being related to local tissue reactions and the fact that it was not eliminated from the body. He therefore recommended the abandonment of its use for ventriculography and encephalography. Of all myelographic agents, Thorotrast was found to be the most irritant to the pia-arachnoid resulting in both systemic and local reactions. The local reactions caused a severe arachnoiditis and a cauda equina syndrome.

Problems could arise following the local injection of Thorotrast with extravasation and the development of a local reaction and mass, the so-called Thorotrastoma (Figure 8.8).[46] Thorotrast was retained in the walls of the vessels and histology showed little balls of Thorotrast in the branches of small cerebral vessels. However, the main danger lay in the permanent retention of a radioactive substance in the body (Figure 8.9). Thorotrast conglomerates emitted radiation as part of the decay of thorium 232. The majority of the radiation was α-radiation and β- and γ-radiation contributed less than 10% to the total dose. Somewhat surprisingly, Almedia Lima concluded as late as 1950 that "the tissue alterations and the radioactivity of Thorotrast are of no importance in the dosage of this substance as used in angiography", and he recommend angiography with Thorotrast (20% colloidal suspension of thorium dioxide) with a dose of 8 ml to either side. Unfortunately, Lima's optimism was to prove unfounded, partly as a result of the work of Hermann Muth and the German Thorotrast Study.[47] Thorotrast was in use in Germany from 1929 to 1955, and the first quantitative biophysical studies to determine the activity concentrations of radionuclides derived from the naturally occurring thorium series were undertaken in Germany from 1946 to 1949. The dose estimates were of such concern that production of Thorotrast was stopped, and it was withdrawn from the market during 1949–1950. This makes it all the more curious that Lima was recommending its use in 1950. A detailed German Thorotrast Study started in 1968 and reported the findings after 20 years in 1988. The results were interesting.[48] The study found an excess

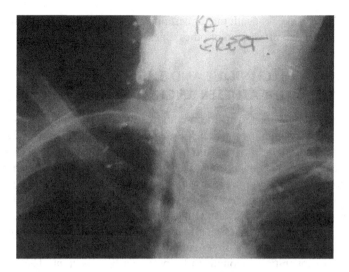

FIGURE 8.8 Thorotrastoma on the right side of the neck. Dense extravasated contrast can be seen following extravasation during a direct carotid artery needle puncture. Author's collection.

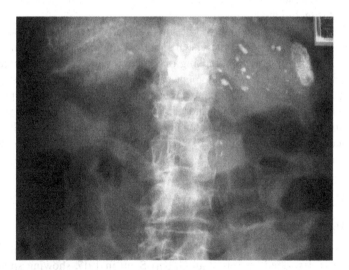

FIGURE 8.9 Retained Thorotrast in the abdominal lymphatic system and spleen (the same patient as in Figure 8.8). Author's collection.

rate of various neoplasms, including malignant liver tumours, myeloid leukaemia, and tumours of bile ducts, pancreas, oesophagus, and larynx. The Thorotrast patients had a statistically significant loss of lifetime as a function of the dose rate, and this could not be accounted for purely by the known Thorotrast-specific diseases. The Study concluded that the long-term irradiation of the reticuloendothelial system not only resulted in an excess death rate of certain neoplasms, but that there was also an acceleration of the manifestation of other illnesses leading to premature death.

THE SELDINGER TECHNIQUE AND THE DEVELOPMENT OF CATHETER ANGIOGRAPHY

A major development related to the method of delivery of contrast medium into vessels and the heart chambers was achieved by Sven Ivar Seldinger (1921–1998) in 1953, working at the Karolinska Clinic in Stockholm.[49] He introduced the needle-guidewire-catheter replacement technique which permits selective catheterisation and injection of most arteries and veins following a simple puncture. The technique might seem obvious with a needle puncture, a wire passed down the needle, and a catheter passed down the wire, with removal of the wire and the catheter left in situ. Prior to Seldinger's technique angiography was either by direct needle injection or by a catheter insertion using a direct cut down. It is remarkable that the pioneering technique was not initially thought to be particularly important. The head of the department of radiology at the Karolinska Institute did not think that his invention could be the basis of a thesis, and Seldinger therefore started a second project on the development of percutaneous cholangiography. Seldinger's thesis of 1966 was entitled 'Percutaneous transhepatic cholangiography' and catheterised the biliary ducts using his technique.

This catheter technique, aided by the low-osmolar contrast agents, permits virtually painless, safe arteriographic visualisation of any arterial or venous territory or cardiac chamber, thus revolutionising diagnostic imaging. The technique is the most important event in the development of angiography. This extremely versatile percutaneous catheterisation technique has been very successfully developed initiating the modern era of intravascular interventional diagnosis and therapy. Seldinger's technique of needle-guidewire-catheter has proven to be extraordinarily versatile enabling access to almost any lumen or cavity and is the mainstay of access for many interventions.

NEURORADIOLOGY

VENTRICULOGRAPHY, ENCEPHALOGRAPHY, AND AIR MYELOGRAPHY

There had been several case reports of patients surviving with intracranial air. One such case was described by Sebastian Gilbert Scott in 1917, showing spontaneous pneumocephaly in a woman who complained of her brain splashing, and Scott could even hear the fluids splashing (Figure 8.10).[50] The patient previously had an orbital tumour removed some years earlier. Walter Dandy (1886–1946) from Johns Hopkins Hospital was aware of such instances, and also of the value of the appearances of abnormal gas collections to diagnose abdominal disease, indicating that the use of air for contrast would be safe. In 1918, Dandy described ventriculography followed by encephalography in 1919. In the latter procedure, air is injected by lumbar puncture

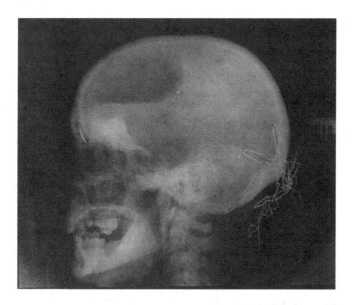

FIGURE 8.10   A patient of Sebastian Gilbert Scott in 1916, showing spontaneous pneumo-cephaly. The fluid level is clearly seen. Author's collection.

in order to fill the ventricular system. Dandy also predicted the development of air myelography for spinal lesions, and subsequently this was performed by Jacobaeus from Stockholm who demonstrated three spinal tumours in 1921. There were disad-vantages to air myelography. Since gas did not mix with the cerebrospinal fluid, the fluid needed to be removed and replaced with gas. This resulted in significant post-myelography headaches. However air, or the more rapidly absorbed oxygen, was the least irritating of all myelographic agents. Unfortunately, the contrast produced using air was relatively poor, even when combined with tomography, and overlying gas-filled organs caused confusing images. In the 1960s, gas myelography was still occasionally performed and survived into the CT era as air meatography when a small quantity of gas was used to outline the acoustic nerve in the internal auditory canal before the universal use of MRI.

Encephalography was not easy to perform, and at the first International Congress of Radiology held in London in 1925, J.W. Pierson, who was a colleague of Dandy, said that the procedure was dangerous and complicated, but in favour said that in competent hands it should not be nearly so dangerous as exploratory craniotomy and could give more information. Dandy had only three deaths in a series of 500 examinations. Following the description of ventriculography by Dandy in 1918, the neurosurgeon Harvey Cushing reproached him for spoiling the intellectual challenge of deducing the site of the brain lesion from the history and physical examination. Currently, the situation is reversed, and it has been quipped that the patient is now referred to the neurologist when the CT or MRI scan is normal!

MYELOGRAPHIC AGENTS

LIPIODOL

Jean-Athanase Sicard (1872–1929) and Jacques Forestier (1890–1978) had been using epidural injections of Lipiodol to treat sciatica and had injected it intrathecally without obvious harm.[51] They described its use in intra-arachnoid, intramuscular, intravenous, intratracheal, and oral locations. They were able to demonstrate the sub-arachnoid space in health and in the presence of disease including tumours. Lipiodol is a viscid, halogenated, poppy seed oil containing 40% iodine as an organic combination. It was a light yellow colour and slowly turned brown due to the release of free iodine. It did not mix with the spinal fluid and was only absorbed slowly. It was viscous, which made removal difficult and tended to break up into globules. Lipiodol was irritant and could cause a late painful arachnoiditis.

Lipiodol was used in many areas, including gynaecological, and was a versatile agent. Lipiodol ultrafluid was used for lymphangiography in the 1980s until the procedure became obsolete, and was also used for sialography.

MYODIL (PANTOPAQUE)

Lipiodol was used for myelography until iopendylate (Pantopaque or Myodil) was introduced in the 1940s. Iopendylate was not water-soluble and was absorbed only slowly. There was again a small risk of adhesive arachnoiditis following its use. Pantopaque is a mixture of ethyl esters of isomeric iodopentyllundecyclic acids containing 30.5% of firmly bound organic iodine. Similar to Lipiodol, the solution became discoloured on exposure to light due to the release of free iodine and needed to be discarded. Pantopaque was less viscous than Lipiodol and so had less of a tendency to break up into globules and was easier to remove. Pantopaque had reactions, including meningitis and a delayed arachnoiditis. Robert Shapiro writing in 1968 said: "All in all, Pantopaque is an eminently satisfactory medium for most problems in the spinal canal, with a low incidence of untoward reactions".[52] In addition to myelography, Myodil was used for ventriculography, and had also been introduced into the amniotic sac to outline the foetus prior to intrauterine blood transfusion.

The practice in the UK was to use a smaller quantity of Myodil for myelography and to aspirate it after the procedure, which is part of the reason for the lower incidence of adhesive arachnoiditis in the UK (Figure 8.11). If the Myodil was left in place, then there would be a prolonged elevation of the serum iodine (Figure 8.12).

The topic of informed consent of patients before radiological procedures is important. By the 1990s, it was good practice to discuss possible side effects and complications with the patient before a radiological procedure, but this did not apply during the period of Myodil use. It was generally believed, somewhat paternalistically, that a patient should not be worried unnecessarily by an overemphasis on side effects since they might then refuse a procedure that the doctor believed would be in their best interests. Judging the actions of one generation by the values of succeeding

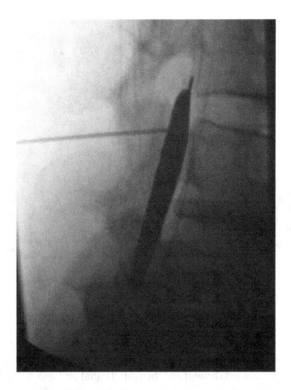

FIGURE 8.11   Needle aspiration of the oily contrast medium Myodil in a patient who had the initial examination abroad, and who had returned to the UK and desired its removal to avoid arachnoiditis. Author's photograph. Author's collection.

generations need to be done with caution. Doctors have generally acted in what they believe to be in the best interests of their patients.

WATER-SOLUBLE AGENTS

Before 1970, only iodinated oils, including Myodil (Pantopaque), were available for myelography. Ionic compounds were generally considered too toxic, although occasionally they were used for lumbosacral radiculography. The possibility of using Conray for myelographic examinations was considered. Conray is sodium iothalamate mixed with a methyl meglumine salt; however, it produced severe local reactions in several patients and its use was abandoned. The advances came when the ionic water-soluble Dimer-X was introduced in 1972, and the non-ionic metrizamide in 1977.

The French company Guerbet developed the ionic compound meglumine iocarmate (Dimer-X), combining two tri-iodinated benzene rings into one large molecule (hence it was a dimer) containing six atoms of iodine and so reducing the osmolality. Dimer-X could only be used in the lower portions of the spinal canal below the

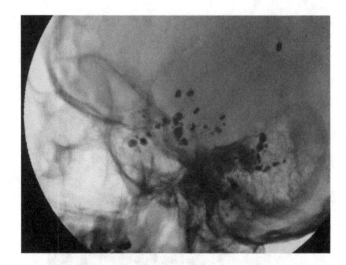

FIGURE 8.12 Globules of Myodil seen on a skull radiograph, once a regular finding. Author's collection.

spinal cord for radiculography, but it produced superb quality radiographs of the lumbosacral nerve roots. Though much less toxic than the previous aqueous contrast media, it had to be used with great care and in a strictly limited dosage. By contrast, metrizamide could be used throughout the spinal canal and was much less toxic than meglumine iocarmate, which it replaced for myelography and radiculography in the late 1970s. Initially, metrizamide was limited in use to the thoracolumbar region, and until 1980 a special licence was needed from the Department of Health to examine the basal cisterns. The water-soluble agents showed the nerve root sheaths better and so Myodil was gradually abandoned.

BRONCHOGRAPHY

The bronchial tree could be opacified with an opaque medium using a variety of techniques. The first experimental bronchogram was performed by Karl Springer from Prague in 1906, which is surprisingly early. In the illustration, a tracheal catheter was used with contrast injected, and showing a normal right bronchogram (Figure 8.13). The examination was unpleasant for the patient and there was therefore a high threshold of referral for performing the examination. The chest physician would be reluctant to submit the patient for the procedure unless there was a degree of confidence about the examination. The introduction of high-resolution computed tomography (HRCT) has considerably changed attitudes to bronchiectasis. Since so many more patients are investigated than was possible with bronchography, it is now known that bronchiectasis is very much more common than had previously been appreciated. Over the years, many contrast agents were used in the bronchial tree, including colloidal silver and bismuth. In the classical technique, Dionosil was introduced by direct tracheal injection or was dripped over the back of the tongue. It

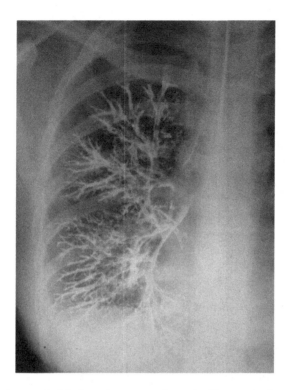

FIGURE 8.13   A normal right bronchogram performed with a tracheal catheter with contrast injected. Author's collection.

could also be introduced using a catheter, or a bronchoscope. It was used not only to diagnose bronchiectasis, but also to investigate lung tumours, cysts, and abscesses in the time before fibre-optic bronchoscopy and CT scanning.

Dionosil (propyliodone) was a contrast agent allied to diodone in a firm organic combination to prevent it from breaking down to iodides or to free iodine. The aqueous form was a 50% aqueous suspension, and the oily form was a 60% suspension in arachis oil.

INTERSTITIAL AIR STUDIES

A gas used as a negative contrast medium could be introduced into the tissues by direct injection. Examples include retroperitoneal air studies and pneumo-mammography to examine breast lumps (Figure 8.14).

RETROPERITONEAL AIR STUDIES

In this technique the retro-peritoneum around the kidney is outlined by gas. Paul Rosenstein from Berlin and Humberto Carelli from Buenos Aires both described the technique independently in 1921. A 10-cm needle was used to make a direct

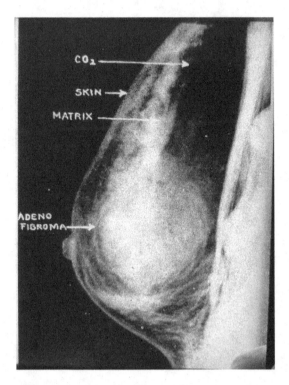

FIGURE 8.14 A labelled pneumo-mammogram performed with carbon dioxide with a fibroadenoma (breast tumour) demonstrated. Author's collection.

retroperitoneal injection. Rosenstein injected 600 ml of oxygen and Carelli injected about 200–400 ml of carbon dioxide. The examination was introduced in the days before the IVU to show the kidneys. Rosenstein emphasised that it was "important that the radiologist became independent of the clinician for these pictures". Rosenstein said that this technique was of value in the following instances:

1. Determining the presence of one or both kidneys. The removal of a solitary kidney would be a disaster.
2. Determining the size of a kidney.
3. Showing the presence of kidney stones more clearly.
4. Diagnosis of displacement of the kidney. This would diagnose the "floating kidney" which was thought to be a cause of symptoms as a part of visceroptosis.
5. Diagnosis of renal tumours and tumours around the kidney.
6. Studying "acute stresses" of the kidney. In unexplained renal colic, the enlarged pelvis could be shown outlined by gas.

In later years, the technique was combined with tomography and was primarily used to show the adrenal glands. The examination illustrated was performed on 14 April

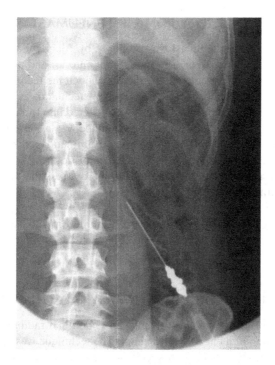

FIGURE 8.15   Direct retroperitoneal interstitial injection to show the kidney. Examination performed in 1943. Author's collection.

1943 by Ernest Rohan Williams (1906–1963) by direct retroperitoneal interstitial injection with a long needle at Hillingdon Hospital in Uxbridge, and would have demanded considerable technical skill (Figure 8.15).

PRESACRAL PERIRENAL PNEUMOGRAPHY

A variant of the technique was presacral perirenal pneumography. This was reviewed by John Laws (1921-1999) from the Hammersmith Hospital in 1958, and it was then almost exclusively used to visualise the adrenal glands.[53] The gas, when injected in front of the sacrum, passes up in the retro-peritoneum and outlines both kidneys and adrenal glands. Laws believed that the technique he described and used was safe and avoided the risk of gas embolism. Laws described the use of pure oxygen on the grounds that its greater solubility in serum makes the risk of inadvertent intravascular injection less serious. The use of carbon dioxide was also described. Carbon dioxide is more than 20 times more soluble in serum than oxygen, and up to 100 ml of carbon dioxide as a gas may be injected intravenously with no serious effects. The preferred gas for presacral injection was therefore carbon dioxide and the quick absorption meant that the procedure caused the patient discomfort for only a short period of time, with the whole examination, including the taking of films being completed in approximately 30 minutes.

PNEUMOMAMMOGRAPHY/RÖNTGEN PNEUMASTASIA

The injection of gas into the breast was proposed by Alberto Baraldi in Argentina in 1933. He initially injected purified air into the anterior and posterior regions of the breast producing emphysema. He further developed the technique by the use of oxygen which was better absorbed. He described the technique as absolutely innocuous, and with minimal symptoms. The gas was injected into three areas: retromammary, retropectoral, and subcutaneous. The technique would show masses within the breast, or evidence of chest wall invasion. Later examinations involved the injection of carbon dioxide. Gas might also be injected into a cyst following aspiration of the fluid contents and followed by mammography to assess the cyst contents in the period before ultrasound.

Contrast studies of the ductal system of the breast were also performed using a wide variety of agents, including Thorotrast, bismuth, Lipiodol, sodium iodide, air, and iodinated water-soluble contrast agents. These studies were performed to investigate bloodstained discharges from the nipple.

DIAGNOSTIC PNEUMOPERITONEUM/GYNAECOGRAPHY

In the technique of pneumogynaecography, an abdominal radiograph is made following the induction of a pneumoperitoneum. The technique was first developed by Eugen Weber from Kiev in 1912. A specifically gynaecological use was described by Otto Goetze from Halle in 1918. Following the induction of a pneumoperitoneum, the patient was placed head down and prone, and a pelvic radiograph was obtained with the pelvic viscera clearly outlined with air. The bowel would fall out of the pelvis, and the uterus, bladder, and ovaries could be identified. Goetze said that he used the technique to diagnose pregnancy in the early months, infantilism, myomata, uterine and adnexal adhesions, pyosalpinx, and ovarian tumours. The technique was reviewed by Marchesi and others in 1955.[54] The authors used between 1,500 and 2,000 ml of gas. The outlines of the uterus and ovaries were demonstrated in this examination; however, like most contrast studies, there is no indication of internal structure. Marchesi was able to show some internal structure by combining gynaecography with hysterosalpingography, that is, by using a negative contrast medium around the uterus and by filling the uterus with a positive contrast medium. This examination was performed as late as the 1970s and before the introduction of ultrasound, which rendered the examination obsolete.

The technique of pneumoperitoneum is now routinely used with the introduction of intraperitoneal gas prior to laparoscopy, or aqueous iodinated contrast may be injected to outline the peritoneal space in the diagnosis of hernia, although this latter technique has been replaced by ultrasound and MRI.

CONCLUSION

In the years since 8 November 1895 when Röntgen discovered X-rays in his laboratory at Würzburg, there have been major developments in all aspects of medical imaging, and many undreamt of by the pioneers. In the forefront has been the synthesis of safer, more effective, more physiological, lower osmolar water-soluble iodinated contrast media, and methods for their delivery.[55] This sequential development has fully realised the most optimistic dreams of Cameron, Binz, Räth, Swick, Wallingford, Hoppe, Almén, and many other researchers.

Many of the examinations that are described in this chapter are invasive and are not without complications. However, it should be noted that these examinations were best practices for their time. Even a relatively short time ago, what for us today would be a simple question to answer, could be surprisingly difficult to answer, and would require an invasive diagnostic procedure. The traditional paradigm of medical care commonly involved invasive diagnostic procedures which were followed by invasive therapy. Modern medicine has replaced this with the model of non-invasive diagnosis and minimally invasive therapy. Medical practice continuously changes and advances are sequentially made, and an advance in one area may facilitate a change in another.[56] The development of safe contrast media had facilitated this change. So successful have been these new imaging technologies that ultrasound systems and magnetic resonance angiography can produce excellent vessel demonstration with additional data on blood flow. Most of the conventional contrast medium examinations are now largely obsolete; however, the newer agents have a firm place in interventional minimally invasive procedures.[57]

NOTES

1. Grainger, R.G., Thomas, A.M.K. 1999. History of intravascular iodinated contrast media. In: *Textbook of Contrast Media*. Eds. P. Dawson, D. Cosgrove, D.J. Allison. Isis Medical Media.
2. de Haën, C. 2019. *X-Ray Contrast Technology: A Revolutionary History*. Boca Raton: CRC Press.
3. Barclay, A.E. 1913. *The Stomach and Oesophagus:. A Radiographic Study*. London: Serrate & Hughes.
4. Bates, S., Beckmann, E., Thomas, A.M.K., Waltham, R. 2012. *Godfrey Hounsfield: Intuitive Genius of CT*. London: The British Institute of Radiology.
5. Felson, B. 1973. *Chest Roentgenology*. Philadelphia: W B Saunders.
6. Williams, F.H. 1901. *The Roentgen Rays in Medicine and Surgery*. New York: The Macmillan Company.
7. Cannon, W.B. 1911. *The Mechanical Factors of Digestion*. London: Edward Arnold.
8. Carman, R.D., Miller, A. 1917. *The Roentgen Diagnosis of Diseases of the Alimentary Canal*. Philadelphia: WB Sanders Company.
9. Barclay, A.E. 1933. *The Digestive Tract: A Radiological Study of Its Anatomy, Physiology, and Pathology*. London: Cambridge University Press.
10. McLaren, J.W., King, J.B., Copland, W.A. 1955. Preliminary observations on veripaque, a colonic actuator for use with barium enemata. *The British Journal of Radiology*, 28(330), 285–294.
11. McAlister, W.H., Anderson, M.S., Bloomberg, G.R., Margulis, A.R. 1963. Lethal effects of tannic acid in the barium enema. *Radiology*, 80(5), 765–775.
12. de Lacey, G. 1978. Retained surgical swabs: possible causes for errors in X-ray detection and an atlas to assist recognition. *The British Journal of Radiology*, 51, 691–698.

13. Voelcker, F., von Lichtenberg, A. 1906. Pyelographie (Röntgenographie des Nierenbeckens nach Kollargolfüllung). *Münchener Medizinische Wochenschrift*, 53, 105–106.
14. Braasch, W.F. 1915. *Pyelography (Pyelo-Ureterography)*. Philadelphia and London: W B Saunders.
15. Braasch, W.F. 1927. *Urography*. Philadelphia: W B Sanders Company.
16. Cameron D. 1917. Aqueous solutions of potassium & sodium iodides as opaque media in roentgenography. *Journal of American Medical Association*, 70, 754–755.
17. Hryntschalk, T. 1929. Studien zur Röntgenologischen Darstellung von Nierenparenchym und Nierenbecken auf Intravenösem Wege. *Zeitschrift für Urologie*, 23, 893–904.
18. Osborne, E.D., Sutherland, C.G., Scholl Jr., A.J., Rowntree, L.G. 1923. Roentgenology of the urinary tract during excretion of sodium iodide. *Journal of the American Medical Association*, 80, 368–373.
19. Swick, M. 1978. Radiographic media in urology, the discovery of excretion urography. *Surgical Clinics of North America*, 58, 977–994.
20. Swick, M. 1929. Darstellung der Niere und Harnwege in Roentgenbild durch intravenose Einbringung eines neuen Kontraststoffes: des Uroselectans. *Klinische Wochenschrift*, 8, 2087–2089.
21. Von Lichtenberg, A., Swick, M. 1929. Klinische Prüfung des Uroselectans. *Klinische Wochenschrift*, 8, 2089–2091.
22. Marshall, V.F. 1977. The controversial history of excretory urography. In: *Clinical Urology*. Vol. 1. Ed. J. Emmett. Philadelphia: Saunders, pp. 2–5.
23. Wallingford, V.H. 1953. The development of organic iodide compounds as X-ray contrast media. *Journal of American Pharmacological Association* (Scientific Edition), 42, 721–728.
24. Hoppe, J.O., Larsen, H.A., Coulston, F.J. 1956. Observations on the toxicity of a new urographic contrast medium, sodium 3,5-diacetamido-2, 4, 6 tri-iodobenzoate (Hypaque sodium) and related compounds. *Journal of Pharmacological and Experimental Therapeutics*, 116, 394–403.
25. Gensini, G., DiGiorgi, S. 1964. Myocardial toxicity of contrast agents used in angiography. *Radiology*, 82, 24–34.
26. Almén, T. 1969. Contrast agent design. Some aspects on the synthesis of water-soluble contrast agents of low osmolality. *Journal of Theoretical Biology*, 24, 216–226.
27. Amdam, R.P., Sogner, K. 1994. *Wealth of Contrasts. Nyegaard & Co.: A Norwegian Pharmaceutical Company 1874–1985*. Oslo: Ad Notam Gyldendal.
28. Dawson, P., Grainger, R.G., Pitfield, J. 1983. The new low-osmolar contrast media: a simple guide. *Clinical Radiology*, 34, 221–226.
29. Holm, M., Præstholm, J. 1979. Ioxaglate, a new low osmolar contrast medium used in femoral angiography. *The British Journal of Radiology*, 52, 169–172.
30. Aspelin, P., Aubry, P., Fransson, S.G., Strasser, R., Willenbrock, R., Berg, K.J. 2003. Nephrotoxic effects in high-risk patients undergoing angiography. *New England Journal of Medicine*, 348, 491–499.
31. Almén, T. 1985. Development of non-ionic contrast media. *Investigative Radiology*, 20, 2–9.
32. Cole, W.H. 1978. Gallbladder disease. *Surgical Clinics of North America*, 58, 917–926.
33. Ross, S.G., et al. 2012. Suddenly death after chest pain: feasibility of virtual autopsy with postmortem CT angiography and biopsy. *Radiology*, 264, 250–259.
34. Thali, M., Dirnhofer, R., Vock, P. (Eds) 2009. *The Virtopsy Approach: 3D Optical and Radiological Scanning and Reconstruction in Forensic Medicine*. Boca Raton: CRC Press.
35. Thomas, A.M.K. 2016. Postmortem imaging: development and historical review. In: *Atlas of Postmortem Angiography*. Eds. Grabherr, S., Grimm J.M., Heinemann, A. Switzerland: Springer International.

36. Anon. 1896. The new photography in Sheffield. *British Medical Journal*, 22 February, 495–496.
37. Rowland, S. 1896. Report on the application of the new photography to medicine and surgery. *British Medical Journal*, February 22, 492–497.
38. Dolby, T. 1976. *Development of Angiography and Cardiac Catheterisation*. Littleton, MA: Publishing Sciences Group.
39. Brooks, B. 1923. Intra-arterial injection of sodium iodide. *Journal of the American Medical Association*, 82, 1016–1019.
40. Berberich, J., Hirsch, S.R. 1923. Die Roentgenographische Darstellung der Arterien und Venen am lebenden Menschen. *Klin Wochenschr*, 2, 2226–2228.
41. Veiga-Pires, J.A., Grainger, R.G. 1982. *The Portuguese School of Angiography*. 2nd ed. Lancaster UK/Boston: MTP Press.
42. Moniz, E. 1931. *Diagnostic des Tumeurs Cérébrales et épreuve de l'Éncephalographie Artérielle*. Paris: Masson et Cie, Éditeurs.
43. Dos Santos, R., Lamas, A.C., Caldas, J.P. 1932. *Artériographie des Membres et de l'Aorte Abdominale*. Paris: Masson et Cie, Éditeurs.
44. Moniz, E., De Carvalho, L. 1932. *A visibilidade dos vasos pulmonares (Angiopneumografia)*. Lisboa: Imprensa Libanio da Silva.
45. Lima, P.A. 1950. *Cerebral Angiography*. London: Oxford University Press.
46. Thomas, A.M.K. 1985. Thorotrast granuloma of the neck. *Journal of the Royal Society of Medicine*, 78, 419.
47. Muth, H. 1989. History of the German Thorotrast study. In: *BIR Report 21: Risks from Radium and Thorotrast*. Eds. Taylor, D.M., Mays, C.W., Gerber, G.B., Thomas, R.G. London: The British Institute of Radiology, pp. 93–97.
48. van Kaick, G., Wesch, H., Lührs, H., Liebermann, D., Kaul, A., Muth, H. 1989. The German Thorotrast study – a report on 20 years follow-up. In: *BIR Report 21: Risks from Radium and Thorotrast*. Eds. Taylor, D.M., Mays, C.W., Gerber, G.B., Thomas, R.G. London: The British Institute of Radiology, pp. 98–104.
49. Seldinger, S.I. 1953. Catheter replacement of the needle in percutaneous arteriography: a new technique. *Acta Radiologica*, 39, 368–376.
50. Scott, S.G. 1917. Two cases of interest. *Archives of Radiology and Electrotherapy*, 21, 237–240.
51. Sicard, J.A. Forestier, J.F. 1928. *Diagnostic et Thérapeutic par le Lipiodol*. Paris: Masson et Cie, Éditeurs.
52. Shapiro, R. 1968. *Myelography*. 2nd ed. Chicago: Year Book Medical Publications.
53. Laws, J. 1958. Radiology in the investigation and management of hypertension. *Postgraduate Medical Journal*, 34, 514–523.
54. Marchesi, F., Olivia, L., Albano, V., Maneschi, M. 1955. *La Pneumoginecografia*. Edizioni Minerva Medica.
55. Grainger, R.G. 1982. Intravascular contrast media – the past, the present and the future. *The British Journal of Radiology*, 55, 1–18.
56. Thomas, A.M.K., Banerjee, A.K. 2013. *The History of Radiology*. Oxford: Oxford University Press.
57. Morris, T.W. 1993. X-ray contrast media: where are we now & where are we going? *Radiology*, 188, 11–16.

9 Radiology and Women

From the earliest days of radiology, women have played a significant role in radiology and the radiological sciences. A number of the first practitioners in the new field of radiology were women, as can sadly be seen by the numbers of women among the pioneer radiation martyrs.[1] There were specific radiological protection considerations for women as patients which became apparent as experience was gained, and innovative radiological examinations were devised for the medical and physical requirements of women.

WOMEN AS RADIOLOGISTS AND PRACTITIONERS

Science and medicine throughout history have been led and dominated by men. What contribution women might have made has been minimised by the failure to report the achievements of the few women who worked in the area of research and in the physical sciences. This had been the case in the West for many centuries and, in ancient Athens, it was forbidden by law for women to either practice or study medicine. The story of Agnodice or Agnodike is therefore remarkable.[2] She was apparently born into a wealthy Athenian family in about the 4th century BC. Agnodice was appalled by the high mortality of women and children in childbirth, and she therefore dressed as a man in order to study medicine. She then moved to Alexandria and became a pupil of the physician Hierophilus of Chalcedon (335–280 BC). Upon returning to Athens, Agnodice practised medicine, treating wealthy Athenian women. Many of the medical problems were related to the shyness of women and to their reluctance to consult male physicians. Agnodice was able to make a significant contribution to a change in the law that would allow women to study medicine and practice it, but solely on women patients. Unfortunately, these issues were still very much at the forefront in the 19th century and laws and regulations again needed to be changed to allow women to study and to practice medicine. Echoing the example of their earlier sister Agnodice, many of the pioneer women doctors were particularly concerned with the healthcare of women and children.

By the time Röntgen made his momentous discovery in 1895, that women could and should be accepted as doctors was already accepted in many countries. Certainly, by the 1890s, battles had been won in the British Isles and women were able to study medicine in Scotland and in England; however, this was not the case in Ireland. In Dublin University, women were not yet admitted to medical degrees. By the end of the nineteenth century, there was a degree of parity in education between men and women in England, and middle-class women had won some battles. The Medical Act of 11 August 1876 prohibited the exclusion of women from medical courses. In 1878, the University of London had accepted women for non-medical degrees, and the first MD taken by a woman was by Mary Scharleib (1845–1930) in 1888. However, in the

DOI: 10.1201/9780429325748-9

mid-1890s, there were still fewer than 180 medical women on the Medical Register in Great Britain.[3] The concern for woman's suffrage and rights in Great Britain was always a part of global movement and was a common cause for women worldwide. Cattie Chapman Catt (1859–1947), the US feminist and suffrage campaigner, said in 1908, that Britain was the storm centre of the women's movement.[4] Attitudes to the higher education varied from country to country, and an example is Maria Salomea Skłodowska (Marie Curie) (1867–1934) who was unable to be educated in her native Poland, and who in 1891 went to France to attend the Sorbonne in Paris where the attitudes were more advanced. Therefore, while the general history of radiology was similar in most countries, the story of the entrance of women into the radiological sciences was very variable reflecting differing social mores in different cultures. In considering the entrance of women into radiology, it is worth considering and comparing the lives of two doctors, Florence Stoney and Evangelia (Lia) Farmakidou, and the radiographer Kathleen Clara Clark.

FLORENCE ADA STONEY (1870–1932)

The first woman radiologist in the United Kingdom was Florence Stoney, and she had a remarkable and productive life (Figure 9.1).[5] Florence had grown up in Dublin and applied to attend the London School of Medicine for Women (LSMW) in 1891, since at that time medical training for women was not possible in Ireland. LSMW was founded in 1874 (Figure 9.2), and became linked to the Royal Free Hospital in nearby Gray's Inn Road in 1877.[6]

FIGURE 9.1 Florence Ada Stoney (1870–1932). Clipping from an unknown magazine. Author's collection.

FIGURE 9.2 The London School of Medicine in Hunter Street, photographed in 2019. Author's photograph.

The life of a female medical student at that time is illustrated in the remarkable novel *Mona Maclean, Medical Student* which was published in 1894 and written by Graham Travers, the pen name of Margaret Todd.[7] Mona notes that whilst the medical woman is presented as being scientific, she is also shown as being feminine, and a friend tells Mona that "the art of dressing one's hair is at least as important as the art of dissecting". Throughout the book, there is a theme of the genus "Medical Woman" not being understood. Mona's aunt spoke of her mingled pride, affection, disgust, and fear. Also, her aunt spoke of her disgust for the lifework that Mona had chosen, and of her fear of her supposed cleverness. Mona's aunt is depicted as despising learned women, and in the society that she inhabited there was no advantage to be gained as an educated woman. A woman was most certainly not to appear too intelligent in society. Mona's uncle said that he was torn asunder on the subject of women doctors. He saw that there was a terrible necessity for women in medicine, and yet saw women taking up medicine as being a sacrifice. Mona's uncle thought that a medical woman becomes hardened and loses something that is central to womanhood. In his opinion, medicine was bad for a man, but a man has some virtues which remain untouched. A woman doctor would therefore lose everything that makes her womanhood fair and attractive and would become hard and blunted. Mona's uncle raised the question of dissection and likened it to human butchery. Mona realised the gulf in viewpoints between herself and her uncle. Mona saw the study of anatomy as an ever-new field for observation, corroboration, and discovery. There was unlimited scope for the keen eye, the skilful hand, the thinking brain, and mature judgement. So, for Mona, to be a true anatomist, one would need to be a mechanician (a mechanic) and a scientist, an artist, and a philosopher. And so, a woman is presented as perhaps especially suited for anatomical studies and dissection. The comparison can certainly be made between Mona and Florence Stoney. Florence was deeply interested in anatomy, and from 1898 to 1903 was demonstrator in anatomy at the LSMW, whilst at the same time she was setting up her general medical practice in London. Florence used to say that her time as an anatomy demonstrator had been of the greatest help in her later work as a radiologist as in a case of an injury, she was able to indicate which muscle was likely to be lying between any bony fragments and was able to give the exact level of a nerve injury and indicate the best way to remove a foreign body.

On 17 November 1899, the Anatomical Society of Great Britain and Ireland (ASGBI), now simply called The Anatomical Society, held its annual meeting at the LSMW. The ASGBI was a new society having been formed in 1887, with the aim to "promote, develop and advance research and education in all aspects of anatomical science". The first annual meeting was held in November 1887 at University College London, which was close to the LSMW. The ASGBI obviously had a positive attitude towards women as members since in May 1894 two ladies from the LSMW were admitted to its membership. It is not surprising that the annual meeting would be held at some point at the LSMW. Florence demonstrated a dissection of an oesophagus which showed "well-marked diverticula". The meeting was written up in the magazine of the LSMW. Florence resigned from her post as anatomy demonstrator in September 1903 in part because she realised that there was no possibility of a woman being appointed to the lectureship.

Florence became interested in the new subject of radiology at a time when the apparatus and facilities were still primitive, and there was very little awareness of either radiological protection or the basic principles of radiobiology. The new field of radiology must have been seen as exciting and novel, and as a member of the BMA, Florence would have read the many articles on "the new photography" in the *British Medical Journal*. The Electrical Department at the Royal Free Hospital was opened in April 1902 with Florence in charge. It should be emphasised that this was a highly unusual department. The radiology department at the Royal Free Hospital was the first that was fully staffed by women, and Florence may well have been the first woman radiologist. Even the two student assistants from the LSMW were both women, and the physics support for the equipment was provided by a woman physicist, Florence's sister Edith Stoney (1869–1938). Amy Shepherd (d. 1936), who became the ophthalmic surgeon to the Elisabeth Garrett Anderson Hospital, commented that much of Florence's work, like many other early radiologists, was done under adverse and even dangerous conditions, and she was "Always ready to help and always resourceful, she gave too freely of a generous spirit in a not over-robust frame, and enlivened even the common task with her flashes of wit and wisdom".[8] There was a certain bravado about radiation that was displayed by many of the pioneers, and Florence never hesitated to take risks in localisation work on the operating table, and following the Great War, she suffered from radiation dermatitis of her left hand.[9]

In the years before the Great War, Florence developed expertise in diagnostic and therapeutic radiology. She paid a visit to the United States in 1914 to observe modern American radiology.[10] This was at a time of rapid development in radiology, and she visited several Eastern towns making a special visit to Schenectady in New York. She found the American doctors to be very helpful and made the interesting observation that "I found the doctors in America, both in the hospitals and in private, very ready to allow me to see the work in their departments – medical women not being kept out of everything so much as in England".

In 1914 when the First World War started, Florence with Mabel St Clair Stobart (1862–1954) and other members of the Women's National Service League (WNSL) approached Sir Frederick Treves (1853–1923) as chairman of the British Red Cross Society offering their medical assistance. Treves rejected the offer saying that there was no work that was fitted for women in the sphere of war. By this time, Florence had 13 years of experience in radiology. It is worth considering why Treves rejected the offer which he should have immediately accepted. Treves was well aware of the value of radiography in military surgery, and he had seen its use at first-hand during the Boer War. Treves also knew from his clinical experience of the bravery that women showed, and said that a woman was as courageous as a man, although might show less resolve in concealing her emotions. Women were also accepted in civilian medical practice by that time. The Association of Registered Medical Women (ARMW), of which Florence was an active member, was expecting that women doctors would primarily be needed on the home front since many men would be serving abroad and vacancies would therefore be created. Certainly, there was a feeling in the War Office that women should be protected from the horror of war, and it was seen as shocking that women would want to be involved. However, since so many women were in foreign service as nurses, why should medical women be treated differently? Female nurses were entirely aware of the horrors of war, taking a major role in caring for the sick and wounded. There was certainly a fear about what would happen should women be given equality with men and any real responsibility, and this is reflected in the difficulty women doctors had in getting both military rank and uniforms.[11] For a professional man, a woman doctor would compete in a way that no female nurse ever could, or indeed would need to.

Florence "wasted no time with arguments or indignation" following the rejection by Treves, and she actively cooperated with Mabel Stobart in organising a voluntary women's medical unit for foreign service with the WNSL. This was typical of Florence who held that all opportunities should be open to women, and that if not then they should make them for themselves. The Belgian Red Cross Society had sent out a request for the assistance of English woman doctors. The unit that was formed, the Stobart Unit, went to serve in Antwerp. Following the fall of Antwerp, Florence worked with the WNSL at the Anglo-French Hospital, No. 2, located at Le Château Tourlaville, near Cherbourg. This was again a hospital entirely staffed by women. On her return to England, Florence was appointed to the Fulham Military Hospital, Hammersmith. She was appointed as Head of the X-ray and Electrical

Department of the Fulham Military Hospital (FMH), and when she took up, her post was the first woman doctor to work under the War Office in England. Florence was the only female member of the medical staff at FMH. Florence personally examined over 15,000 cases, and the use of fluoroscope added considerably to her radiation exposure.

Early in 1929, Florence and Edith Stoney made a visit to India, following in the footsteps of a number of early graduates of the LSMW, where there had been an interest in women's health in India since the early days. There was a high death rate in India in childbirth and many Indian women were reluctant to see male doctors. Florence was a strong believer in the therapeutic value of sunlight. By the 1920s, a scientific rationale for the value of ultraviolet radiation in the treatment of bone disease, including rickets, was developing. Vitamin D was discovered in 1918, and by the end of the 1920s, the part played by sunlight in its synthesis was understood. Florence spent some time investigating the causes and effects of osteomalacia (bone softening) during their visit to India in 1929, which she attributed to a limited exposure to ultraviolet rays. She gave advice to various hospitals on the installation of ultraviolet light for the treatment of rickets, osteomalacia, "and other diseases of darkness".

During the Stoneys' visits to Cape Town in 1930, Florence gathered further data and used it to write her final published paper, on "The Pelvis and Maternal Mortality" presented in March 1930 to the BIR.[12] This paper discussed pelvic deformities and showed both historical examples and cases from Africa, India, and locally in Bournemouth where she was now working. Florence described the desirable rounded or gynaecoid shape of the normal female pelvis, and stated that the oval pelvic shape could cause difficulties in childbirth. She associated the rounded pelvic shape with women who "had plenty of ultra-violet light (vitamin D)", which was an assertion at the cutting-edge of knowledge at that time. The importance of this paper was recognised by many, and was quoted by Noel Hypher in 1931.[13]

One patient she looked after was a 32-year-old married, but never pregnant, woman in Bournemouth who seldom went out into the daylight and became bedridden as a result. Florence wrote that she attended the hospital by ambulance for nine months before a skiagraph (radiograph) was taken which led to the correct diagnosis (of osteomalacia or bone softening). Florence treated her with ultraviolet light for 12 months and she recovered her power of walking, and her fractures all healed. In this paper from the end of her career, Florence's concern for the health of mothers and young girls is obvious. Florence emphasised the need for sunlight and good bone health in growing girls, and she ended by saying "Maternal mortality in this country is partly due to want of sunlight on the whole skin surface (or its equivalent in other forms of vitamin D.)" and exclaims "If preventable – why not prevented?"

During the visit to South Africa, a reporter from the *Cape Times* observed that Florence was suffering from the effects of overexposure to X-rays. This visit was two years after she had retired from radiological practice, and her hands were still showing signs of radiation dermatitis.

During her residence in Bournemouth, Florence was treated by Sydney Watson Smith (d. 1950), who was honorary consulting physician and dermatologist to the

Royal Victoria and West Hampshire Hospital. There is a funded Sydney Watson Smith Lecture of the Royal College of Physicians of Edinburgh. He was chairman of the Bournemouth Division of the BMA of which Florence was a member, and wrote

> It is now notorious that those pioneers in X-ray work and those exposed unduly to unprotected rays usually die very painful deaths, and Dr Stoney was no exception: she suffered greatly and bravely, and she knew quite well what would be the manner of her death.[14]

Florence had personally known many of the radiology pioneers, and she was fully aware of the manner of their deaths, knowing that she would most probably develop aplastic anaemia or cancer. Florence suffered from the harmful effects of radiation which was to be expected of someone practising radiology in the early period, but she lived as long as she did is a testimony to her care when she gave radiological treatments and diagnostic examinations.

Florence developed a cancer of the spine, and moved back to London for treatment, at the Cancer Hospital in Fulham Road. She died on 7 October 1932 at the age of 63. On her death certificate, the cause of death was given as fibrosarcoma. Although Florence exhibited the effects of chronic radiation exposure, there is no clear evidence that her death was a result of radiation, and so her name was not included as a radiation martyr on the Martyr's Memorial in Hamburg.

Florence was mourned by both friends and colleagues, and many obituaries were written, including *The British Journal of Radiology*,[15] and *The Bournemouth Daily Echo*.[16] Helen Chambers, who was the pathologist at the Marie Curie Hospital in London, remarked that Florence was blessed with an exceptional ability and intellect. Agnes Savill wrote of her "knowledge of suffering humanity, wide sympathy, and rare gift of whimsical humour in conversation". Savill added: "Her charming and welcome smile, her eyes with the bright twinkle behind the glasses, her amusing and entertaining comments – at radiological gatherings we shall all miss that quiet pioneer, that little figure with a great heart". Mabel Ramsay, who had accompanied Florence to Antwerp, remembered her in *The Times* as "British to the core, she will yet always remain in my memory as a very courteous and kindly Irish lady"[17]. Her health, "good though never robust ... she accomplished much by steady application and conserving her strength". Of her personality, "She had a good deal of quiet humour. She had gentle kindliness and rich sympathy for suffering, and showed courage in her last, long and painful illness".

EVANGELIA (LIA) FARMAKIDOU (1890–1982)

Whilst the story of the introduction of radiology itself is similar from one country to another, the story of how women became radiologists shows considerable variation and the situation in Greece was very different from the British Isles.[18] In particular upper-class women gradually took on more typically male roles which, in many cases, required a university education. Even bolder women with more dynamic personalities and stronger wills would hesitate to compete against their male counterparts and to take on roles that were revolutionary for that period. In the first half of the 20th century, it was a very bold undertaking for a woman, even for those of the upper class, to complete her studies in the predominantly male field of medicine.

It was considered even more remarkable to undertake postgraduate studies abroad, especially in a speciality such as radiology which required a thorough knowledge of complex equipment and its use. One can only imagine the conditions faced by women when they returned to the Greek towns to practice medicine in their chosen speciality. At the time, there was only a limited freedom of thought, combined with a fear of both new ideas and the unknown which existed in the minds not only of the patients but also of others who worked in the field of medicine. All these factors surely served as serious obstacles for young female physicians.

The first Greek female radiologist was Evangelia (Lia) Farmakidou (1890–1982) who was born in Athens in 1890 (Figure 9.3).

She enrolled in the Medical School of the National and Kapodistrian University of Athens obtaining her licence to practice medicine in July 1913 (Figure 9.4). In that same year, she was awarded a Doctoral Degree from the Medical School of the National and Kapodistrian University of Athens. In 1916, during the Great War, Lia went to Paris with the intended purpose of studying under the great French Electro-Radiologist Joseph Belot (1876–1953). Whilst it was unusual at his time for women to become doctors in Greece, it was quite unprecedented for women to obtain post-graduate education abroad. Lia was impressed by the possibilities of new technology

FIGURE 9.3 Evangelia (Lia) Farmakidou (1890–1982) as an adolescent with her pet dog. Image courtesy of Christos S. Baltas.

FIGURE 9.4 Evangelia (Lia) Farmakidou's ID card, with photograph, for the Athens Medical Association. Image courtesy of Christos S. Baltas.

and this influenced her choice of a speciality. After returning to Greece, she worked at a trauma care centre in Athens. It was not possible to train in radiology in Greece and therefore in 1925 she went and specialised in radiology at the University of Munich under Rudolph Grashey (d. 1950), who was one of the most prominent of the pioneer radiologists.

After her return from Munich, Lia opened a private radiological laboratory building at 13 Gravias Street and practised both radiodiagnosis and radiotherapy.

Lia was interested in the role of women in Greece and was a member of the Hellenic Federation of University Women and represented Greece in 1932 at the Sixth Conference of the International Federation of University Women (IFUW), organised by the International Federation of University Women in Edinburgh, Scotland.

Her love for radiology and the breadth of her contacts led to the foundation of the Hellenic Radiological Society in 1933. Lia Farmakidou was one of the 20 founding members of Hellenic Radiological Society. She actively fought for the recognition of the speciality in Greece, as well as for the foundation of the first Chair of Radiology at the Medical School of the National and Kapodistrian University of Athens.

Lia took an active role in the Hellenic Radiology Society acting as curator of the library and special secretary. She was elected adviser to the Board of Directors of the Hellenic Radiology Society in 1955. In the 1950s, she moved her radiological laboratory to 22 Mavromichalis Street using radiotherapy in the treatment of skin diseases and aesthetics.

Lia Farmakidou was successful during a time when the speciality was dominated by men. She was highly esteemed by her colleagues. Her approach to practice was pioneering and innovative.

During her scientific career, she gave interviews in scientific journals and in the daily press, as well as lectures to the "Parnassos" Literary Association.

In the last years of her life, she remained active, full of energy, and lively. She died in Athens on the 10th of November 1982, after a long and full life, at the age of 92.

KATHLEEN CLARA CLARK (1898–1968)

Kathleen Clara Clark, known as Katie or Kitty, was a pioneer in radiography and was responsible for many advances (Figure 9.5).[19] At the age of 23, she began her training at Guy's Hospital in London, the only civilian radiography training school in the country at that time. In 1921, she passed the first-ever qualifying examination held by the Society of Radiographers (SoR). As might be imagined, her family was opposed to her choice of profession, particularly since there had been the recent high-profile death of William Ironside Bruce (1876–1921) from the effects of radiation. Her family finally acquiesced, and they gave her an entire outfit of clothes made from silk since it was believed that these were impenetrable to radiation!

The SoR in the United Kingdom had been set up in 1920. Those who had been in active radiographic practice for over ten years were given membership without

FIGURE 9.5 Kathleen Clara Clark (1898–1968), passport photograph. Courtesy of the late Marion Frank.

examination. All other applicants had to take a new examination, the MSR or Member of the Society of Radiographers, which took place in January 1922. Katie worked initially at the Princess Mary's Hospital in Margate before moving in 1927 to the Royal Northern Hospital in London as a junior radiographer. A few months after starting at Margate, she had undertaken the training of student radiographers, and teaching was to be the major feature of her career.

She was aware of the lack of adequate training for radiographers and founded a School of Radiography at the Royal Northern Hospital in 1929 which became a model for schools elsewhere.[20] Kathleen was by then senior radiographer and she became the first tutor of the School of Radiography, remaining until 1935. Kathleen proved herself to be a gifted teacher. The School of Radiography was one of the first in the country and soon made a name for itself, attracting many pupils.

In 1935, with M.R. Bell at the Royal Northern Hospital, she wrote a paper on the technique for femoral neck radiography.[21] This was an important paper and described the technique that was necessary in cases of fractured neck of femur. Adequate imaging of the hip was needed for surgical pinning of the fractured femoral neck.

In 1935, she left the Royal Northern Hospital to become the co-founder and Principal of the Ilford Radiographic Department at Tavistock House, where she was involved in instruction and research into radiography and medical photography (Figure 9.6). Ilford was a large company selling all that was needed for photography, including X-ray photography.[22] Ilford saw their interest as not ending with the manufacture of the film and wanted to collaborate with the users to help in the promotion of science and development of the photographic technique. In many hospitals,

FIGURE 9.6 The bust of Miss K.C. Clark which was placed in a treasured position in the reception area of Tavistock House. It was made by Kathleen Parbury ARBS after she had attended an Ilford refresher course as a radiographer. The bust is currently displayed at the Society of Radiographers. Author's photograph.

clinical photography was part of the X-ray department. The department was to serve as a meeting place for those interested in radiography. Under her guidance, the Ilford Department of Radiography and Medical Photography at Tavistock House developed a worldwide reputation.

Within a year of starting at Tavistock House, Katie began her work on *Positioning in Radiography*, which was destined to become a world-famous textbook. It was published in 1939 and sold for 3 guineas (£3.15 p), which was a significant sum of money.[23] It had 500 pages and 1,400 illustrations and diagrams, and by 1942 had sold about 6,500 copies. The book became the standard work of reference for radiographers and has been through many editions. The book was accompanied by a teaching slide collection and this has been preserved, and even today it remains a useful resource.

Positioning in Radiography is a very interesting book for several reasons. Firstly, it standardised the radiographic projections, and so similar projections were able to be made in all hospitals. This had been a concern of Ironside Bruce. Katie Clark was keen to standardise both positioning and exposure. Secondly, the book is very artistic. The illustrations do not come across as cold and entirely objective scientific images. It is therefore not surprising to read that the artist Francis Bacon acknowledged *Positioning in Radiography* as a crucial source and commented that it was his favourite medical textbook. Lawrence Gowling indicated that Bacon repeatedly borrowed from the photographs in the book for his artistic work. The images of the body that Francis Bacon made have an almost radiographic quality and there is the impression that multiple layers of the body are seen at the same time and that one is not just looking at the skin surface.

The second edition of *Positioning in Radiography* was published in January 1941. The third edition appeared in June 1942 and the fourth edition in April 1945 and further editions appeared at regular intervals. The book became the most important book on radiography ever written and was hugely influential. The great theme throughout the life of Katie Clark was the standardisation of radiographic projections and techniques. Whilst standardisation might be seen in a negative manner for some occasions, for Katie standardisation was about promoting high quality and on advising as to the best technique that should be used to obtain optimal results. She was President of the Society of Radiographers from 1935 to 1937 (and the first woman President).

During the Second World War, Kathleen worked with the RAMC to perfect radiographic techniques for use in casualty clearing stations. From 1937 to 1939, Kathleen worked with the Chief Radiologist of the RAMC on the localisation of foreign bodies, which was to prove valuable when the war started in 1939. In the June of 1939, she had made a visit to Sweden and Denmark to observe progress in radiography.[24] She was able to meet the radiologist Eric Lysholm (1891–1947) and the neurologist Herbert Olivecrona (1891–1980). Eric Lysholm developed his widely used skull table for radiography, which was almost universally in utilisation. She found that in Sweden the term "radiographer" was unknown, the technical work being performed by radiologists and trained nurses.

Kathleen continued with a programme of technical demonstrations and exhibitions at Tavistock House, and the department was greatly valued. The normal commercial activities of Ilford in photography were greatly reduced and activities were expanded due to the high wartime demand for X-ray materials. As a company, Ilford worked with the RAF in the production of films needed for aerial reconnaissance, and as a spin-off the new high-speed emulsions resulted in improvements in X-Ray films. This was marketed as "Red Seal" film, which not only had a finer definition, but also the improved speed allowed for a lower X-ray dosage for the patient.

Katie was involved with experiments in mass miniature radiography (MMR), which was in part stimulated by the shortage of X-ray film. MMR was used for the early detection of pulmonary tuberculosis, which was increasing in prevalence partly as a result of the poor wartime living conditions. The camera used 35-mm film instead of the full X-ray plate of 14 × 17 inches. The image on a fluorescent screen was recorded on the recently popularised 35-mm film, and Ilford promoted the special camera. Patients could be screened quickly, and if the MMR study was abnormal, a full-sized chest film was obtained. In 1942, she was put in charge by the Medical Research Council of a technical team to assess the value of the technique in examining office and factory workers. In 1943, she was to direct the training of 40 such teams that had been recruited by the Ministry of Health.

Following the German invasion of Norway, the Norwegian Medical Services relocated to London and were housed next to Kathleen's department at Tavistock House. Ilford offered the Norwegians radiographic facilities, and as a result an experiment was undertaken whereby 23,000 Norwegian servicemen were radiographed. The experience proved the value of the technique. For her wartime services, the Norwegian Government awarded Katie the Norwegian Liberty Cross.

Mass miniature radiography of the chest was being developed in the 1940s, and there was considerable interest in mass chest radiography, and it was thought to be useful in examining military recruits, for workers in the war industry, and for the examination of season ticket holders for the public air raid shelters. The main object of mass chest radiography was to detect early pulmonary disease. By 1943, 70 mass miniature radiography units had been established throughout the country. Her book *Mass Miniature Radiography of Civilians* (MRC special report series No. 251), written jointly with P. D'Arcy Hart, Peter Kerley, and Brian Thompson appeared in 1945, and it became the standard text on the topic.[25] The technique was to become more important in the 1950s when effective treatments for pulmonary tuberculosis with drug therapies became more available following clinical trials. In 1945, she was awarded the MBE for her services to radiography, particularly for her work on mass miniature radiography of the chest.

The post-war period was a time of rapid technological advancement in film technology. The X-ray group made a study of the materials that were available for use as a base material for X-ray film, and these included triacetate, polycarbonate, and polyester. There were also investigations into automatic processing such as "dunking", which replicated manual techniques, or the use of rollers for transporting films through the processing cycle. Other areas investigated included packaging of the X-ray film without interleaving, the use of automatic film changing, and the use and future of silver emulsions as opposed to other techniques.

In 1948, she presented the 11th Stanley Melville Memorial Lecture, and perhaps unsurprisingly her subject was "Chest radiography: an investigation into the possibilities of standardisation"[26]

Her view was that there was no standardisation of X-ray equipment to ensure uniform output from one unit to another; different units from different makers with the same setting of the controls may give outputs which may vary by more than 100%. There was therefore an urgent need for the manufacturers of X-ray equipment to agree upon the calibration of their sets, so that a radiographer may be able to ensure a given output with a given setting of the controls apart from variation in the mains supply.

She was committed to fostering cooperation and contact between radiographers throughout the world and was a driving spirit behind the formation of the International Society of Radiographers and Radiological Technicians (ISRRT), which was formally founded in 1962 as a non-profit organisation (Figure 9.7).[27] Kathleen took part in preliminary meetings in Munich in 1959. The ISRRT is an organisation composed of 71 national radiographic societies from 68 countries representing more than 200,000 radiographers and radiological technologists. Katie was particularly influential in formulating the educational policies of the ISRTT.

Katie remained as Principal at Ilford until her retirement in 1958. From the 1st of May 1958, she was appointed as Consultant on Radiography to Ilford Limited, which post she held until 1964.

From September 1958 to February 1959, she undertook a lecture tour of Australia and New Zealand. In New Zealand, she was awarded honorary membership of the

FIGURE 9.7 Three radiographers involved with the ISRRT: Kathleen "Katie" or "Kitty" Clara Clark, Ernest Raymond "Ray" Hutchinson, and Marion Frank (1964). Courtesy of the late Marion Frank.

New Zealand Society of Radiographers and also the Watvic Jubilee Award for Merit in Radiography. In Australia she was made an Honorary Fellow of the Australasian Institute of Radiography.

Sadly, she suffered an incapacitating stroke in September 1964 and was forced to stop working. On 20 October 1968, she died at the age of 70, and her passing left a great gap in the world of radiography and was the end of an era.

Much has changed in the radiographic world since 1968. We now have ultrasound, CT scanning, and MRI. However, standardisation of technique remains as important in the world of MRI sequences or CT protocols as it did in the world of MMR and plain films. A 13th edition of Clark's *Positioning in Radiography* was published in 2015 and her legacy has been maintained.[28]

RADIOLOGICAL PROTECTION CONSIDERATIONS FOR WOMEN

The topic of radiation protection in women is a large one; however, two topics stand out as worthy of deeper consideration. These are the relationship of chest fluoroscopy using the thorascope to breast cancer, and the work of Alice Stewart and antenatal radiography.

THE THORASCOPE AND BREAST CANCER

The thorascope was a common item that was used in the mid-20th century and was found in physician's offices. It consisted of an upright cabinet with an attached simple fluorescent screen and was used for upright fluoroscopy. The examination would be performed in a darkened room following dark adaptation of the eyes because of the dim image on the fluorescent screen. The thorascope was designed as a compact and completely self-contained unit and was recommended for the consulting room of the private physician or in tuberculosis centres. The apparatus was easy to use and gave up to 5 mA with a kV control from 55 to 85 kV. However, its very ease of use could result in unnecessary dosage with untoward biological effects. In 1991, Boice and others[29] reviewed the incidence of breast cancer in 4,940 women who had been treated for tuberculosis between the years of 1925 and 1954. They identified 2,573 women who had been examined using X-ray fluoroscopy for an average of 88 times when they had lung collapse therapy for pulmonary tuberculosis. This was a useful treatment and involved creating a controlled pneumothorax, which rested the lung. In this group of 2,573 women, there were 147 breast cancers identified, when only 113.6 were to be expected. More importantly, there was no excess of breast cancer in 2,367 women who had been treated by other methods. The increased rates for breast cancer were seen in women after an interval of 10–15 years after the initial fluoroscopy. The age at exposure strongly influenced the risk of radiation-induced breast cancer, and it was found that young women were at the highest risk and that women over the age of 40 were at lowest risk. The mean radiation dose to the breast was estimated to be 79 cGy, and there was strong evidence for a linear relationship between the dose received and the breast cancer risk. The young and glandular breast is more sensitive to radiation than the older and more fatty organ. This has implications for the use of ionising radiation in young women. It is unfortunate that such an elegant piece of apparatus was associated with harmful effects.

ALICE STEWART AND THE OXFORD CHILDHOOD CANCER SURVEY

The Oxford Childhood Cancer Survey was initiated by Alice Stewart (1906–2002) in 1954 and led to a series of reports that were published between 1956 and 1975.[30] There had been an increasing concern about the effects of low doses of radiation, and in his 1958 presidential address to the British Institute of Radiology, L.F. Lamerton reviewed the relation of the available clinical and experimental data on radiation effects to the problem of the possible hazards of small doses of radiation.[31] Radiography of the abdomen in pregnancy was performed for suspected renal disease, pelvimetry, to diagnose possible pelvic disproportion, to locate the placenta, and to diagnose twins (Figure 9.8). Alice Stewart was a British doctor and epidemiologist, who was interested in the effects of low-dose radiation. The Oxford study was important for its time since it involved the collaboration of many doctors. The aim of the study was to find an explanation for the increase in the death rate from leukaemia that had taken place in children in England and Wales between the First and Second World Wars, and ionising radiation was under suspicion as an etiological factor. The question became whether children with leukaemia had been exposed to radiography and other environmental factors to a greater extent than other children of the same age. Since leukaemia was rare, the study needed to cover the entire

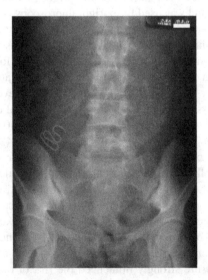

FIGURE 9.8 Abdominal radiograph for pregnancy, taken on 11 August 1967. The head is in the maternal pelvis, the spine is to the left, and an intrauterine contraceptive device (IUD) is seen on the right. The IUD used was a Lippes loop which was first introduced in 1962. It consisted of a plastic double-S loop, and was used from the 1960s to the 1980s. Author's collection.

country, and Alice Stewart received copies of the death certificates of all children under 10 years of age. This was later extended up to 15 years of age. The death certificates included children who had died of malignant disease of all types, so a comparison could be made between children with leukaemia and children with other malignant diseases. Alice Stewart could not do it all herself, so she obtained the assistance of the Medical Officers of Health who were then asked to interview the mothers for her using a standard questionnaire. To provide a control group, the Medical Officers of Health were asked to select from the local birth registers surviving children of the same sex, who were born about the same time, and were resident in the same region as that of the child with malignant disease with whom they were matched. A preliminary report in 1956 showed that a greater proportion of mothers of children who died of leukaemia or other type of malignant disease when compared to mothers of children without malignant disease reported that their abdomens had been radiographed during pregnancy, and this observation continued for subsequent years. The study was retrospective, and this meant that the data needed to be verified. Stewart had to show that the mother's statement that her abdomen had been X-rayed in pregnancy was correct. This was done by checking radiology department records and checking antenatal records. Stewart showed to her satisfaction that association was not distorted by memory. There was always the possibility that the cause of the malignancy also caused the mother to be radiographed. It is difficult to exclude this sort of hypothesis entirely from observational studies; however, no evidence for this could be demonstrated. Richard Doll concluded that if these and other similar data are accepted as indicating that exposure to diagnostic X-ray procedures *in utero* caused an increased risk of malignant disease in the child, it could then be inferred that doses to the foetus of the order of 10 mGy would cause a malignant tumour to develop in approximately one in every 2,000 children.[32]

However, the topic is controversial. In 1962, Brian MacMahon from the Epidemiology Department of Harvard School of Public Health published the results of a study to test the hypothesis that *in utero* exposure to diagnostic radiography increases childhood mortality from neoplasia.[33] He noted that there had been seven previous studies in which statistically significant excess cancer mortality in prenatally radiographed children was not demonstrated; however, they were of a smaller size than his current study. This study considered all children born in 37 large maternity hospitals in the Northeast Region of the United States from 1947 to 1954. MacMahon found a higher incidence of prenatal radiography in children with cancer that was statistically significant. Cancer mortality was 40% higher in the radiographed than non-radiographed members of the study population. The excess mortality was most marked in ages 5–7, with the excess risk exhausted by the age of 8. MacMahon concluded that no significant variation with stage of pregnancy with exposure was evident. He also noted that the association of intrauterine radiography did not explain the high incidence of leukaemia in first births noted previously. MacMahon's results showed that the relative risk was lower than that suggested by the original data of Alice Stewart and her colleagues.

These studies were being performed at the height of the Cold War and the public was concerned about radiation. Alice Stewart's final report was published in 1958, which was the same year that the Campaign for Nuclear Armament was formed. The medical profession was slow to change its practice and although there was some fall off in requests for radiography in pregnancy, it was not until major American medical groups ceased to recommend the practice that it ceased. Requests for X-ray pelvimetry continued for a considerable period of time and only slowly diminished. The London course for radiologists in the 1980s had obstetrical radiology in the syllabus, and the standard textbook of the period had a detailed chapter illustrating fetal abnormalities.[34]

NOTES

1. Meyer, H. 1937. *Ehrenbuch der Röntgenologen und Radiologen aller Nationen*. (Band XXII) Sonderbände zur Strahlentherapie. Berlin und Wien: Urban & Schwarzenberg.
2. Saldarriaga, N. 2015. Agnodice: The first female physician ... maybe. Classical Wisdom Limited, https://classicalwisdom.com/people/agnodice-first-female-physician -maybe/ (accessed 21 March 2021).
3. McIntyre, N. 2014. *How British Women Became Doctors: The Story of the Royal Free Hospital and Its Medical School*. London: Wenrograve Press.
4. Mukerjee, S. 2018. Sisters in arms. *History Today*, 68, 72–83.
5. Thomas, A.M.K., Duck, F.A. 2019. *Edith and Florence Stoney, Sisters in Radiology* (Springer Biographies). Switzerland: Springer Nature.; 1st ed. 2019 edition (1 July 2019).
6. Todd, M. (Graham Travers). 1918. *The Life of Sophia Jex-Blake*. London: Macmillan & Co.
7. Travers, G. (Margaret Georgina Todd). 1897. *Mona Maclean, Medical Student: A Novel*. 12th ed. Edinburgh: William Blackwood.
8. Sheppard, A. 1932. Obituary. Dr. Florence Stoney, OBE. *The Magazine of the London (Royal Free Hospital) School of Medicine for Women*, 27, 128–13.
9. Obituary. 1932. Florence Ada Stoney, O.B.E., M.D. (Lond.), D.M.R.E. (Camb.). *The British Journal of Radiology*, 5(59), 853–858.
10. Stoney, F.A. 1914. X-ray notes from the United States. *Archives of the Roentgen Ray*, 19, 181–184.
11. Leneman, L. 1993. Medical women in the First World War, ranking nowhere. *British Medical Journal*, 307, 1592–1594.
12. Stoney, F. 1930. The pelvis and maternal mortality. *British Journal of Radiology*, 3(33), 426–429.
13. Hypher, N. 1931. The diagnostic value of radiology in obstetric practice. *British Journal of Radiology*, 4(40), 171–177.
14. Watson Smith, S. 1932. The late Dr. Florence Stoney. *British Medical Journal*, ii, 777.
15. Obituary. 1932. Florence Ada Stoney, O.B.E., M.D. (Lond), D.M.R.E. (Camb). *British Journal of Radiology*, 5, 853–858.
16. Obituary. 1932. Dr Florence Stoney O.B.E. *Bournemouth Daily Echo*, 2 October.
17. Ramsay, M. 1932. *The Times*. 12 October.
18. Baltas, C.S., Balanika, A.P., Graeme, H., Fezoulidis, J.V. 2014. Pioneer Greek female radiologists (1895–1950): modern daughters of Asclepius? *The Invisible Light (The Journal of the British Society for the History of Radiology)*, 38, 10–24.
19. Thomas, A.M.K. 2020. "Our Katie": Kathleen Clara Clark MBE, MSR, Hon. FSR, Hon. MNZSR, FRPS (1898–1968). *The Invisible Light (The Journal of the British Society for the History of Radiology)*, 47, 26–38.

20. Jewesbury, E.C.O. 1956. *The Royal Northern Hospital, 1856–1956: The Story of a Hundred Years' Work in North London.* London: HK Lewis & Co. Ltd.
21. Clark, K.C., Bell, M.R. 1935. Neck of femur technique. *Radiography*, 1, 74–81.
22. Hercock, R.J., Jones, G.A. 1979. *Silver by the Ton. The History of Ilford Limited, 1879–1979.* London: McGraw-Hill Book Company (UK) Limited.
23. Clark, K.C. 1939. *Positioning in Radiography.* 1st ed. Messrs. London: Ilford: W. Heinemann.
24. Clark, K.C. 1940. A visit to Sweden and Denmark. *Radiography (The Journal of the Society of Radiographers)*, 6, 21–27.
25. Clark, K.C., D'Arcy Hart, P., Kerley, P., Thompson, B.C. 1945. *Mass Miniature Radiography of Civilians for the Detection of Pulmonary Tuberculosis (Guide to Administration and Technique with a Mobile Apparatus Using 35-mm. Film: And Results of a Survey).* Medical Research Council Special Report Series No. 251. His Majesty's Stationery Office, London.
26. Clark, K.C. 1949. Chest radiography: an investigation into the possibilities of standardization (The 11th Stanley Melville Memorial Lecture). *Radiography (The Journal of the Society of Radiographers)*, 15, 97–107.
27. ISRRT web site. http://www.isrrt.org (accessed 22 January 2020).
28. Whitley, A.S., Jefferson, G., Ken Holmes, K., Sloane, C., Anderson, C., Hoadley, G. 2015. *Clark's Positioning in Radiography.* 13th ed. Boca Raton: CRC Press.
29. Boice, J.D., Preston, D., Davis, F.G., Monson, R.R. 1991. Frequent chest X-ray fluoroscopy and breast cancer incidence among tuberculosis patients in Massachusetts. *Radiation Research*, 125, 214–222.
30. Greene, G. 1999. *The Woman Who Knew Too Much: Alice Stewart and the Secrets of Radiation.* Ann Arbor: University of Michigan Press.
31. Lamerton, L.F. 1958. An examination of the clinical and experimental data relating to the possible hazard to the individual of small doses of radiation. *The British Journal of Radiology*, 31, 229–239.
32. Doll, R. 1981. Radiation hazards: 25 years of collaborative research. *British Journal of Radiology*, 54, 179–186.
33. MacMahon, B. 1962. Prenatal X-ray exposure and childhood cancer. *Journal of the National Cancer Institute*, 28, 1173–1191.
34. Highman, J.H. 1980. Obstetric radiology. In: *A Textbook of Radiology.* Ed. D. Sutton. London: Churchill Livingstone, pp. 918–938.

10 Tomography
Mechanical to Computed

A PREMONITION OF TOMOGRAPHY

One of the strengths of simple radiography may also be seen as one of its weaknesses. Whilst we are fascinated by the image of Bertha Röntgen's hand with its ring, we have to remember that it is a simple shadow, and that the three dimensions of the body are reduced to a two-dimensional photograph. However, what is so remarkable is that the deeper implications of Röntgen's discovery were appreciated immediately as is so often the case with new developments.

Wilhelm Röntgen had sent the off-print of his First Communication and a series of photographs to the Austrian physicist Franz Exner (1849–1926), and Exner had displayed them at a meeting in Vienna on 4 January 1896. This was brought to the attention of Zacharias Lecher (1829–1905) who was a reporter at the Viennese daily newspaper *Die Presse*. In the Sunday edition of 5 January 1896, an unsigned article (presumably written by Lecher) was published on the front page, and entitled "A sensational discovery". The last paragraph observed that if one lets phantasy run free, then one can envision that it will become possible to improve the method of the photographic process using the rays emitted by a Crookes' tube in such a way that only a part of the soft tissues of the human body remains transparent, that a deeper layer can be fixed on the plate and thus an inestimable aid would be gained for the diagnosis of numerous other groups of diseases than those of the bones. Such an achievement, such a progress on the opened track, would not be out of reach provided the preceding premise proved correct. The author then admitted that all this was an over-audacious fantasy of the future. Robert Dondelinger[1] has pointed out that despite the fact that this article is the first landmark publication in the history of X-rays, and was subsequently abundantly quoted in the literature, the prophecy of tomography that was printed in the last paragraph on the second page of the newspaper was simply never spotted, in spite of being in clear view.

A further article, and similar to the piece in *Die Presse*, appeared in the feature supplement of the influential German newspaper, the *Frankfurter Zeitung* of 7 January 1896. Willi Kalender,[2] the inventor of multislice spiral CT, notes that the unknown author of the article "expressed truly prophetic thoughts only a few days after the first reports were published on Roentgen's discovery of X-rays and prior to their first medical use". However, Kalender indicated that we really do not know what the author has in mind, but that perhaps he "wished to obtain a view similar to that with an anatomical preparation after superimposed layers of tissue have been removed". However, the author's prediction that X-rays would be of an inestimable

DOI: 10.1201/9780429325748-10

aid for the diagnosis of diseases than those of the bones proved to be entirely accurate. These two predictions of 1896 should not be overinterpreted; however, it does seem to be a prefiguring of tomography, and the two newspaper articles were not referred to by any of the developers of classical tomography.

CLASSICAL TOMOGRAPHY

Classical tomography was a highly successful and much utilised technique. It came into use in the 1930s and lasted for over 60 years, passing out of use only when there was widespread availability of high-quality body CT. As is the case in so many areas, the technique reached its zenith as it was passing out of use. It is poignant that in December 1973 in *The British Journal of Radiology*, and immediately following the three papers describing CT, a paper was published from Harvard Medical School and the Massachusetts General Hospital in Boston[3] describing the use of planar tomography in the diagnosis of abdominal aortic aneurysm. The authors were pleased to report that tomography revealed the aneurysm in 36% of cases when plain films had been unhelpful. The images in the article are not particularly easy to interpret; however, the paper does illustrate the difficulties of traditional radiology when making what are reasonably straightforward diagnoses using modern techniques.

Radiography involves an X-ray source, a patient, and an object. The problem to be solved is the presence of confusing superimposed images when a complex body is displayed on a two-dimensional X-ray film. There were a number of solutions and these included plain films in multiple projections, stereoscopy, bi-plane fluoroscopy with rotation of the body or using the X-ray machine in right angles, and tomography. The principle behind conventional tomography is a simple one. For the majority of tomography devices the patient remains still and the X-ray tube and the film in its cassette both move. Everything then gets blurred out of vision apart from the chosen plane in the patient which does not move relative to the X-ray tube and the film. The development of conventional tomography is a complex topic, with many individuals contributing.[4] Perhaps, as in other areas, what is more important is to note the contributions that individuals made rather than to award a prize to a winner. As is often the case, an idea occurred to several individuals at a similar time. This was the case for tomography, and as will be seen was also the case for MRI and CT. The techniques used for tomography varied depending on the inventor and this is part of the reason that so many names were used. Names used for classical tomography include body section, planogram, laminogram, stratigram, and tomogram. There were approximately ten inventors and, as Dewing noted, these different investigators were each fired with zeal and pride of authorship.[5] As Dewing further recounts, these various workers put on an unseemly, and even acrimonious, wrangle over priority during the early 1930s. So, any historical account of the development of tomography would have received objections from one quarter or another. There were similar wrangles and animosities in other areas, including priorities in intravenous contrast agents, and most especially in the development of MRI. The history of science is filled with the need for recognition and jealousies as detailed by Morton Meyers in his well-known study of the topic.[6] It's perhaps more of a case that human nature does not change rather than that of history repeating itself.

The earliest attempt at tomography was made by Karol Mayer (1882–1946) of Poznań University in Poland, who in 1916 tried to show the heart without it being obscured by the overlying ribs. Mayer moved the X-ray tube during the exposure and by so doing blurred the image. The heart was blurred less since it was closer to the film.

It was between 1917 and 1921 that the French physician André-Edmund-Marie Bocage (1892–1953) developed the theoretical basis for tomography, and he applied for a French patent in 1921. In his method, images of surrounding tissues were blurred out whilst those of interest were clearly shown. His basic principles were found in later apparatus and he also suggested pluri-directional tomography. In linear tomography the apparatus moves in a straight line, whereas in pluri-directional the movement is complex with more effective blurring. The subsequent development of tomography was delayed by the concentration on linear devices and pluri-directional tomography was not developed until the late 1940s. Linear tomography had significant technical limitations. Bocage himself did not develop a working apparatus until 1938.

The story of tomography is complex and only the major developments can be described. In 1930, Alessandro Vallebona (1899–1987) from Genoa initially used a very simple apparatus in which the tube and film remained stationary and the object was rotated in the beam, and this is the principle from which autotomography was developed which was used in air encephalography. Bernard Georg Ziedses des Plantes (1902–1933) published in 1931, and Gustav Grossmann (1879–1957) in 1935. Whilst he was a medical student, Ziedses des Plantes commented on the fact that the image seen when using a microscope is restricted to one plane of the object. This suggested to him that a similar effect could be obtained during an X-ray exposure by moving the film and tube. This should not be particularly surprising since both light microscopy and radiography are optical processes. By the mid-1930s, it is recorded that the principles of tomography had been independently invented (or discovered) by at least eight inventors in four countries, and had been patented five times in four countries. In spite of this, it was invented once more and again independently of the previous workers! Much of the apparatus used for tomography was expensive and complex, and it was left to Edward Wing Twining (1887–1939) to devise a simple attachment to the Potter-Bucky X-ray couch and this was widely used (Figure 10.1).[7]

Linear tomography, in spite of its limitations, was a highly useful technique. Tomography was used extensively in the chest to show cavities or mass lesions (Figure 10.2), and could be combined with retroperitoneal air studies to show the adrenal glands, and was routinely used in urography. Linear tomography as a technique lasted well into the era of CT. There were a number of reasons for this, including the limited availability of the CT scanner and also the limitations of early body CT. To investigate a possibly hilar abnormality in the chest, linear tomography was superior to CT and both examinations were sometimes performed. Tomography also persisted as part of the IVU when it was performed almost routinely apart from

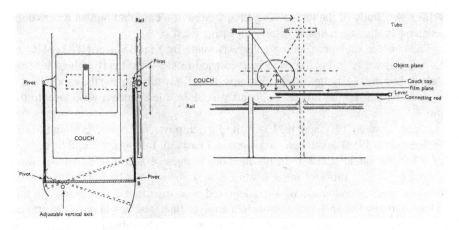

FIGURE 10.1 The simple principle behind conventional tomography as illustrated by Edward Wing Twining. The patient stays still and the X-ray tube and the film plate both move. From Twining, 1937, see Note 7. Courtesy of the *British Journal of Radiology*.

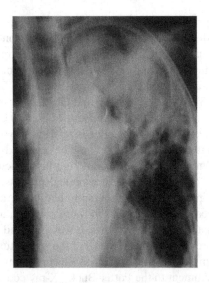

FIGURE 10.2 Linear tomogram of complex disease in the left upper lobe, showing consolidation, cavitation, and calcification. Author's collection.

the situation of emergency studies. The introduction of high-quality CT enabled the introduction of the CT KUB (kidneys, ureters, bladder) study and CT urography. In the emergency situation the CT KUB gave a result in a matter of minutes, whereas the emergency IVU required an intravenous contrast injection, and obtaining the result might take several hours. In addition, as the older conventional X-ray rooms were upgraded to digital radiography, there was usually no tomographic attachment on the table.

AXIAL TRANSVERSE TOMOGRAPHY

The technique of axial transverse was developed by the British radiographer William "Willie" Watson (1895–1966) who patented the method in 1937. The resultant image is of a section of the body transverse to the vertical axis of the body. For the technique to work and project the image on to the film, the focus needs to be above the film.[8] The apparatus was commercially marketed as the Sectograph. In 1938, American radiologic technologist Jean Kieffer (1897–1972), working independently of Watson, devised a similar apparatus.[9] Whilst the Sectograph was lent to various hospitals, Watson had to admit in 1962 that no useful work had been done with it. Post-war further work was done by Alessandro Vallebona and he produced his Zuder apparatus in 1947. Whilst the idea for axial transverse tomography might have been ingenuous there was no widespread uptake of the technique.

PLURI-DIRECTIONAL TOMOGRAPHY

It is not obvious why it took so long for pluri-directional tomography to be developed with all of its advantages, and particularly since it had been suggested by many of the early workers. The simple linear method was dominant for many years, and whilst linear tomography was superior to plain films, it produced confusing artefacts, inefficient blurring, and incomplete imaging. The first useful pluri-directional device appeared in 1949 and was developed by Raymond Sans (1902–1996) and Jean Porcher (1910–1971) being built by the laboratory workshop at the Pitié-Salpêtrière University Hospital in Paris. The first model had an independent hand-tilting table and a two-motor drive for the mechanism. It was commercially marketed by the Massiot Company and marketed as the Polytome and announced in 1951. There were various pluri-directional movements employed, with tube and film moving in circular, elliptical, hypocycloidal, spiral, sinusoidal, or a random path. Different companies preferred different movements; for example, the French company CGR preferred the spiral movement. It was not always apparent which movement was preferable for a given indication.

Pluri-directional tomography was to prove of particular value in ear, nose, and throat imaging. In his 1952 book on the subject,[10] Eric Samuel (1914-1997) did not even mention tomography; however, by the second edition of 1978,[11] he noted that for the middle ear and para-nasal sinuses tomography was essential. Linear tomography was difficult to interpret with relatively thick 2-mm cuts obtained in practice and artefacts (so-called parasitic shadows). The complex hypocycloidal cuts were of the order of 1 mm thickness. Whilst the 1978 book mentioned CT, there were no CT illustrated and it would be some years before the resolution of CT improved for accurate assessment of the petrous bones. Hypocycloidal tomography was still being used into the 1980s.

It is rather poignant to read Littleton who in 1976 stated that with the recent introduction of multiple pluri-directional devices, it would appear that the history of tomography was just beginning. Certainly, he thought that the best tomographic device and the optimal tomographic procedures have not yet been developed, and that future is brilliant. This is another example of a procedure being optimised and perfected only to become obsolete. Other examples include air encephalography and contrast media for intravenous cholangiography. Computerised tomography was announced in 1972, with whole-body scanning being possible from February 1974. Again, and perhaps more surprisingly, Littleton thought that CT scanning would not supplant conventional tomography, but saw it rather as an augmentation to the currently available technology. It was perhaps the case that he was so committed to one technique, and had invested so much time on it that he could not envisage it being supplanted.

TOMOSYNTHESIS

In 1969, Grant and Garrison[12] described a three-dimensional radiographic technique, consisting of 20 conventional radiographs that were reduced and projected into a display volume according to a geometry which matches the X-ray source geometry. The images were integrated on the viewing screen, and produced a truly three-dimensional image which allowed viewing behind opacities. The technique could also be applied to electron microscopy. Tomosynthesis has been particularly valuable in the breast when the digital technique has resulted in a considerable dose reduction and superior diagnostic qualities when compared to two view mammography.

PANORAMIC DENTAL RADIOGRAPHY

Panoramic dental radiography is essentially a modified form of tomography. As with linear tomography, the theory was developed independently by several workers: by Numata in Japan in the 1930s[13] and by Yrjö Paatero (1901–1963) in Finland in the 1940s. In 1954, Sydney Blackman (1898–1971)[14] from the Eastman Dental Hospital in London visited Paatero in Helsinki and persuaded the company Watson & Sons to develop a prototype. Paatero came to England and the initial prototype was demonstrated at the Royal Dental Hospital in 1954 to an enthusiastic audience.[15] Watson then developed a production model, the rotograph. The rotograph used a single centre of rotation and was the first to produce a continuous image of the jaws from condyle to condyle. Blackman wanted to use the rotograph to introduce a form of mass dental radiography, similar to mass chest radiography, to locate and control dental disease, and especially in the growing child. In reality, the rotograph was immediately taken up as an easy alternative to a set of "full mouth" radiographs.[16] The rotograph was superseded by the orthopantomograph, which was developed by Paatero working with Siemens, and which used a three-centre rotation.

COMPUTED TOMOGRAPHY

The Nobel Prize for Physiology or Medicine for 1979 was awarded jointly to Allan MacLeod Cormack (1924–1998) and Godfrey Newbold Hounsfield (1919–2004) for their development of computer-assisted tomography. James Bull, the pioneer British neuroradiologist, reviewed the history of computed tomography and said that seldom in the history of medicine had a new discovery swept the world quite so quickly as did computed tomography.[17] The CT scanner has not only transformed medical care but can also be said to have changed the way we look at the body. The same might also be said for the original discovery of X-rays by Wilhelm Conrad Röntgen in 1895. The discovery of X-rays swept the world even more rapidly than the invention of CT scanning and also changed the way we look at ourselves with its demonstration of living anatomy. Both the discovery of X-rays and the invention of computed computer-assisted tomography resulted in the award of a Nobel Prize, with Wilhelm Röntgen receiving the first Nobel Prize for Physics in 1901.[18] The history of the development of computerised axial tomography is complex.[19,20,21,22] The need for tomography, either mechanical or computerised, lies in the fact that in conventional radiography a three-dimensional object is displayed as a two-dimensional image with no depth data. The X-rays come from a point source and the object radiographed appears on photographic film as a shadow of varying densities depending on differences in absorption.

ALLAN CORMACK

Allan Cormack was born in February 1924 in Johannesburg, South Africa, and matriculated from Rondebosch Boys High School in Cape Town in 1941.[23] He studied electrical engineering at the University of Cape Town and switching to physics and graduating in 1944. He completed his master's thesis on the subject of X-ray crystallography in 1945. He then moved to Cambridge working at the Cavendish Laboratory as a research student under the nuclear physicist Otto Frisch (1904–1979). Cormack returned to Cape Town as a lecturer in physics, and described his work as being rather lonely because of the small numbers of nuclear physicists in South Africa.

In 1956, Cormack was working at Groote Schuur Hospital, and became interested in the absorption of radiation in the body and in the basis of what became computed tomography. In 1955, the Hospital Physicist at Groote Schuur Hospital had resigned, and the law in South Africa required that a qualified physicist should supervise the use of radioactive isotopes. As he was the only nuclear physicist in Cape Town, Cormack therefore spent a day and a half at the hospital working with the radiation therapist J. Muir Grieve. He became interested in radiotherapy treatment planning and in the use of isodose charts. The isodose charts that were in use at that time described the passage of radiation as through homogeneous materials and showed the distribution of the dose of radiation. The body is, of course, not homogeneous and contains tissues of varying densities, including gas-filled tissues, solid organs, and bone. Cormack realised that treatment planning could be facilitated if the distribution of attenuation coefficients in the body could be determined. This distribution of attenuation coefficients could be determined by external measurements, and at some point later, it occurred to Cormack that this data might have diagnostic significance. How to determine this distribution of attenuation coefficients was the problem that Cormack then set himself.

In 1958, he took a sabbatical from his post in South Africa and worked at the Harvard University cyclotron. Cormack was then offered a position as assistant professor in the department of physics at Tufts University, Massachusetts, USA, which he accepted. Cormack moved to the United States and became a US citizen in 1966. He continued to work on CT; however, his main interest for most of this time was in nuclear and particle physics and he pursued the CT scanning problem only intermittently. By 1963, he had developed his ideas on CT to such an extent that they were published in the *Journal of Applied Physics*.[24] Cormack had tested his ideas using phantoms. However, when his paper appeared, there was little response from the scientific community, but he continued his normal course of research and teaching. He was appointed chairman of the physics department at Tufts University from 1968 to 1976 and retained his interest in nuclear physics and the interaction of subatomic particles. In the period 1970–1972, Cormack became aware of developments related to CT scanning and following this he devoted much of his time to these problems, retiring in 1980.

Allan Cormack developed the mathematical basis of what became CT scanning and worked quite independently from other workers. Cormack developed a mathematical approach to looking at the problems of variations in body tissues that are important in radiotherapy. The exponential attenuation of radiation through material had been known and used for over 60 years when Cormack considered the problem; however, the attenuation was determined for parallel-sided blocks of material. Cormack found out rather to his surprise that the generalisation had not been transferred to inhomogeneous materials. Cormack initially considered a fine beam of radiation passing through a body with circular symmetry. In 1957, he made experiments using a phantom that had circular symmetry. By 1963, Cormack was ready to make experiments on a phantom that did not have circular symmetry. The apparatus consisted of two cylinders containing a detector and a gamma ray source. The phantom lay between the two cylinders and the work was done in the summer of 1963. Cormack then considered how many measurements need to be made since only a finite number of measurements can be made with beams of a finite width. The results were presented for publication in graphical form and were published in 1964. Cormack developed the mathematics of line integrals independently from other workers. There was almost no response to the publications, although Cormack related that the most interesting reprint request was from the Swiss Centre for Avalanche Research since the method he had described could be used for examining snow on a mountain assuming that the source or detector could be placed in the mountain under the snow. There was certainly no commercial interest in the work of Cormack.

Cormack became aware of the pioneering work of Radon only in the late 1970s.[25] Cormack had thought that this problem would have been a standard part of the 19th century mathematical repertoire, but perhaps surprisingly had found no reference to it. The mathematical basis for reconstruction had been developed by Johann Radon (1887–1956). Between 1912 and 1919, Radon was Assistant Professor of Mathematics in the Department of Mathematics at the Technische Hochschule Wien (Technical University Vienna, Austria). Johann Radon applied the calculus of variations to differential geometry, and this led to various applications in number theory. In 1917, he published his fundamental work on what was termed the "Radon transform", which mathematically transforms two-dimensional images with lines into a domain of possible line parameters, where each line in the image will give a peak positioned at the corresponding line parameters. This idea of reconstructing a function from a set of projections therefore plays a significant role in the development of computed tomography. Cormack commented that if he had known of Radon's work that it would have saved him a lot of work.

What is interesting is that in 1980, Cormack discovered that Radon had been anticipated in his work by the Dutch physicist Hendrik Lorentz (1853–1928). Lorentz had first solved the three-dimensional version of the problem before 1905. Cormack said that in three dimensions, the plane is replaced by a volume, and the straight lines are replaced by the planes. Cormack commented that it's not known why or when Lorentz solved the problem because he never published his results. Lorentz found a solution to the three-dimensional problem where a function is to be recovered from its integrals over planes. Cormack noted that we have no idea why Lorentz thought of the problem, or what his method of proof was, and that the only reason that we know of his work is that in a 1906 paper a student of Lorentz recalled that it had been mentioned in a lecture as something that had been dealt with.[26] This student was Hermann Bernard Arnold Bockwinkel (1881–1960) and the paper was on the propagation of light in biaxial crystals.[27] The Radon transform should therefore be the Lorentz transform, and Cormack commented that there was a joke that Lorentz should have a transform as well as a transformation! Cormack said that working on Radon's problem gave him a great deal of pleasure and, that as was usual, each question that was answered raised new questions.[28] So why wasn't the mathematical problem solved in the 19th century, why didn't Lorentz publish his work, and why wasn't the work of Radon better known? There are so many instances when workers are independently working in the same area, and in his 1983 paper, Cormack discusses the independent discoveries of Radon's problem.

OTHER WORKERS

There were a number of workers over the years looking at developing tomography from its mechanical form. In the mid-1940s, Shinji Takahashi (1912–1985) in Japan had worked on the principles underlying rotational radiography and developed what came to be called sinograms. In 1957, Korenblyum and his co-workers build a medical CT scanner in Kiev in the then USSR. Korenblyum used a television screen-equipped analog computing device to solve the respective integral equation, converting the information contained in the X-ray sinogram into an image of the layer in the body being studied.[29] In 1960, William Oldendorf (1925–1992) made experiments to demonstrate the feasibility of CT scanning using a rotating phantom made of nails and mounted on a track.[30] In the early 1960s, David Kuhl (1929–2017) and Roy Edwards developed an apparatus for emission CT scanning (radioisotope section imaging).[31] Their Mark III scanner had four detectors viewing four different aspects of the brain simultaneously, with a self-contained computer. Kuhl and Edwards used their Mark III scanner routinely for brain-scanning, and preferred the transverse section for detailed examination of suspect regions shown on the rectilinear nuclear medicine scan.

GODFREY HOUNSFIELD

Godfrey Hounsfield (Figure 10.3) was born in Newark in Nottinghamshire in the UK on 28 August 1919, going to school at the Magnus Grammar School in Newark.[32,33] He grew up on a farm and from his earliest days was interested in how things work, performing many practical experiments. He was always interested in aeroplanes and flying, and when the Second World War started, he joined the Royal Air Force (RAF) as a volunteer reservist. He became interested in radar and radio communications and worked as a radar mechanic instructor. In the RAF, he attended the Royal College of Science in South Kensington, which was then occupied by the RAF, and then he went to the Cranwell Radar School. He passed the City and Guilds examination in Radio Communications. After the war, Hounsfield attended the Faraday House Electrical Engineering College in London, and received their diploma. Hounsfield was never an undergraduate at a university and the only university degrees he received, and there were six of them, were honorary.

Hounsfield had joined EMI in Middlesex in 1951 where he worked on guided weapons and radar. He was involved with computers which were still in their infancy, and from 1958 he led a design team that built the first all-transistor computer constructed in Britain, the EMIDEC 1100. Following this, his next project was to design a thin-film computer store. When this project was abandoned, he started to work that was to result in the CT scanner. By 1972, Hounsfield was Head of the Medical Systems Research Department.

And so, in the 1960s Godfrey Hounsfield was working at EMI Ltd. in Hayes, Middlesex.[34,35,36] When the EMI project to design a thin-film computer store was abandoned, Hounsfield was not immediately assigned another task. Instead, he was

FIGURE 10.3 Godfrey N. Hounsfield (1919–2004) in the old library of the British Institute of Radiology. Author's photograph.

allowed to suggest his own ideas for research. Hounsfield put forward a project for automatic pattern recognition which had no apparent medical implications. It was in 1967 that it occurred to Hounsfield that there were medical implications and the EMI scanner and computed tomography were born. This illustrates the value of research that is undertaken with no immediate commercial benefit. Hounsfield had two problems. The first problem was similar to that of Cormack and was one of automatic pattern recognition using external measurements. The second problem was to build a functioning and practical scanner having realised the medical possibilities. The project was undertaken with limited resources and there is a persistent urban legend that money from a well-known pop group, the Beatles, who were on the EMI label funded the research. There is no truth to this and yet this keeps on being proposed. A recent example is in a presidential message to the American Osler Society in 2021 by Robert Mennel who categorically states that the Beatles were very important in the development of CT. Mennel says that the Beatles made up 30% of EMI's profits and that much of that money from the Beatles was ploughed into the development of the CT scanner.[37] There is no evidence to support Mennel's assertion and yet it is curiously persistent, as are many other urban legends.

Hounsfield was looking at internal structure and considered a closed box with an unknown number of items inside. The box could be looked at from multiple directions using an X-ray source and a radiation detector. The results of the transmission readings could then be analysed by the computer and then presented slice in a single plane. Hounsfield developed a mathematical approach to determine the nature of the objects in the box in a process of reconstruction. The original apparatus was very simple and resembled that used by Cormack. The basis of the apparatus was a simple lathe holding the object to be examined (Figure 10.4). On opposite sides

FIGURE 10.4   The lathe bed that Hounsfield used for his initial work displayed at the UK Annual Radiology Congress in 2005 (UKRC 2005). Author's photograph.

FIGURE 10.5   Perspex phantoms of varying complexity (UKRC 2005). Author's photograph.

were a radiation source (initially an americium radioisotope source) and a radiation detector. The early experiments were made using Perspex phantoms of varying complexity (Figure 10.5). The readings were taken using a scintillation counter which counted the gamma ray photons. It took nine days to take the picture and 15 minutes of computing time to reconstruct the picture. Following the use of Perspex phantoms, a section of the human brain in formalin in a Perspex box was used as a phantom. Most of the pictures taken on the lathe bed were scanned in 1969 and 1970

Hounsfield looked at practical applications of the technique and approached the Department of Health in London in 1968. The Department of Health became aware of the possibility of CT scanning in 1968 when Cliff Gregory was visited by Godfrey Hounsfield.[38] If a medical use could be found, then this would stimulate the development of the project. Hounsfield met Cliff Gregory and Gordon Higson, who were scientific advisers at the Department of Health and Social Security (DHSS). Hounsfield was then introduced to Evan Lennon, a radiological adviser to the DHSS. In January 1969, Gregory, Higson, and Lennon visited EMI at Hayes to see Hounsfield's apparatus. Hounsfield was making proposals in 1968–1969 for the development of the scanner. On 7 October 1969, Hounsfield made detailed proposals, and these were remarkable in that Hounsfield foresaw the potentials of CT for the next few decades. Lennon knew that Frank Doyle (1926–1999) (Figure 10.6) from the Hammersmith Hospital was working on the problem of bone density measurements. At this period, the scanning time was between three and four minutes and it was therefore necessary to scan organs that had no respiratory movement. Lennon thought that scanning would involve scanning the brain and measuring the bone density of the spine.

FIGURE 10.6   Frank Doyle (1926–1999). Author's collection. Unknown photographer.

Lennon introduced Godfrey to Doyle, and Doyle was asked to provide samples of bone for scanning. Doyle was measuring ulnar bone density by immersing the fore-arm in a water bath and using conventional radiography. In order to evaluate the proposal, there would be a need to expand the group beyond the Scientific and Technical Branch and so Frank Doyle was added to the circulation list.[39] Frank Doyle gave Hounsfield two lumbar vertebrae of different densities. Hounsfield examined the vertebrae and returned to Doyle with computer printouts of numbers in the coronal plane of the vertebral bodies. Hounsfield had already worked out a scale of numbers and Doyle was impressed with the result. It was a recurring theme with Hounsfield, even after the development of the EMI scanner, that he preferred the computer print-out of numbers to a pictorial presentation of the data. Lennon also made contact with two other radiologists. These were James Ambrose (1923–2006) (Figure 10.7) and Louis Kreel (1925–2019) (Figure 10.8). James Ambrose was a neuroradiologist from Atkinson Morley's Hospital in South London, and Louis Kreel was from the Royal Free Hospital, moving to Northwick Park Hospital in Harrow when the body scanner was installed. Louis Kreel was a remarkable man and talented in many areas. He was a particularly fine teacher and his book for radiology trainees was instantly popular.[40] Ambrose later discovered that Hounsfield had been previously dismissed by an eminent radiologist as a crank.

Doyle had some concerns about Hounsfield's methods, and later recounted them in a letter to James Bull, professor of neuroradiology at the National Hospital for Nervous Diseases in London's Queen Square (now the National Hospital for Neurology and Neurosurgery).[41] Doyle was concerned about the use of heterogeneous X-rays that this might introduce a variation in the absorption coefficients, that collimation could not produce a completely parallel beam of X-rays, and that scatter

FIGURE 10.7   James Ambrose (1923–2006). Image courtesy of Lucy Ambrose.

FIGURE 10.8   Louis Kreel (1925–2019). Author's photograph.

would be a problem. Doyle acknowledged that Hounsfield was a knight's move ahead of him at every turn. Every objection that Doyle raised had already been considered and Hounsfield's calculations had demonstrated that there was not a serious problem. Doyle concluded that Hounsfield's idea of computed tomography was worth supporting.

Frank Doyle gave Hounsfield two lumbar vertebrae of different densities. Hounsfield examined the vertebrae and returned to Doyle with computer printouts of numbers in the coronal plane of the vertebral body. Hounsfield had already worked out a scale of numbers and Doyle was impressed with the result. Hounsfield drew a histogram, and at the top wrote "air = −630". It was a recurring theme with Hounsfield that even after the development of the EMI scanner, he preferred the computer printout of numbers to a pictorial presentation of the data. Doyle and Hounsfield compared the histogram with radiographs taken in different projections and also with microdensitometer tracings of the radiographs. There was a meaningful correlation between the techniques, and in time Doyle realised the superiority of CT to his own methods of measuring bone mineral density.

It became apparent that EMI would not spend any more money on developing the new technique without the support of, and a contribution from, the Department of Health. Lennon and Higson reported to the Department of Health, who agreed to give the necessary support. The department agreed to share the costs with EMI on a 50:50 basis. The three radiologists then worked more closely with EMI. Frank Doyle supplied bone specimens, James Ambrose supplied brain specimens, and Louis Kreel supplied abdominal specimens. The collaboration was to prove fruitful.

It became apparent that EMI would not spend any more money on developing the new technique without the support and a contribution from the DHSS. Lennon and Higson reported to the Department of Health, who agreed to the necessary support. The three radiologists then worked more closely with EMI. Frank Doyle supplied bone specimens, James Ambrose supplied brain specimens, and Louis Kreel supplied abdominal specimens. Work continued on specimen radiography and then on 14 January 1970, there was a meeting at the Department of Health between the three radiologists: Dr. Lennon, Mr. Gregory, and Mr. Higson. The initial results were very promising and it was agreed that a prototype machine should be produced. Because of the difficulties of abdominal scanning, it was agreed upon that the prototype would be a brain machine and that this was to be at Atkinson Morley's Hospital. Atkinson Morley's Hospital had developed as a convalescent hospital in the countryside in Wimbledon with links with St. George's Hospital in central London at Hyde Park Corner. There had been an expansion of the hospital during the Second World War as London became increasingly subject to air attacks.[42] Atkinson Morley's Hospital had several advantages as a location for the scanner. The hospital was quite close to EMI, the scanner could be placed in a discreet location, patients could be examined without too much advertisement, and the department had an innovative approach to radiological practice. During this period, James Ambrose developed a close relationship with Godfrey Hounsfield.

Jamie Ambrose[43] is a key figure in the development of CT scanning and in neuroradiology. Ambrose was born in Pretoria, South Africa, on 5 April 1923. He initially studied science at Johannesburg and during the Second World War, he joined the Royal Air Force. After the war, he studied medicine at Cape Town and came to the United Kingdom in 1954 to study radiology, initially at the Middlesex Hospital and then at Guy's Hospital. Ambrose became interested in neuroradiology and in 1959 went to work at Atkinson Morley's Hospital, where he spent his working life. Atkinson Morley's Hospital was in Wimbledon in South London and by 1948, it had become the busiest neurosurgical unit in London. Neurosurgery at Atkinson Morley's Hospital had been developed by Wylie McKissock. McKissock had visited Stockholm and had been very impressed by the close collaboration between the surgeon Herbert Olivecrona and the radiologist Eric Lysholm. McKissock disliked the current invasive neuroradiological techniques of angiography and pneumoencephalography and James Ambrose shared his concerns. The department at Atkinson Morley's Hospital actively investigated alternative imaging techniques, including cranial ultrasound and nuclear medicine brain scans with the support of the physics department at St George's Hospital. Ambrose presented his work on cranial ultrasound in 1969 to the meeting of the British Medical Association in Leicester and although the paper was well received, Ambrose would admit that the technique was not generally useful. Ambrose was therefore well prepared to respond positively to Godfrey Hounsfield and to his novel ideas regarding cranial imaging. In 1974, James Ambrose and Godfrey Hounsfield jointly received the Barclay Prize of the British Institute of Radiology, and James Ambrose received honorary membership in 1993.

The meeting on 14 January 1970 concluded that a prototype machine should be built. In February 1970, the provisional specifications for such a clinical prototype were submitted by Godfrey Hounsfield, and was worked into its final form over the next three months. This was decided to be a head scanner and was to be installed at Atkinson Morley's Hospital.

In June 1971, Hounsfield went to Amsterdam to attend the Second Congress of the European Association of Radiology, and presented a lecture on experimental CT of animals. The importance of the presentation was not appreciated at that time.

The prototype scanner was installed at Atkinson Morley's Hospital on 1 October 1971. It is quite remarkable that Hounsfield went in one move from the primitive lathe bed apparatus to the prototype CT scanner. This prototype CT scanner looks very similar to modern CT scanners and is on permanent display in the Science Museum in South Kensington, London. The scanning time was four minutes per slice with a slice thickness of a little over 1 cm. There was no computer attached to the machine and the data had to be taken by car on magnetic tape to be analysed by EMI at Hayes. The data was reconstructed using an ICL 1905 mainframe computer and a picture with an 80 × 80 matrix took 20 minutes to reconstruct. The software was written by Stephen Bates who made major contributions to the early CT computing at EMI. It would have been possible to reconstruct the data using a 160 × 160 matrix, but that would have taken considerably longer. Ambrose had felt that at least six months of work would be needed to build up an appreciation of the normal and abnormal. The first patient scanned on the new machine was a 41-year-old lady with a suspected frontal lobe tumour. The data was acquired and the tapes were sent to EMI. The results were returned after two days. The cystic tumour in the left frontal lobe was clearly shown, and Ambrose said the result caused Hounsfield and himself to jump up and down like football players who had just scored a winning goal. Radiology was changed forever. The scan is presented in the opposite direction from modern scans and is viewed as a neurosurgeon wood look, which is from above. Modern scans are viewed as looking from below upwards. The reverse of the original Polaroid has the words of Hounsfield "original 1st PATIENT SCANNED" with a request for it to be returned to him.

The preliminary results were presented at the 32nd Annual Congress of the British Institute of Radiology, which was held at Imperial College in April 1972. The session was held on the afternoon of Thursday 20 April, chaired by the neuroradiologist George du Boulay, and was entitled "New Techniques for Diagnostic Radiology" (Figure 10.9). The paper presented by Ambrose and Hounsfield was entitled "Computerised axial tomography (a new means of demonstrating some of the soft tissue structures of the brain without the use of contrast media)". As might be expected, the paper produced a sensation and the first press announcement was in *The Times* on 21 April 1972. The presentation appeared as an abstract in the *British Journal of Radiology*[44] with three papers appearing later that year. The first paper by Hounsfield[45] described the technical background, the second paper was by Ambrose[46] describing the clinical findings, and the third paper by Perry

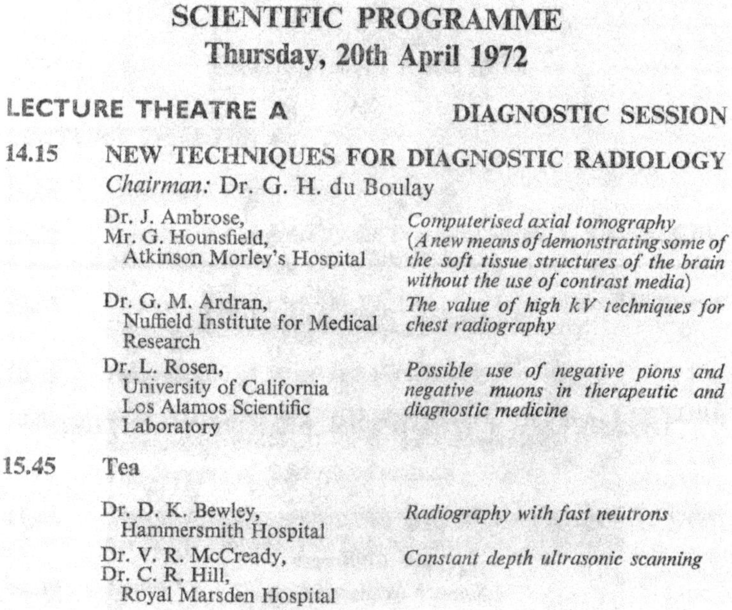

SCIENTIFIC PROGRAMME
Thursday, 20th April 1972

LECTURE THEATRE A   **DIAGNOSTIC SESSION**

14.15   NEW TECHNIQUES FOR DIAGNOSTIC RADIOLOGY
         Chairman: Dr. G. H. du Boulay

Dr. J. Ambrose, Mr. G. Hounsfield, Atkinson Morley's Hospital	*Computerised axial tomography* *(A new means of demonstrating some of* *the soft tissue structures of the brain* *without the use of contrast media)*
Dr. G. M. Ardran, Nuffield Institute for Medical Research	*The value of high kV techniques for* *chest radiography*
Dr. L. Rosen, University of California Los Alamos Scientific Laboratory	*Possible use of negative pions and* *negative muons in therapeutic and* *diagnostic medicine*

15.45   Tea

Dr. D. K. Bewley, Hammersmith Hospital	*Radiography with fast neutrons*
Dr. V. R. McCready, Dr. C. R. Hill, Royal Marsden Hospital	*Constant depth ultrasonic scanning*

FIGURE 10.9 The programme for the session "New Techniques for Diagnostic Radiology" at the 32nd Annual Congress of the British Institute of Radiology, held on 20 April 1972. Courtesy of the British Institute of Radiology.

and Bridges[47] looking into radiation doses. EMI then started the production of a brain machine and made five: one for the National Hospital, Queen Square; one for Manchester (Figure 10.10); one for Glasgow; and two for the United States: one for the Mayo Clinic and the other for the Massachusetts General Hospital. All machines were installed in the summer of 1973. James Bull described the new machine to Fred Plum (1924–2010), a leading American neurologist. Plum commented that the United States would need at least 170 brain machines to cover the neurology departments and that it would soon become unethical for a neurologist to practice without access to a brain machine since it saved many patients from unnecessary suffering from the currently available techniques (Figure 10.11). Ambrose also looked at the use of the new scanner in orbital lesions which were previously difficult to image.[48]

Godfrey Hounsfield received many honours for his work on CT scanning. In 1972, he received the prestigious MacRobert Award and the referee said that "no comparable discovery has been made in this field since Röntgen discovered X-rays in 1895". He was elected to the Fellowship of the Royal Society. In 1974, with James Ambrose, Godfrey Hounsfield received the Barclay Prize of the British Institute of Radiology, followed by the honorary membership of the Institute in 1980. In 1979, he received the Nobel Prize for Physiology or Medicine jointly with Allan Cormack.

FIGURE 10.10   Ian Isherwood (1931–2018), the pioneer of CT and MRI in Manchester, at the UK Annual Radiology Congress in 2005, looking at the lathe bed that Hounsfield used to develop CT. Author's photograph.

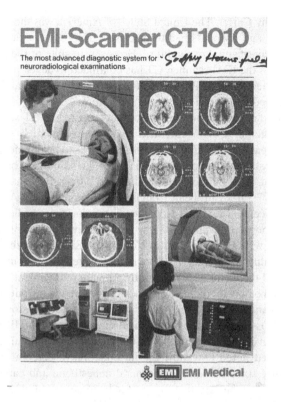

FIGURE 10.11 The brochure for the EMI scanner CT1010 "the most advanced diagnostic system for neuroradiological examinations", signed by Sir Godfrey Hounsfield. Author's collection.

Godfrey Hounsfield was knighted by Her Majesty Queen Elizabeth in 1981. Until his death after a long illness on 21 August 2004, Hounsfield remained interested in the development of CT and in MRI. He was involved with the Department of Radiology at the Royal Brompton Hospital and regularly attended meetings at the British Institute of Radiology, including the annual Hounsfield Lecture, the eponymous lecture given each year in his honour.

Whilst the new scanner was obviously effective, it was also very expensive. EMI had various problems, which Melvyn Marcus outlined in the *Sunday Telegraph* of 30 July 1978 in an article entitled "It's crisis time for scanners". The difficulties that EMI experienced were from two directions. Firstly, competition from other companies: there were several court cases for patent infringements, including a suit filed by EMI against Ohio-Nuclear in 1976 and Pfizer in 1977. In 1974, Siemens became the first of the traditional X-ray companies to announce a CT scanner, and other companies followed suit. By the late 1970s, Kalender noted that 18 companies were marketing a CT scanner in what became a technology boom. However, more significant was the clampdown on hospital expenditure in the United States under the administration of President Jimmy Carter. The United States of America was the largest market for the EMI scanner and hospitals had to meet very rigorous conditions to be able to undertake any major capital expenditure. However, in spite of the problems experienced by EMI and the EMI scanner, CT has continued to develop. The award of the Nobel Prize for Medicine and Physiology to Godfrey Hounsfield and Allan Cormack in 1979 emphasised the arrival of the new technique.

It is difficult to overestimate the impact that the EMI or CT scanner has had on investigative medicine. A *British Medical Journal* editorial of 1975 saw the benefits of the scanner to patients as being so overwhelming as to be obvious,[49] and there was no need for any randomised controlled trial to make sure that what was being seen could be believed. The London neuroradiologist James Bull introduced the book based on the 1977 First European Congress on Computerised Axial Tomography in Clinical Practice, and said that Hounsfield's revolutionary radiological technique was the most important advance in X-ray photography since Wilhelm Röntgen had radiographed his wife Bertha's hand in November 1895 in Würzburg.[50] Within five years of the introduction of the CT scanner, there were machines in major centres on all continents and the technique was rapidly adopted into clinical practice. It is now difficult to imagine how medicine was practised before the introduction of CT scanning. The CT scanner followed by the widespread use of ultrasound and MRI has resulted in the blossoming of the radiological sciences since the 1970s.

In 1984, Bryan Jennett reviewed both the benefits and the burdens of high-tech medicine.[51] Jennett noted how slow the introduction of cerebral angiography had been, particularly since it required a general anaesthetic and carried a significant morbidity. Jennett emphasised that for a test to become accepted as a normal investigation not only must it produce aesthetically pleasing pictures or satisfying complete data, but it must also benefit a significant number of patients. In 1978, a thoughtful editorial in the *American Journal of Roentgenology* took a measured approach to the introduction of the scanner, stating that one of the difficulties in evaluating a diagnostic test is its remoteness from health outcomes, and also noting that diagnosis is not an end in itself, and that technological triumphs are confronted by critical questions about their worth.[52] Jennett noted that different centres had different experiences following the introduction of the scanner, and that in one neurological/neurosurgical centre, there was no change in morbidity or mortality, and although fewer invasive tests were performed, the total costs for diagnosis had increased. Jennett, working in Glasgow, wrote that the CT scanner often revealed a clinically unsuspected haematoma in the brain following trauma, and that at that time his unit was still using long established criteria for transfer in which clinical deterioration needed to have taken place before transfer from the local hospital to the neurosurgical centre would take place. Teasdale and colleagues from Glasgow reviewed their admissions from 1974 to 1978 and wrote that if the potential benefits of CT scanning in the management of head injuries were to be realised, then patients must be scanned sooner than they had been in the past. This would therefore mean that more patients should be moved to a neurosurgical unit where the scanner was located, and that simple criteria for transfer should be established.[53] In 1978, Bartlett and others, from the neurosurgical department at the Brook General Hospital in South London, wrote on the evaluation of the cost-effectiveness of diagnostic equipment, "the brain scanner case", and made important observations.[54] An EMI brain scanner had been introduced into the Brook Hospital in February 1976 and there had been a significant reduction in angiography and air encephalography. Bartlett demonstrated that the efficient use of conventional equipment required the centralisation of neuroradiological services, which would then result in major cash savings. In contrast, the pattern of demand for CT scanning, in addition to the acknowledged clinical efficiency of the scanner and its unique role in the head-injured patient, only emphasised the need for improved access to scanners. Bartlett concluded that in the interest of the patient the pattern of service must change, and in their "Option 3" they suggested that scanners would be introduced into selected district hospitals where all head injury services would be centred. This option accepted the pattern of local centralisation of major accident services, and that the number of cases of cranio-cerebral trauma requiring transfer to neurosurgical units would fall by about half. This is what happened in the decades following 1978, and can be illustrated by the experience of the Bromley group of hospitals. In 1978, there were accident and emergency services on the four hospital sites in Bromley. The accident and emergency services were gradually centralised onto one site, and a CT scanner was placed on this site. The four hospital sites, all with accident units and inpatient beds, were reduced to three. One site became the acute site with inpatient beds, one site became a non-acute inpatient site, one site became purely outpatient site, and one site was closed. All of Bartlett's recommendations were realised.

The CT scanner has led to the modern paradigm shift in medical investigation and treatment. Traditional radiological investigation was often invasive and the resulting treatment was also invasive. The modern paradigm is that of non-invasive diagnosis and minimally invasive treatment. Minimally invasive treatment has at its basis an accurate pretreatment diagnosis. The CT scanner also changes the way radiological images were interpreted. In traditional radiology, the presence of an abnormality was commonly inferred by the distortion of normal anatomy. In the head, the presence of a brain tumour could be inferred by a displaced blood vessel when filled with contrast media at angiography or from a distorted cerebral ventricle which had been filled with air at pneumoencephalography. Both angiography and pneumoencephalography required the patient to be an inpatient and anaesthesia needed to be used since both techniques were physically unpleasant for the patient. As a contrast, the CT scan could be performed as an outpatient and the patient only had to keep still in the scanner gantry. Unlike angiography or pneumoencephalography, the CT scanner showed the abnormality directly as in the case of the first patient with the left-sided frontal lobe tumour. This ability to see the abnormality directly has had several effects. The abnormality is shown more clearly and a likely diagnosis is easier to make. The extent of the abnormality is easier to define and so the radiological–pathological correlation is enhanced. Because the CT scan is non-invasive, it can be repeated easily unlike invasive techniques and so the response of the disease to treatment can be assessed quite easily. This has greatly facilitated the monitoring of the effects of medical treatments in clinical trials. Since the CT scan is non-invasive, the threshold for performing a radiological examination is reduced since doctors are understandably reluctant to perform potentially hazardous examinations without a very strong clinical indication. Since the introduction of the CT scanner, the threshold for performing radiological investigations has been reduced. Whilst this has had the effect of considerably increasing the work of X-ray departments, it has also meant that abnormalities may be diagnosed at an earlier stage in their natural history and earlier diagnosis means that treatment is facilitated. The CT scanner can be used to guide the radiologist in interventional radiology and facilitate the biopsy of tumours (Figure 10.10), the drainage of fluid collections, and radiofrequency ablation.

By 1975, the first whole-body CT scanner had been introduced into clinical use. By 1975, EMI were marketing a body scanner, the CT5000, the first of which was installed at Northwick Park Hospital (NPH) in Harrow, North West London (Figure 10.13). The first body scanner in the United States was installed at the Mallinckrodt Institute and had its first clinical use in October 1975. By this time, scan time had been reduced to 20 seconds, for a 320 × 320 image matrix. By building a scanner with a larger aperture, it became possible to make images of cross sections of any part of the body. Hounsfield announced the existence of this scanner at the First International Conference on CT Scanning, which was held in Bermuda in 1975, where he showed images of the abdomen of a pig. This announcement was greeted by a standing ovation. A whole-body machine, the CT5000, was then installed at NPH. Thirty detectors were used and the image acquisition took about 20 seconds – short enough for the patients to hold their breath. It was also unlike the body CT where Louis Kreel had diagnosed carcinoma of the pancreas on his first case in 1975.[55] Graeme Bydder who worked with Kreel at Northwick Park Hospital describes how Kreel reported the cases at three levels. These were one for the radiologist, one for the referring clinician, and interestingly one for the patient. This was so to enable all three to know what the imaging findings were, and this was at a time when patients were not meant to even see their own medical records. Kreel also provided annotated prints of the scans so that the images could be understood at the three different levels. The reports were also accompanied by personal letters to the clinician, often including hand-drawn diagrams, as well as a relevant publication of his to clinicians who were referring patients for the first time.[56] In 1977, Kreel summarised his experience of computerised tomography using the EMI general-purpose scanner in a paper that is worth reading today.[57] Whilst the role of CT in the brain was assured by 1977, the use of CT in the body was still to be determined. It is difficult to overestimate the major achievements of Kreel and his team at Northwick Park Hospital in putting body CT on to a firm footing (Figure 10.11). Kreel showed many pathologies for the first time on CT, and his work on asbestos and the lung is particularly interesting. Chest disease following asbestos exposure could be imperceptible on a plain film, and Kreel was able to demonstrate CT changes.[58] In December 1977, Kreel ran a course on medical imaging at Northwick Park Hospital for the British Council (Course No. 733). This course towards the end of the 1970s can be seen as summarising the changes in radiology that had taken and were taking place transforming traditional diagnostic radiology into contemporary medical imaging. A book of the course was produced entitled *CT, U/S (Ultrasound), IS (Isotope Scanning/Nuclear Medicine) and NMR (Nuclear Magnetic Resonance)*.[59] It was a vision of the future of radiology, and remarkably included a section on NMR, and as Bydder notes, this was long before a single patient was scanned in the UK, and the prediction is made that there would be radiology systems that combined different forms of imaging such as CT and IS. When the first body images from NPH were shown, EMI were inundated with orders. However, EMI did not have the capacity to manufacture scanners in large numbers and developmental work was still needed.

When the whole-body CT scanner was being developed, it became apparent to Godfrey Hounsfield and the team at EMI that a cross section of the body would have a significant role in radiotherapy treatment planning.[60] The availability of an accurate cross-sectional picture of the body, as a CT slice, would have a profound impact on the precision and implementation of radiotherapy treatment planning. Prior to the introduction of CT scanning, radiotherapy treatment planning was imprecise and time-consuming. There was an imbalance between the accuracy of the treatment that could be delivered by the linear accelerator and the then available treatment plans. The CT scan allowed computer programs to guide the treatment beam in a process requiring only a few minutes. The radiotherapy planning system is linked to the CT diagnostic display console and the radiation isodose distribution curves can be overlaid onto the CT image. The CT density numbers could be used to calculate the effect of inhomogeneities in the tissues in the path of the radiation beam. The areas to be irradiated at therapy can be marked as could sensitive areas to be avoided could be located precisely. This was the very problem that Allan Cormack had been considering back in 1956 when he had observed the radiotherapy treatment planning when a technician would superpose isodose charts generated with isodose contours, which the physician would then examine and adjust as necessary until a satisfactory dose-distribution was found.[61]

The impact of CT scanning on the X-ray equipment industry is interesting. The CT scanner (or EMI scanner) developed by EMI Medical Inc. was quite unexpected by the large X-ray equipment companies and it took them some time to catch up. The X-ray industry had been concentrating on increasingly sensitive X-ray film-intensifying screen combinations with ever finer resolution. The EMI/CT scanner was initially of lower resolution and used an entirely different physical principle. Philips, who were a major manufacturer, approached several radiologists who had experienced the EMI scanner, and had been told that the CT scanner would never amount to anything because of its poor spatial resolution. The radiologists had completely ignored the intrinsic advantage of CT with its high contrast resolution. This resulted in in a delay in Philips developing their CT scanner. There are stories told of Hounsfield approaching several eminent radiologists with his early ideas and not being greeted with any enthusiasm. It is salutary to consider how the research that led to the CT scanner did not come from the major X-ray companies in the same way that the research that led to our modern low-osmolar non-ionic contrast media did not develop in one of the large contrast-media companies but rather in a small Norwegian company.[62] In the same way that the first paper by Torsten Almén[63] on the idea of a non-ionic contrast agent had no immediate impact in the radiological community, the significance of the papers of Allan Cormack was not recognised. In the mid-1960s, EMI had only a minor involvement in medicine by way of its subsidiary

SE Labs; however, by the mid-1970s, EMI was a world leader in advanced medical technology with a full order book and a significant market potential at that time was valued at £100 million per year. EMI had initiated an entirely new business, based on innovative technology, and most unusually with almost no start-up losses. Pandit in his history of Thorn-EMI states that it was actually the revolutionary importance of the EMI scanner in medical diagnosis that was in part responsible for the failure of EMI Medical as a business.[64] EMI in developing the scanner went through a period of remarkable growth; however, this resulted in major pressures on the organisation. The CT scanner rapidly became an essential part of radiological diagnostic equipment and the traditional radiology equipment companies had to develop their own scanner or lose significant business. Pandit presents the situation as one of life and death for other manufacturers. The patents held by EMI were comprehensive; however, there were soon to be many patent cases in the courts upon which Hounsfield was to spend considerable time and effort. This is discussed by Pandit who presents many American companies as having a cavalier disregard for the rights of the patent holder. EMI was used to a more gentlemanly approach in the United Kingdom, and in the United States encountered companies who would make a commercial scanner first and then discuss terms with the patent holder "on the courthouse steps". This story of the rise and decline of EMI is salutatory, and today EMI has no medical connection. Although EMI Medical made significant losses in its last two years, during its lifetime the EMI scanner made significant profits. EMI decided to merge with Thorn in October/November 1979 and the merged company (Thorn-EMI) very soon decided to divest itself of its CT interests. EMI scanner production, including the innovative new third-generation CT7070 scanner, was therefore terminated. EMI Medical had been unable to appropriately manage its very rapid international growth, and perhaps the answer would have been for EMI to have originally bought a pre-existing radiology equipment company in the United States with an established infrastructure and experience.

ELECTRON BEAM CT

The story of the use of X-rays to produce digital tomographic images is complex. The concept of computed tomography had resulted in many developments, including electron beam CT, and has been reviewed by Kulkarni and colleagues.[65] Kulkarni noted that the early scanners were not able to provide diagnostic images of the moving heart, and likened the scanner to a heavy X-ray source mounted on a gantry, and that being like a large camera on a moving tripod. In 1976, Douglas Boyd from the University of California San Francisco realised that an entirely new mechanism for scanning was needed and used an electron beam. The electron beam CT (EBCT) that Boyd developed did not use any moving parts and significantly improved temporal and spatial resolution. From the 1980s to the early 2000s, EBCT became a dominant modality for cardiac imaging. EBCT resulted in two significant developments: coronary artery calcium scoring (CAC) and CT coronary angiography. CT coronary angiography was part of the progressive development of non-invasive imaging. CAC using CT was part of a shift from symptomatic to asymptomatic imaging. Whilst

Kulkarni stated that this was radical for that time because it meant that the purpose of examination of the heart was for prognostic and not diagnostic information, it should be remembered that this was intrinsic to medical imaging from the earliest days. A patient may feel well and have no physical signs and yet the radiograph may show an abnormality. This concept was as difficult for physicians as it was for their patients. EBCT became obsolete when single-detector CT evolved into the modern multislice scanners. The high temporal resolution of EBCT, which was far superior to that of contemporaneous CT scanners, was diminished by the superior spatial resolution of MDCT. Perhaps the most interesting aspect of EBCT is that it demonstrates how a technique could be developed to meet a particular medical problem.

Conventional CT scanning continued to develop, and spiral CT represented a significant advance in the technology of CT scanning, significantly increasing its clinical value. The first clinical cases and performance measurements were presented as work in progress by Willi Kalender (Figure 10.12), Peter Vock, and Wolfgang Seissler at the 75th anniversary meeting of the Radiological Society of North America in 1989.[66,67] The technique was then fully described in a paper in *Radiology* in 1990.[68] The development of spiral CT and multislice scanning was primarily made possible by advances in computing. Data could be reconstructed almost instantaneously. In spiral CT, the rotation of the X-ray source and the movement of the table and patient are made simultaneously, ensuring that the time for acquisition of the raw data is markedly reduced. Willi Kalender has reviewed the use of spiral CT and describes the remarkable results that can be achieved by modern multislice scanners (Figure 10.14).[69] The improved spatial resolution allows for virtual endoscopy and faster scanning enables complex dynamic studies. The most dramatic improvement

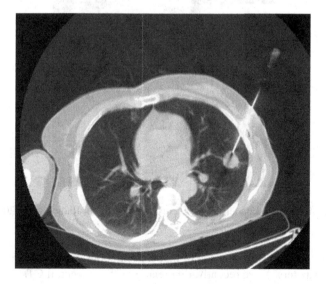

FIGURE 10.12 CT used for intervention, in this case a guided lung biopsy from 2004. The needle is seen on the right of the image. Author's collection.

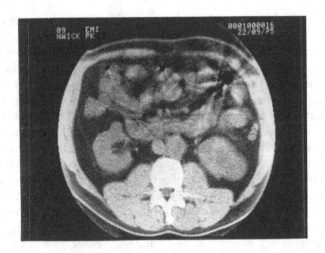

FIGURE 10.13 Abdominal CT scan at the level of the kidney's taken at Northwick Park Hospital on 20 October 1975. Author's collection.

FIGURE 10.14 Willi Kalender, at the European Congress of Radiology in 2004. Author's photograph.

was made because of the provision of higher continuous X-ray power and improvements in computers. More recently, dual-energy CT, or spectral CT, uses two separate X-ray photon energy spectra, which enables the examination of materials that have different attenuation properties at different X-ray energies. Kalender finishes his article by stating that predictions are particularly difficult when they are concerned with the future. However, some problems remain the same, and the very speed of acquisition and high resolution of modern scans produces significant reporting pressures on departments. In 1979, the Working Party on CT Scanning of the Royal College of Radiologists reported that at national level, the problems of providing whole-body CT scanners must be considered with the general problems which beset contemporary radiology, i.e. shortage of radiologists and shortage of money.[70] Although these problems persist, considering the clinical utilisation of modern scanners they should not be seen as unexpected.

The CT scanner is in clinical use in all parts of the body and it has transformed the lives of countless patients. Godfrey Hounsfield was told by many patients and relatives how very grateful they were for the CT scanner and he felt humbled and pleased.

NOTES

1. Dondelinger, R.F. 2020. The idea of tomography was published together with the news of the discovery of Röntgen rays: a historical note. *European Radiology*, 30, 4141–4142.
2. Kalender, W.A. 2000. *Computed Tomography, Fundamentals, System Technology, Image Quality, Applications.* Munich: Publicis MCD Werbeagentur GmbH.
3. Newhouser, J.H., Fleischli, D.J. 1973. Tomography of abdominal aortic aneurysms. *British Journal of Radiology*, 46, 1952–1062.
4. Littleton, J.T. 1976. *Tomography: Physical Principals and Clinical Applications.* Baltimore: William & Wilkins Co.
5. Dewing, S.B. 1962. *Modern Radiology in Historical Perspective.* Springfield: Charles C. Thomas.
6. Meyers, M.A. 2012. *Prize Fight: The Race and Rivalry to be First in Science.* New York: Palgrave, Macmillan.
7. Twining, E.W. 1937. Tomography, by means of a simple attachment to the Potter-Bucky Couch. *The British Journal of Radiology*, 10, 332–347.
8. Watson, W. 1940. Differential radiography (II). *Radiography*, 6, 161–172.
9. Watson, W. 1962. Axial transverse tomography. *Radiography*, 18, 179–189.
10. Samuel, E. 1952. *Clinical Radiology of the Ear, Nose and Throat.* London: H.K. Lewis & Co.
11. Samuen, E., Lloyd, G.A.S. 1978. *Clinical Radiology of the Ear, Nose and Throat.* 2nd ed. London: H.K. Lewis &Co.
12. Garrison, J.B., Grant, D.G., Guier, W.H., Johns, R.J. 1969. Three dimensional roentgenology. *American Journal of Roentgenology*, 105, 903–908.
13. Horner, K. 1995. Dental radiology. In: *The Invisible Light, 100 Years of Medical Radiology.* Ed. A.M.K.Thomas. Oxford: Blackwell Science.
14. Mason, R.A. 1998. Sydney Blackman 1898–1971: a pioneer of panoramic radiography. *Dentomaxillofacial Radiology*, 27, 371–375.
15. Tammisalo, E.H. 1975. Professor Yrjö V. Paatero: the pioneer of panoramic oral tomography. *Dentomaxillofacial Radiology*, 4, 53–56.
16. Mason, R. 2002. A family affair. *The Invisible Light, The Journal of the Radiology History and Heritage Charitable Trust*, 18, 17–28.
17. Bull, J. 1981. History of computed tomography. In: *Radiology of the Skull and Brain: Technical Aspects of Computed Tomography.* Eds. T.H. Newton, D.G. Potts. St Louis: Mosby.

18. Thomas, A.M.K., Banerjee, A.K, Busch, U. 2004. *Classic Papers in Modern Diagnostic Radiology*. Berlin: Springer Verlag.
19. Friedland, W., Thurber, B.D. 1996. The birth of CT. *American Journal of Radiology*, 167, 1365–1370.
20. Webb, S. 1990. *From the Watching of Shadows. The Origins of Radiological Tomography*. Bristol and New York: Adam Hilger.
21. Webb, S. 1992. Historical experiments predating commercially available computed tomography. *British Journal of Radiology*, 65, 835–837.
22. Webb, S. 1995. The invention of classical tomography and computed tomography. In: *The Invisible Light: 100 Years of Medical Radiology*. Ed. A.M.K. Thomas. Oxford: Blackwell Science.
23. Vaughan, C.L. 2008. *Imaging the Elephant: A Biography of Allan MacLeod Cormack*. London: Imperial College Press.
24. Cormack, A.M. 1964. Representation of a function by its line integrals, with some radiological applications. *Journal of Applied Physics*, 34, 2722–2727.
25. Cormack, A.M. 1978. Sampling the radon transform with beams of finite width. *Physics in Medicine and Biology*, 23, 1141–1148.
26. Cormack, A.M. 1992. 75 years of radon transform. *Journal of Computer Assisted Tomography*, 16, 673.
27. Bockwinkel, H.B.A. 1906. Physics: Mr. LORENTZ offers a communication by Mr. H.B.A. BOCKWINKEL: "About the propagation of light in a biaxial crystal around an oscillating center point". *Verh. Kon. Akad. Wet. Wissen. Naturk.*, 14, 636.
28. Cormack, A.M. 1982. Computed tomography: some history and recent developments. *Proceedings of Symposia in Applied Mathematics*, 27, 35–42.
29. https://arxiv.org/pdf/2004.03750.pdf (accessed 1 July 2021).
30. Oldendorf, W.H. 1978. The quest for an image of the brain, a brief historical and technical review of brain imaging techniques. *Neurology*, 28, 517–533.
31. Kuhl, D.E., Edwards, R.Q. 1970. The Mark III scanner: a compact device for multiple-view and section scanning of the brain. *Radiology*, 96, 563–570.
32. Bates, S., Beckmann, E., Thomas, A.M.K., Waltham, R. 2012. *Godfrey Hounsfield: Intuitive Genius of CT*. London: The British Institute of Radiology.
33. Wells, P.N.T. 2005. Sir Godfrey Newbold Hounsfield KT CBE. *Biographical Memoirs Fellows Royal Society*, 51, 221–235.
34. Hounsfield, G.N. 1976. Historical notes on computerised axial tomography. *Journal of the Canadian Association of Radiology*, 27, 135–142.
35. Hounsfield, G.N. 1979. Computer reconstructed X-ray imaging. *Philosophical Transactions of Royal Society London A*, 292, 223–232.
36. Hounsfield, G.N. 1977. The EMI scanner. *Proceedings of Royal Society London B*, 195, 281–289.
37. Mennel, R. 2021. A message from the president. *The Oslerian*, 22, Issue 1, 1–3.
38. Higson, G.R. 1987. Seeing things more clearly. *The British Journal of Radiology*, 60, 1049–1057.
39. Bates, S., Beckmann, E., Thomas, A.M.K., Waltham, R. 2012. *Godfrey Hounsfield: Intuitive Genius of CT*. London: The British Institute of Radiology.
40. Kreel, L. 1971. *Outline of Radiology*. London: William Heinemann Medical Books.
41. Bull, J. 1981. History of computed tomography. In: *Radiology of the Skull and Brain: Technical Aspects of Computed Tomography, Volume 5*. Eds. T.H. Newton., D.G. Potts. St. Louis: The C.V. Mosby Company, 3835–3849.
42. Gould, T., Uttley, D. 1996. *A History of the Atkinson Morley's Hospital 1869–1995*. London: The Athlone Press.

43. Ambrose, J. 1996. You never know what is just around the corner. *Rivista di Neuroradiologia*, 9, 399–404.
44. Ambrose, J., Hounsfield, G. 1973. Computerised transverse axial tomography. *British Journal of Radiology*, 46, 148–149.
45. Hounsfield, G.N. 1973. Computerized transverse axial scanning (tomography). Part 1. Description of system. *British Journal of Radiology*, 46, 1016–1022.
46. Ambrose, J. 1973. Computerized transverse axial scanning (tomography). Part 2. Clinical application. *British Journal of Radiology*, 46, 1023–1047.
47. Perry, B.J., Bridges, C. 1973. Computerized transverse axial scanning (tomography). Part 3. Radiation dose considerations. *British Journal of Radiology*, 46, 1048–1051.
48. Ambrose, J.A.E., Lloyd, G.A.S., Wright, J.E. 1974. A preliminary evaluation of fine matrix computerized axial tomography (Emiscan) in the diagnosis of orbital space-occupying lesions. *British Journal of Radiology*, 47, 747–751.
49. Editorial. 1975. Non-invasive investigations of the brain. *British Medical Journal*, 2, 295–6.
50. Bull, J.W.D. 1977. Foreword. In: *Computerised Axial Tomography in Clinical Practice*. Eds. G.H. du Boulay, I.F. Moseley. Berlin: Springer-Verlag.
51. Jennett, B. 1984. *High Technology Medicine, Benefits and Burdens*. London: Nuffield Provincial Hospital Trust.
52. Fineberg, H.V. 1978. Evaluation of computed tomography: achievement and challenge. *American Journal of Roentgenology*, 131, 1–4.
53. Teasdale, G., Galbraith, S., Murray, L., Ward, P., Gentleman, D., McKean, M. 1982. Management of traumatic intracranial haematoma. *British Medical Journal*, ii, 1695–1697.
54. Bartlett, J.R., Neil-Dwyer, G., Banham, J.M.M., Cruikshank, D.G. 1978. Evaluating cost-effectiveness of diagnostic equipment: the brain scanner case. *British Medical Journal*, 2, 815–820.
55. Kreel, L. 1975. Computed tomography in the evaluation of malignant disease. *Transactions of the Medical Society of London*, 92–93, 139–144.
56. Bydder, G. MRIS history UK. https://mrishistory.org.uk/wp-content/uploads/2019/05/gmbchapter-v9.pdf (accessed 19 July 2021).
57. Kreel, L. 1977. Computerized tomography using the EMI general purpose scanner. *The British Journal of Radiology*, 50, 2–14.
58. Kreel, L. 1976. Computer tomography in the evaluation of pulmonary asbestosis. *Acta Radiologicia Diagnosis*, 17, 405–412.
59. Kreel, L. 1979. *Medical Imaging: CT, U/S, IS, NMR*. Chicago: Year Book Medical Publishers.
60. Hounsfield, G.N. 1979. Nobel lecture. https://www.nobelprize.org/prizes/medicine/1979/hounsfield/lecture/ (accessed 22 August 2021).
61. Cormack, A.M. 1979. Nobel lecture. https://www.nobelprize.org/prizes/medicine/1979/cormack/lecture/ (accessed 22 August 2021).
62. Grainger, R.G., Thomas, A.M.K. 1999. History of intravascular iodinated contrast media. In: *A Textbook of Contrast Media*. Eds. P. Dawson, D. Cosgrove, D.J. Allison. Oxford: Isis Medical Media.
63. Almén, T. 1985. Development of non-ionic contrast media. *Investigative Radiology*, 20, 2–9.
64. Pandit, S.A. 1996. *From Making to Music: The History of THORN EMI*. London: Hodder & Stoughton.
65. Kulkarni, S., Rumberger, J.A., Jha, S. 2021. Electron beam CT: a historical review. *American Journal of Roentgenology*, 216, 1–7.

66. Vock, P., Jung, H., Kalender, W.A. 1989. Single-breathhold voluminetric CT of the hepatobiliary system. *Radiology*, 173(P), 377.
67. Kalender, W.A., Seissler, W., Vock, P. 1989. Single-breath-hold spiral volumetric CT by continuous patient translation and scanner rotation. *Radiology*, 173 (P), 414.
68. Kalender, W.A., Seissler, W., Klotz, E., Vock, P. 1990. Spiral volumetric CT with single-breathhold technique, continuous transport, and continuous scanner rotation. *Radiology*, 176, 181–183.
69. Kalender, W.A. 1996. Spiral CT in the year 2000. In: *Spiral CT of the Chest*. Eds. M. Rémy-Jardin, J. Rémy. New York/Berlin, Heidelberg: Springer, pp. 322–329.
70. *Report on the Working Party on CT Scanning*. 12th November 1979. The Royal College of Radiologists.

11 NMR to MRI

MAGNETISM AND ELECTROMAGNETISM

Magnetism is a mysterious phenomenon. When we bring two bar magnets together and can feel their attraction and repulsion, and when we can observe iron filings arranged along a magnetic field, then this seems almost magical, and must have seemed as such to our ancestors. We now think that we understand, and yet do we really? We are like children splashing on the sea shore and thinking that we experience the ocean. What the study of natural phenomena has shown since the earliest times to us is that there is always more to understand; there are always more depths to explore. The understanding of what seem to be apparently simple physical phenomena may lead to profound conclusions. Similarly, apparently separate phenomena such as electricity and magnetism may be related in unsuspected ways.

Magnets have been known since earliest times in the form of magnetic rocks or lodestones, which are the naturally occurring forms of magnetite.[1] In the cartoon "The Lodestone Rock", a knight in ferrous armour is attracted to the lodestone, as illustrated on a menu for the *Sette of Odd Volumes* from 26 November 1901 to illustrate a talk by Silvanus P. Thompson (1851–1916), the first president of the British Institute of Radiology (Figure 11.1). Titus Lucretius Carus (c.99BC–c.55BC) in *De Rerum Natura* (On the Nature of Things) described how iron can be attracted to a stone which the Greeks call the magnet, named because it came from the land of the Magnetes.[2] Lucretius noted that the stones can make a chain of rings all hanging to itself and that each can feel the binding force of the stone. Pliny the Elder (23–79) described in his *Naturalis Historia* (Natural History) of AD77 the sympathy that the lodestone has with iron, and how a series of magnetic rings could be attracted to a lodestone. The lodestone was also known to the ancient Chinese and to the Arabians.

Peter Peregrinus, or Pierre de Maricourt, was a French military engineer, and he described the use of the floating compass needle for navigation in his text *Epistola Petri Peregrini de Maricourt ad Sygerum de Foucaucourt, militem, de magnete* (Letter of Peter Peregrinus of Maricourt to Sygerus of Foucaucourt, Soldier, on the Magnet) of 1269.[3] His compass needle was set into a divided circle with a reference line. The earliest compass card marked up with the names of the four winds is credited to Andrea Bianco of Venice in 1426. In ancient times, and with basic instruments, the fact that the compass needle did not point exactly to the true north was not appreciated. Then, as time passed, the compass makers learnt to vary the setting of the needle on the card according to the geographical location of the card. There were a variety of explanations as to why the compass needle did not point true north, such as the action of the stars, the influence of spirits, or to the existence of lodestone mountains. Fanciful maps giving the locations of lodestone rocks or magnetic islands were made by cartographers, and so in Geraldus Mercator's (1512–1594) great chart

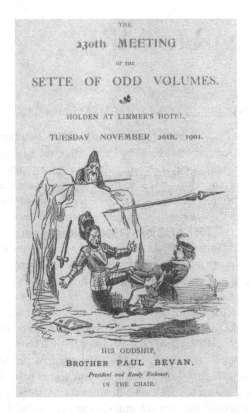

FIGURE 11.1 "The Lodestone Rock": a knight in armour is attracted to the lodestone. A menu for the Sette of Odd Volumes from 26 November 1901 to illustrate a talk by Silvanus P. Thompson. Author's collection.

two rocks are shown in the sea and located north of Eastern Siberia. The navigator Christopher Columbus (c.1451–1506) also recognised that the declination of the needle varied, and in his early voyage this had nearly caused a mutiny of his sailors.

A new observation was made by Georg Hartmann of Nürnberg (1489–1564) in 1544 (although not immediately published), and Robert Norman who was a compass-maker from Limehouse in London. In 1581 Norman published his little book *The Newe Attractive. Containing a Short Discourse of the Magnes or Loadstone.* Hartman and Norman observed that in a magnetised needle, the northern end pointed downwards giving it a dip or inclination. Norman therefore devised a dipping needle, and was able to measure the dip at London, and developed a theory of point perspective. As exploration of the world continued, the navigators were able to make many observations, and bring in additional data, and Norman's point perspective theory could not be supported.

However, it was the great William Gilbert of Colchester (1540–1603)[4] who transformed the understanding of magnets. Gilbert was appointed as personal physician to Queen Elizabeth the First in February 1601, and attended her in her last illness.

Gilbert had been President of the Royal College of Physicians of London in 1599. Gilbert made a detailed study of magnets in both Italy and England for a period of about 20 years. In 1600, Gilbert published his monumental book *De Magnete*[5] with the subheading of *magneticisque corporibus, et de magno magnete tellure, physiologia nova*. Gilbert's astounding conclusion, which no one had suspected previously, was that the compass needle works because the earth itself is a giant lodestone or magnet. So, in his title page, Gilbert proclaims, "the great magnet, the Earth". The title of the 17th chapter of the first book of *De Magnete* expresses Gilbert's views well when he states "That the globe of the earth is magnetick, & a magnet; & how in our hands the magnet stone has all the primary forces of the earth, while the earth by the same powers remains constant in a fixed direction in the universe". It was Gilbert who coined the term "terrestrial magnetism", and this expression had never been used previously. Therefore, the needle points in a polar direction is because of the globe of the earth acting as a whole. Gilbert demonstrated many proofs of his discovery in his book, and his work is a masterpiece of practical experimental science. He constructed lodestone globes, or terellas, to demonstrate his theories (Figure 11.2). Local variations could be explained by variations in the earth's crust. Gilbert also discussed acquired magnetism, and illustrated how this can be produced by hammering an iron bar lying in a north–south direction on an anvil. However, in his explanation as to how the lodestone actually works, Gilbert uses classical neoplatonic ideas following those of the pre-Socratic philosopher Thales of Miletus and of Aristotle. The universe itself is seen as animate, and so the lodestone is also animate and is "the choice offspring of its animate mother the Earth". Gilbert's treatise is a major literary and scientific achievement, and was translated into English for the Gilbert Club in 1900 at the 300th anniversary of its publication.[6] Silvanus

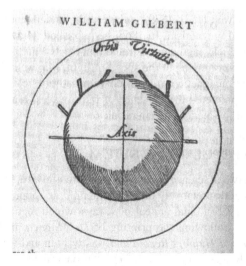

FIGURE 11.2   Gilbert's Terrella (magnetic globe) with compass needles from *De Magnete* of 1600. Public domain.

Thompson documented Gilbert's scientific attitude, and pointed out that "He poured out the vials of his wrath upon the empty-headed and inert philosophers who merely copy from one another and invent high sounding Greek words wherewith to cloak their ignorance".

The intimate relationship between electricity and magnetism was demonstrated in April 1820 in an elegant experiment by Hans Christian Ørsted (1777–1851) of Copenhagen. Ørsted showed that a compass needle was deflected when an electrical current was started or stopped in an adjacent voltaic pile. This discovery resulted in considerable interest in electromagnetism, and further work was undertaken by François Arago (1786–1853) and André-Marie Ampère (1775–1836). Ampère made major contributions, particularly applying mathematics to the study of electromag-netism. In his Bakerian Lecture of 1832, Michael Faraday (1791–1867) discussed terrestrial magnetoelectric induction and magnetism, and notes that "it is a striking thing to observe the revolving copper plate become thus a new electric machine". Faraday investigated electromagnetic induction and his electromagnet may be viewed at the Royal Institution in London. Essentially, Faraday realised that any body that has charge and velocity will have a magnetic field. The earth's magnetic field is created by its rotation and the creation of an angular magnetic moment, and in a similar manner a magnetic field is produced by charged nuclear particles spin-ning within tissues. A similar comparison can be made in that the earth's magnetic field strength is not uniform, being weaker at the equator than the North Pole. If the strength is measured, then the location between the Pole and equator can be determined. This is not dissimilar to Gilbert's suggestion that the dip of a compass needle might be used to determine latitude, although this method proved not to be successful (Figure 11.3). However, it is the case that gradient magnetic fields used in magnetic resonance imaging (MRI) can be used to determine spatial localisation in bodily tissues.[7]

James Clerk Maxwell (1831–1879)[8] enlarged the understanding of the phenom-ena of electricity and magnetism. In 1865, he published "A Dynamical Theory of the Electromagnetic Field" and demonstrated that electric and magnetic fields travel through space as waves moving at the speed of light. Maxwell formulated the clas-sical theory of electromagnetic radiation, and unified both electrical and magnetic phenomena, which predicted the presence of radio waves. Maxwell's work can be seen as the origin of the modern age in physics.

MEDICAL AND POPULAR MAGNETISM

The relationship between science and popular culture is always interesting. It is note-worthy that MRI features relatively little in contemporary popular culture compared to the excitement and continual appeal of X-rays, which regularly feature in films and advertisements. Following the printing of *De Magnete* in 1600, a stage play *The Magnetic Lady, or Humors Reconciled* was written and was the final comedy of the playwright Ben Jonson (1532–1637).[9] Jonson was well acquainted with the contemporary scientific literature on magnetism and that the lodestone "expels gross humors". The allegorical images in the play relate to the attraction of opposites, and

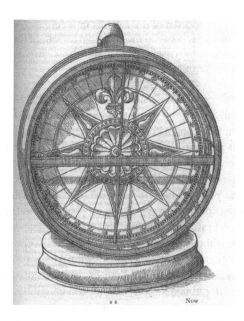

FIGURE 11.3 "Instrument of the Declination": the declination of a compass needle from the fifth book of Gilbert's *De Magnete* of 1600. Public domain.

then to their reconciliation. The play was premiered in 1632 by the King's Men and performed at the Blackfriars Theatre in London. The plot is somewhat convoluted as is often the case in Caroline drama. The characters are Lady Lodestone who is guided by the advice of Master Compass. Parson Palate tells Lady Lodestone that Master Compass is "the perfect instrument your Lady should sail by". Other characters are her niece Placentia Steel, the foppish Sir Diaphanous Silkworm, and the gruff soldier Captain Ironside. The play was last staged in 2010 at the White Bear Theatre in Kennington, South London, and updated in period to the 1930s. MRI has featured in fiction on many occasions, including a story involving the teenage spy Alex Rider, an episode of the television series *Dr. Who*, and the action-thriller film *Unknown* of 2011 directed by Jaume Collet-Serra. None approach the interest and fascination of *The Magnetic Lady*.

The relationship between magnetism and medicine is complex. By the end of the 18th century, better quality and more powerful magnets were being manufactured, and there had been reports of the treatment of stomach pains and toothache using such magnets. The story of Franz Anton Mesmer (1734–1815) is remarkable.[10] It is too easy to represent what transpired with Mesmer and his followers as an example of either ignorance or quackery. This would be a mistake, and Mesmer and his followers were sincere and believed in the truth of their observations. Mesmer practised as a doctor in Vienna, publishing his doctorate in 1765 on the subject of the influence of celestial bodies on disease. Mesmer believed that the universe was filled with a universal fluid, which could influence humans. Mesmer wrote:

I never gave up close observation of my patients in accordance with my theory, and became so involved that I experienced the condition myself, in the rise and fall of the illness. In the end I came to the conclusion that the ebb and flow in the body of the sick person was similar to those of magnets.

Mesmer wrote to Dr. Unzer about a patient called Franzl that he had treated in 1774, that in the month of July when the patient had another attack, he fixed two magnets of horseshoe type to her feet and a heart-shaped magnet to her breast. Mesmer noted that suddenly she felt a burning sensation spreading from her feet through all her joints like a glowing coal, with severe pains at the hips, and likewise from both sides of the breast to the crown of the head. This continued throughout the night, causing copious sweating. The pains gradually went off, and she became insensitive to the magnets. The symptoms disappeared and she recovered from the seizure.

News of the cure of Franzl by Mesmer spread throughout Vienna and there was much public interest, although his medical colleagues were sceptical. A further remarkable cure was that of Wilhelm Bauer, who was a professor of mathematics in Vienna. Bauer had been a sleepwalker, or somnambulist since his childhood and would wake up with pains and convulsions and coughing up blood. Bauer was completely cured by magnetic treatments, and he made a public announcement of Mesmer's magnetic cure on 15 March 1775. Mesmer based his views on those of Isaac Newton (1643–1727) who had postulated the presence of the ether through which light, gravity, and magnetism were transmitted. The idea of the ether was popular in the 19th century, and one of the last proponents was Sir Oliver Lodge (1851–1940), although the concept remained controversial. Mesmer believed that passes with the hand could influence the fluid and work under the action of animal magnetism. There would therefore be no need for pharmacological or surgical interventions. Mesmer's historic treatise *Mémoire sur la découverte du Magnétisme Animal* was published in 1779.[11] Mesmer was obliged to leave Vienna and set up a salon in Paris which became popular with the aristocracy. A commission was eventually set up, headed by no less a person than Benjamin Franklin (1706–1790), who had himself made major contributions to the understanding of electrical phenomena. The commission concluded that Mesmer's results were due to the action of the imagination. We would perhaps now say the results were due to a combination of suggestion, hypnosis, and the placebo effect.

Perhaps animal magnetism would have disappeared if it had not been for the work of John Elliotson (1791–1868).[12] In 1832, Elliotson was appointed to the Chair of Clinical Medicine at University College in London, and became Senior Physician to University College Hospital. Elliotson was a popular teacher and one of the most influential doctors of his time, and in 1839 his influential textbook *The Principles and Practice of Medicine* was published.[13] In 1837, Elliotson had met Du Potet when he visited London. Baron Jules Denis Du Potet de Sennevoy (1796–1881) was a physician at the Pitié Salpêtrière Hospital in Paris, and was a leading exponent of animal magnetism in France, and a successor of Mesmer. Mesmerism had gone into decline at the time of the French Revolution but was currently undergoing a revival. Du Potet saw in animal magnetism the presence of "the magic of antiquity", and was also interested in many curious phenomena, including levitation and spirit communications. Elliotson was interested and started his own investigations into animal magnetism. Elliotson attracted many patients and practised mesmeric treatment both in his house and for his hospital patients at University College Hospital, and held publicly attended séances in the hospital's lecture theatre. There was considerable opposition from both his colleagues in the hospital and also the medical establishment.

Elliotson's casebooks were preserved and one of the most remarkable instances involved the Okey sisters. Elliotson examined them with the help of Du Potet when they were admitted to Elliotson's wards with fits. The sisters Elizabeth and Jane Okey's symptoms were abolished when in a magnetic trance. As the magnetic sessions progressed, Miller noted "the sisters began to acquire some of the transcendent abilities associated with clairvoyant mesmerism". The sisters became able to see into their own bodies, and in a deeper trance they could see inside the bodies of other patients. Elliotson then took the patients around the wards where, in a magnetic trance, they claimed that they could diagnose the state of other patients' internal organs. As Wendy Moore points out in her recent study of Elliotson, the patients became doctors and the doctors became passive observers.[14] It is never good when doctors are passive observers, whether of mesmeric trances or of the findings of contemporary medical imaging. As a result of Elliotson's activities, the hospital experienced chaotic scenes, and he was forced by the hospital authorities to resign his post in 1838.

However, Elliotson continued his research, and in 1848 he wrote his well-known study of performing of painless surgery using animal magnetism as anaesthesia.[15] He also published a journal *The Zoist*,[16] which contains many examples of animal magnetism, used for anaesthesia or for medical cures. It is noteworthy that the title page of *The Zoist* quotes Galileo "This is TRUTH though opposed to philosophy of ages". Mesmerism and electrical psychology were considered by Dods in 1886 who noted that all motion and power originated in mind, and that the human spirit comes into contact with matter through an electromagnetic medium.[17] In the 19th century, electrical phenomena in the body were increasingly appreciated both in popular culture and in science. The poet Walt Whitman (1819–1892) in his collection *Leaves of Grass* of 1855 celebrated the physicality of the body in his poem "I sing of the body electric" and describes the intimate relationship of body and soul. Certainly, modern functional MRI elegantly demonstrates the intimate relationships between cerebral function and human actions, and is perhaps a modern electrical psychology.

Knowledge of both the mind and the brain has developed since Elliotson's time. The Okey sister's conditions resembled those of the patients with hysteria seen by Jean-Martin Charcot (1825–1893) at the Salpêtrière in the 1880s. The magnetic trance itself was put on to a firmer base by James Braid (1795–1860) who has been called the Father of Hypnosis. Braid developed the idea of hypnosis, separating it "from the occult shadows of mesmerism", with insights into the nature of trance and by coining the word hypnosis.[18] However, the performance of surgery under hypnosis is well recognised, and Gilbert Frankau and others have viewed Mesmer as a forerunner of modern psychotherapy.

Magnetism has continued to fascinate with a belief in its healing power. There are many links between magnetism and medicine, both in mainstream practice and in quackery; however, at times the distinction between the two has been blurred. Indeed, the distinction between medicine and quackery must be blurred. In popular medicine, both now and in Victorian times, magnets have been worn as bracelets and in corsets,[19] and this shows no sign of ceasing with contemporary promotion of medicinal lodestones.[20] Finally, magnetism was to have a role in medical imaging in an entirely unsuspected manner.

NMR TO MRI

The story of the application of NMR (nuclear magnetic resonance) to the biomedical sciences is complex, and throughout its history we find groups working simultaneously and arriving at equivalent conclusions. This can make the assignment of priority difficult, if not impossible.

Isidor Isaac Rabi (1898–1988) in the early 1930s first measured the magnetic properties of atomic nuclei in two classic papers.[21,22] Rabi was Jewish and Austrian and emigrated to the United States, as did so many others. His PhD thesis was on the magnetic properties of crystals. Rabi was able to first demonstrate and measure single states of rotation of atoms and molecules and determine the mechanical and magnetic movements of the nuclei, and this might be viewed as prefiguring the principle of magnetic resonance. Rabi was successful in his experiments having been helped in September 1937 by a visit and advice from a Dutchman, Cornelis Jacobus Gorter (1907–1980), who had tried similar experiments unsuccessfully. Russian scientists were also working on magnetic resonance, and electron paramagnetic resonance was discovered at the University of Kazan in 1944 by Yevgeni K. Zavoisky (1907–1976). Zavoisky had worked on NMR in 1941; however, the signal he obtained was unstable and was difficult to reproduce, and he did not develop his research.

Although Rabi had measured the magnetic properties of nuclei, it was Felix Bloch (1905–1983), working at Stanford University with Hansen and Packard, and Edward Purcell (1912–1997), working at Harvard University with Torrey and Pound, who independently discovered nuclear magnetic resonance in 1946. The results were published in consecutive issues of *Physical Review*.[23] Bloch was of Swiss origin, moving to the United States where, in the late 1930s, he worked at the Berkley cyclotron in California, measuring magnetic moments of the neutron. Bloch, whilst working with Luis Alvarez (1911–1988), had discovered nuclear induction in 1940 whilst endeavouring to find the proton resonance in water so as to measure the magnetic field in another experiment.[24] Bloch extended the work of Rabi and his collaborators on the magnetic resonance method of determining nuclear magnetic moments in molecular beams allowing the determination of the neutron moment. In 1946, Bloch's Stanford Group had finally obtained the predicted resonance signal in water by adiabatic fast passage.[25] Purcell and his group discovered the principle of magnetic resonance simultaneously working at Harvard University,[26] observing the proton resonance of paraffin in a waveguide.

The importance of the discoveries of Bloch and Purcell were apparent immediately. It is remarkable that these two groups, both working independently, came up with essentially identical discoveries, although describing them using different terminologies.[27] There are similarities to the invention of the CT scanner by Godfrey Hounsfield, in which there were multiple groups working independently on the problem of image reconstruction.[28] Hounsfield's group at EMI was poorly resourced and used an old lathe bed for their experiments, and similarly in NMR both the Harvard and Stamford groups used borrowed electromagnets. These three NMR pioneers received the Nobel Prize for Physics: Rabi in 1944 and then Bloch and Purcell receiving theirs jointly in 1952.

Since the principles of NMR were understood, it might be wondered why there was such a long time interval before the various practical applications were realised. Part of the reason was related to the need for the development of sufficiently powerful computers for analysis of the complex data sets.

NMR IMAGING

Raymond Damadian (1936–) is a physician who qualified from Albert Einstein College of Medicine in New York, and after medical qualification pursued interests in biophysics, physical chemistry, and mathematics. Damadian can be credited as inventing the MRI (Magnetic Resonance Imaging) scanner and published a seminal paper in the journal *Science* in 1971 on the detection of tumours by NMR.[29] Damadian then suggested in 1972 that NMR could be of value in medical diagnosis. Damadian showed that different tissues emitted signals, which varied in length, and that malignant tissues emitted signals that lasted longer than those from benign tissues. He filed a patent for this human NMR scanner, which was granted on 5 February 1974. By 1977, Damadian had completed construction of the first whole-body NMR scanner,[30] and this is described in the classic book *A Machine Named Indomitable* of 1985.[31] Damadian founded the FONAR Corporation in 1978 to produce and market his MRI scanner, FONAR – "field focused nuclear magnetic resonance".

In 1973, Paul Lauterbur (1929–2007) in New York found that it was possible to make two-dimensional images by adding a gradient to a magnetic field. His important paper was published on 16 March 1973 in the journal *Nature*.[32] Lauterbur coined the term "zeugmatography" (from the Greek zeugmo, meaning yoked or joined together) for this process. Lauterbur was particularly important for his work in transforming the NMR signals into images, and is said to have had an unexpected epiphany while eating a hamburger. Lauterbur then went out and bought a notebook, and worked out his idea in a few days.[33] His ideas enabled the single-dimensional NMR spectroscopy to move into the spatial orientation, which is the basis of MRI. Lauterbur went on to publish over a hundred papers in the field of MRI and many patents.

The Physics Department at Nottingham University had an interest in the applications of NMR, and in the January of 1974, Moore and Hinshaw attended an international meeting in Bombay where they heard Lauterbur discuss the possibility that NMR could be used to produce images. On returning to Nottingham, a spin mapping technique was used to image water phantoms. Peter Mansfield (1933–2017)

FIGURE 11.4   Peter Mansfield (1933–2017). Author's photograph.

(Figure 11.4) and Andrew Maudsley developed the line technique, which in 1977 led to an *in vivo* cross-sectional image through a finger.[34] In 1977, the Nottingham team succeeded in producing an image of a wrist, and in April 1978 the Nottingham group produced high-definition images of a rabbit. They emphasised that NMR would be able to provide both functional and anatomical information. By January 1979, they demonstrated recognisable anatomical detail in the living human forearm, although their first cranial scan of 28 May 1979 from their prototype body scanner, whilst being recognisable as a head, showed less anatomical detail than was seen in the experimental head CT scans. They saw MRI as more than an alternative to CT scanning since it would become possible to achieve tissue discrimination and characterisation through the analysis of the complex NMR signal. In 1982, Brian Worthington (1938–2007) (Figure 11.5) from the NMR group in Nottingham gave the George Simon Memorial Lecture to the Royal College of Radiologists and reviewed the current state of NMR imaging.[35] Worthington was the first radiologist to be elected to the Royal Society of London. In his lecture, Worthington reflected on the rapid progress that had been made in the field, and noted that several whole-body systems were being evaluated in clinical trials. It was noteworthy that much of the work had been carried out in Great Britain, with units at Hammersmith Hospital in London, at Aberdeen University, and at Nottingham University.

In 1976, after the success of the EMI CT scanner, and again with the assistance of the Department of Health and Social Security (DHSS), an NMR research team was established by Thorn-EMI.[36] The NMR group began with a Walker 0.1 tesla resistive magnet and, under the direction of Hugh Clow and Ian Young (1932–2019)

FIGURE 11.5 Brian Worthington (1938–2007). Author's photograph.

FIGURE 11.6 Ian Young (1932–2019). Image courtesy of Graeme Bydder.

(Figure 11.6), produced the first published human NMR image of the brain in November 1978. This was followed by an inversion recovery (IR) image of Ian Young's brain in the autumn of 1979 that showed a very high grey-white matter contrast (Figure 11.7). The image was shown by Godfrey Hounsfield in his Nobel Prize lecture on computed medical imaging on 8 December 1979, together with his explanation of the image contrast in terms of differences in T1. Ian Young added to the gradient echo and spin echo (SE) sequences in 1980. With funding from the

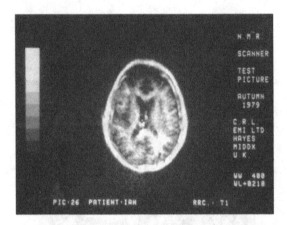

FIGURE 11.7   The inversion recovery image of Ian Young's brain whose grey-white matter contrast he was explaining. The image was taken in Autumn 1979 at the Central Research Laboratories at EMI Ltd. in Hayes. Image courtesy of Graeme Bydder.

scientific and technical services branch of the DHSS, headed by Gordon Higson (who had worked so fruitfully with Godfrey Hounsfield), his group built a 0.15 tesla system based on the world's first commercial large-bore cryomagnet. This was built by Oxford Instruments under the direction of Martin Wood. The MR system was shifted to Hammersmith Hospital in London and imaging of patients began in March 1981. Frank Doyle was actively involved with the research and development, with the team working on the MRI appearances of the liver and brain. Doyle and his co-workers' paper on MRI of the brain was published in *The Lancet* in 1981, and the use of an inversion recovery technique resulted in a remarkable differentiation between grey and white matter. The basal ganglia were seen clearly and better than on CT, as was the posterior fossa and the basal ganglia.[37]

There was a tremendous sense of excitement in the radiology department of Hammersmith Hospital at that time as Graeme Bydder brought new MR images to the departmental lunchtime meetings, fascinating the junior radiologists. In November 1981, Young published a study of 10 patients with multiple sclerosis, in which 112 lesions were detected with MRI and only 19 were seen with CT.[38] It was a quantifiable, decisive, clinical advantage for MRI over CT, and lead to a flurry of activity. Oxford Instruments' order book for magnets increased from 1 million pounds (1.12 million euros) in 1981 to 25 million pounds (28 million euros) in 1982.

Oxford Instruments had opened their factory in West Oxford in 1971, and in 1980 manufactured the first commercial superconducting magnet to be used for MRI scanning. Oxford Instruments had previously worked from a series of sites, including a garden shed in Oxford's Northmoor Road, before opening their purpose-built factory (Figure 11.8). Once it had been decided to develop a whole-body MRI scanner, Oxford Instruments were approached with the proposal to develop a superconducting magnet with a 1-metre inner diameter and a field strength of 0.3 tesla. This was a significantly larger magnet than Oxford Instruments had previously constructed. This was during

FIGURE 11.8 The Osney Mead site of Oxford Instruments in 2019. The 1971 factory is on the right, and the 1982 MRI factory is on the left. Author's photograph.

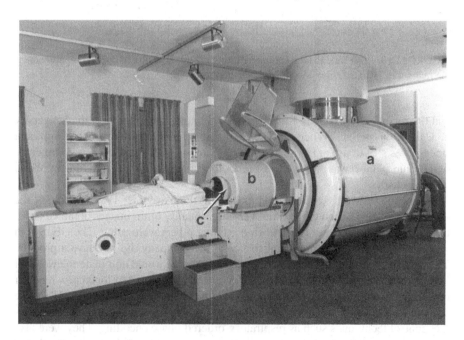

FIGURE 11.9 The NMR/MRI scanner with the Oxford Magnets cryogenic magnet at Hammersmith Hospital in the early 1980s. Image courtesy of Graeme Bydder. Public domain.

a period of great interest in MRI, since by 1980 the majority of companies that were selling CT scanners were either planning or were developing prototype MRI systems.

Oxford Instruments delivered their first two superconducting magnets in 1980. One magnet went to Ian Young and EMI at Hammersmith Hospital (Figure 11.9),

FIGURE 11.10 The Blue Plaque commemorating the manufacture of the first superconducting magnet in 1980 at Osney Mead. Author's photograph.

and the second went to Leon Kaufman at the Imaging Laboratory of the University of California at San Francisco. The quality of the supplied magnet is apparent since it was still in use at Hammersmith Hospital over 20 years later. A commemorative blue plaque was unveiled at the factory site in 2007 by Sir Martin Wood (Figure 11.10).

The world of MRI was turned on its head at the November 1982 meeting of the RSNA in Chicago, when GE showed very-high-quality brain images obtained at 1.5 tesla, which was ten times the field strength that others were using. The GE system was heavily marketed, although at that time the company did not have a product. This was the beginning of the so-called field wars that would dominate the world of MRI for the next four to five years. Ian Young and his group developed new receiver coils, they used low-bandwidth acquisitions, the short inversion time (STIR) pulse sequences, heavily T2-weighted gradient echo sequences, and various motion artefact control techniques such as respiratory-ordered phase encoding. They were also an early user of the contrast agent gadolinium-DTPA which had been developed by Hanns-Joachim Weinmann of Schering (Figure 11.11) and showed it was not a simple issue, and that it was perfectly possible to perform good clinical examinations at a low-field strength and using less expensive systems.

Sadly, Frank Doyle's illness ended his work with NMR at Hammersmith Hospital, and his colleague Robert Steiner (1918–2013) (Figure 11.12) took up the lead of the project at the Hammersmith MRI Centre that now bears his name.

FIGURE 11.11 Hanns-Joachim Weinmann. Author's photograph.

FIGURE 11.12 Robert Steiner (1918–2013). Author's collection, unknown photographer.

MRI machines were gradually introduced in several European and American centres. Although there were many enthusiastic claims being made for this new technique by the media, there were also critical articles being published that deprecated the premature exploitation of what was still an unproven technique. This had also been the case when CT scanning had been introduced.

In the early 1980s when NMR was being introduced into hospitals there was considerable public concern about the use of nuclear power and nuclear weapons. There was also a more critical public attitude to the applications of scientific discoveries, and an awareness that outcomes were not always benign, and this concern has continued and, indeed, has intensified. This point was made by the Aberdeen MRI pioneer John Mallard (1927–2001) (Figure 11.13) in an interview in 2000 when he stated that "Nuclear was associated with bombs and wars and God knows what".[39] The change in terminology from NMR to MRI is therefore significant. In the public mind, the word nuclear has negative connotations, whereas magnetic seems natural and "green" and has traditional associations with healing as has been seen. The word resonance sounds like a musical tone and also has positive connotations. It was the US pioneer of NMR imaging Alexander Margulis (1921–2018) who promoted the new name, and the term MRI was generally accepted. There were, however, scientists outside the United States who objected to the name change calling it "American silliness", and Kelly believes that these scientists expressed impatience with the way the rest of the world is expected to adjust to what is happening in the United States. Criticism of the name change on NMR to MRI draws attention to the technoscientific forms of imperialism that accompany the development of new technologies, but there is nothing new in this. However, there is perhaps another significance to the change in terminology. The term NMR belongs to the physicist, whereas MRI belongs to the domain of the doctor. In the early days of NMR imaging, the images were in colour, and there was an emphasis on the value of NMR spectroscopy in

FIGURE 11.13   John Mallard (1927–2001). Author's photograph.

diagnosis.[40,41] When MRI entered the world of medical radiology, the images were now viewed in a grey scale and presented in a similar manner to the CT scans. Indeed, for a non-specialist, the CT and MRI images can be confused.

Magnetic resonance spectroscopy has resulted in the award of two Nobel Prizes for Chemistry: Richard Ernst (1933–2021) in 1991 for the development of NMR spectroscopy, and Kurt Wuthrich in 2002 for the use of NMR spectroscopy in determination of the three-dimensional structure of biological macromolecules. In 2003, Paul Lauterbur and Peter Mansfield were awarded the Nobel Prize for Physiology or Medicine for their pioneering research in MRI. Peter Mansfield has written an interesting autobiography on his "long road to Stockholm", with the subheading "The Story of Magnetic Resonance Imaging".[42] It is noteworthy that Raymond Damadian was not included with Lauterbur and Mansfield for the Nobel Prize and this aroused considerable controversy and bad feelings at the time. In a chapter entitled "Antagonisms to MRI", Mansfield discusses the growing rivalry that occurred between Lauterbur and himself, and also his meetings with Damadian. Mansfield gives an account of his three meetings with Damadian and the problems that Damadian was having.

The decision to exclude Damadian raises two questions. Firstly, why are the conclusions of the Nobel Committee, which at times have been more than a little curious, seen as the signifier of human excellence; and secondly, why was Damadian, when he is seen to have contributed so much, excluded? The controversy has been discussed by the radiologist Morton Meyers in a book perceptively entitled *Prize Fight: The Race and Rivalry to Be First in Science*.[43] Meyers shows that scientists are not simply disinterested individuals working for the good of humanity, but that recognition is desired, both in personal reputation and financially, and that competitiveness and resultant acrimony are not limited to one research group or nationality but rather are inherent components of the race to scientific fame. Meyers notes that Lauterbur worked closely with a patent attorney, and that Mansfield benefitted so much from his patents that he was able to buy a new MRI scanner for Nottingham. There is cooperation in science, but also considerable competition for grant money and recognition. Meyers noted that in his 1973 *Nature* article, Lauterbur did not give as a reference the 1971 *Science* article by Damadian. In later years, Lauterbur did give various explanations for this significant omission; however, it is recognised that the paper in *Nature* resulted in a lifelong enmity between Damadian and Lauterbur. Meyers, however, indicated that Lauterbur had another major reason for excluding Damadian's paper. In order to get a patent, Lauterbur had to show that his work was novel and that there was no significant prior art, and this would not be the case if Damadian had been referenced. It is notable that Stuart Young in his historical review of NMR imaging in his 1984 book on basic principles of NMR does not mention Damadian and references to Lauterbur and his 1973 *Nature* paper several times.[44] While Damadian received many awards in his long career, his exclusion from the 2003 Nobel Prize resulted his initiating in a major newspaper campaign costing an estimated $1,200,000. Damadian was determined to prove the justice of his position, yet in spite of his and the FONAR Corporation's ability to win major awards in patent infringement cases, he was prevented from winning the coveted Nobel Prize.

Whilst the acrimony between Damadian and Lauterbur was well known within the scientific community, the viewing of the full-page newspaper advertisements would leave the public with few illusions about the nature of scientific rivalry. The controversy resurfaced at a "dramatic plenary session" at the 2014 European Congress of Radiology (ECR) when the "bitter conflict" resurfaced.[45] Following the presentation by Morton Meyers when he stated that Damadian should have been the third recipient for the Nobel Prize, there was an audience comment by Peter Rinck, the chairman of the European Magnetic Resonance Forum, who accused Damadian of plagiarising the work of Erik Odeblad. Indeed, Damadian never referenced Odeblad original findings. In 1955, Erik Odeblad (1922–2019) and Gunnar Lindström from Stockholm published their first NMR studies, including relaxation time measurements of both living cells and excised animal tissue.[46] Xia and Stilbs have shown that it is quite certain that it was Odeblad and Lindström who published the first biomedical study using NMR in 1955. They used a primitive NMR instrument that Lindström had built for his graduate research and using it Odeblad and Lindström studied the characteristics of the NMR signal in calf cartilage. They speculated that the signal differences between water and biological tissues could be attributed to the absorption and organisation of the water molecules to the proteins in the tissue, and Xia and Stilbs show that this was remarkably accurate.[47] Odeblad was a major pioneer of NMR in medicine, and in recognition of this, in 2012 he received European Magnetic Resonance Award in a special ceremony in Umeå, Sweden. On this occasion, the two prize categories for Basic Science and Medical Sciences were combined into a single award. In spite of this award, it is worth asking why the work of Odeblad has been largely forgotten. Following the ECR lecture, Hans Ringertz, who had been ECR president in 1997 and who had chaired the voting committee for the 2003 Nobel Prize, expressed his frustration that there were still those who were campaigning for Damadian. The Nobel Prizes have often been controversial, including the first Nobel Prize for Physics in 1901 when Philip Lenard was resentful about the award being made solely to Wilhelm Conrad Röntgen for the discovery of X-rays.[48] For many controversies, it is only with the passage of time that the dust is allowed to settle.

MRI AND ANATOMY

William Morton in 1896 had seen that the radiological examination of the body could produce more information than could be found in either the anatomy theatre or in the pathology department.[49] If this was the case for plain film radiography, it was even more the case for MRI. The issues that were encountered in early MRI were similar to those seen by early radiologists[50] and included a new understanding of anatomy, the presence of anatomical variants and congenital anomalies, and the correlation of abnormalities to clinical findings. This was fully appreciated by Paul Lauterbur who wrote a perceptive foreword to *Nuclear Magnetic Resonance Imaging in Medicine* in 1981 when NMR was transforming into MRI.[51] Lauterbur wrote that to peer within the human body, examining internal organs with the same ease and delicate discrimination with which we can look at the skin, and with the same freedom from hazard, had not until recently been even a dream. To see the internal equivalents of a flush, a discolouration, a slight puffiness, or a quiver had seemed obviously impossible, although radiologists long ago had learnt to discern many subtle differences in their X-ray shadows. Lauterbur realised that everything was now different; new techniques, born of scientific cleverness and digital computers, had stimulated the physicians' dreams and were rapidly changing the procedures, the ambitions, and the economics of diagnostic medicine. Lauterbur is seen to complement the words of Morton. William Morton would have been delighted and enchanted by the applications of NMR to medical imaging, and whilst the early NMR images are of low resolution, the images now achieved routinely are of anatomical quality.

CONCLUSION

The development of NMR and MRI has taken place over a considerable period of time with many individuals advancing knowledge. It was in 1159 that Johannes Parvus, known as John of Salisbury (1120–1180), wrote in his *Metalogicon*:

> Bernard of Chartres used to say that we are like dwarfs on the shoulders of giants, so that we can see more than they, and things at a greater distance, not by virtue of any sharpness of sight on our part, or any physical distinction, but because we are carried high and raised up by their giant size.

This truth is well illustrated in the development of MRI, and we may thank the pioneers for their contributions. And as Paul Dreizen said, it is sad that this important scientific discovery with direct human benefit is marred by controversy, and sadder yet that the Nobel award has exacerbated rather than settled an unnecessary controversy.[52]

NOTES

1. Thompson, S.P. 1891. *Gilbert of Colchester, and Elizabethan Mag*netizer. London: The Chiswick Press.
2. Bailey, C. (Trans.) 1929. *Lucretius on the Nature of Things*. Oxford: The Clarendon Press.
3. Peter Peregrinus of Maricourt. 1902. *The Epistle of Peter Peregrinus, on the Magnet*. Trans. S.P. Thompson. London: The Chiswick Press.
4. Thompson, S.P. 1903. William Gilbert and terrestrial magnetism. *Geographical Journal*, 21, 611–618.
5. Gilbert, W. 1958. *De Magnete*. Trans. P. Fleury Mottelay. New York: Dover.
6. Gilbert, W. 1900. *William Gilbert of Colchester, Physician of London, on the Magnet, Magnetick Bodies also, and on the Great Magnet the Earth*. Trans. S.P. Thompson. London: The Chiswick Press.
7. Young, S.W. 1984. *Nuclear Magnetic Imaging: Basic Principles*. New York: Raven Press.
8. Tolstoy, I. 1981. *James Clerk Maxwell: A Biography*. Edinburgh: Canongate.
9. Jonson, B. 2000. *The Magnetic Lady*. Ed. P. Happé. Manchester: Manchester University Press.

10. Walmsley, D.M. 1967. *Anton Mesmer*. London: Robert Hale.
11. Frankau, G. 1948. *Mesmerism by Doctor Mesmer (1779)*. London: Macdonald.
12. Miller, J. 1983. A Gower street scandal. *Journal of the Royal College of Physicians of London*, 17, 181–191.
13. Elliotson, J. 1839. *The Principles and Practice of Medicine*. London: Joseph Butler.
14. Moore, W. 2017. *The Mesmerist*. London: Weidenfield & Nicolson.
15. Elliotson, J. 1843. *Numerous Cases of Surgical Operations without Pain in the Mesmeric State*. London: H. Baillière.
16. The Zoist. *A Journal of Cerebral Physiology and Mesmerism and Their Application to Human Welfare*. London: Hippolyte Baillère.
17. Dods, J.B. 1886. *The Philosophy of Mesmerism and Electrical Psychology*. Ed. J. Burns. London: James Burns.
18. https://www.historyofhypnosis.org/james-braid/ (accessed 16 January 2019).
19. MacFarlane, R. (Ed.). 2017. *A Practical Course in Personal Magnetism*. Wellcome Collection/Profile Books, London.
20. https://crystal-information.com/encyclopedia/magnetite-lodestone-properties-and -meaning/ (accessed 9 December 2018).
21. Rabi, I., Cohen, V.W. 1933. The nuclear spin of sodium. *Physical Review*, 43, 582.
22. Rabi, I., Cohen, V.W. 1934. Measurement of nuclear spin by the method of molecular beams: the nuclear spin of sodium. *Physical Review*, 46, 707–712.
23. Hahn, E.L. 1990. NMR and MRI in retrospect. In: *NMR Imaging*. Eds. P. Mansfield and E.L. Hahn. London: The Royal Society.
24. Alvarez, L.W., Bloch, F. 1940. A quantitative determination of the neutron moment in absolute nuclear magneton. *Physical Review*, 57, 111–122.
25. Bloch, F., Hansen, W.W., Packard, M.E. 1946. Nuclear induction. *Physical Review*, 69, 127–129.
26. Purcell, E.M., Torrey, H.C., Pound, R.V. 1946. Resonance absorption by nuclear magnetic moments in a solid. *Physical Review*, 69, 37–38.
27. Andrew, E.R. 1984. A historical review of NMR and its clinical implications. *British Medical Bulletin*, 40, 115–119.
28. Bates, S., Beckmann, E., Thomas, A.M.K., Waltham, R. 2012. *Godfrey Hounsfield: Intuitive Genius of CT*. London: The British Institute of Radiology.
29. Damadian, R. 1971. Tumor detection by nuclear magnetic resonance. *Science*, 171, 1151–1153.
30. Damadian, R., Goldsmith, M., Minkhoff, L. 1977. NMR cancer. XVI FONAR image of the living human body. *Physiological Chemistry and Physics*, 9(9), 7–108.
31. Kleinfield, S. 1985. *A Machine Called Indomitable*. New York: Times Books.
32. Lauterbur, P.C. 1973. Image Foundation by induced local interactions: example employing nuclear magnetic resonance. *Nature*, 242, 190–191.
33. Dawson, M.J. 2013. *Paul Lauterbur and the Invention of MRI*. Cambridge: The MIT Press.
34. Mansfield, P., Maudsley, A.A. 1976. Planar and line-scan spin imaging by NMR. *Proceedings of the XIXth Congress Ampère*, Heidelberg, 247–252.
35. Worthington, B.S. 1983. Clinical prospects for nuclear magnetic resonance. *Clinical Radiology*, 34, 3–12.
36. Bydder, G. M. 1983. Clinical aspects of NMR imaging. In: *Recent Advances in Radiology and Medical Imaging 7*. Ed. R.E. Steiner. Edinburgh: Churchill Livingstone, pp. 15–33.
37. Doyle, F.H., Gore, J.C., Pennock, J.M., Bydder, J.M., Orr, J.S., Steiner, R.E., Young, I.R., Burl, M., Clow, H., Gilderdale, D.J., Bailes, D.R., Walters, P.E. 1981. Imaging of the brain by nuclear magnetic resonance. *The Lancet*, ii, 53–57.

38. Young, I.R., Hall, A.S., Pallis, C.A., Legg, N.J., Bydder, G.M., Steiner, R.E. 1981. NMR imaging of the brain in multiple sclerosis. *The Lancet*, ii, 1063–1066.
39. Kelly, J.A. 2008. *Magnetic Appeal. MRI and the Myth of Transparency*. Ithaca: Cornell University Press.
40. Radda, G.K., Bore, P.J., Rajagopalan, B. 1983. Clinical aspects of 31-P NMR spectroscopy. *British Medical Bulletin*, 40, 155–159.
41. Alger, J.R., Shulman, R.G. 1983. Metabolic applications of high-resolution 13-C nuclear magnetic resonance spectroscopy. *British Medical Bulletin*, 40, 160–164.
42. Mansfield, P. 2013. *The Long Road to Stockholm*. Oxford: Oxford University Press.
43. Meyers, M.A. 2012. *Prize Fight: The Race and the Rivalry to Be First in Science*. New York: Palgrave Macmillan.
44. Young, S.W. 1984. *Nuclear Magnetic Imaging, Basic Principles*. New York: Raven Press.
45. Ward, P. 2014. Bitter conflict over recognition for discovery of MRI resurfaces in dramatic plenary session. *ECR Today*, March 10, 1.
46. Rinck, P.A. 2012. Europe celebrates the forgotten pioneer of MRI – Dr. Erik Odeblad. *Rinckside*, 23(6), 11–12.
47. Xia, Y., Stilbs, P. 2016. The first study of cartilage by magnetic resonance: a historical account. *Cartilage*, 7, 293–297.
48. Thomas, A.M.K., Banerjee, A.K. 2013. *The History of Radiology*. Oxford: Oxford University Press.
49. Morton, W.J. 1896. *The X-Ray or Photography of the Invisible and Its Value in Surgery*. London: Simpkin, Marshall, Hamilton, Kent & Co. Ltd.
50. Thomas, A.M.K. 2016. Vesalius, Röntgen and the origins of modern anatomy. *Vesalius*, 22, 79–91.
51. Lauterbur, P.C. 1981. Foreword. In: *Nuclear Magnetic Resonance Imaging in Medicine*. Eds. L. Kaufman, L.E. Crooks, A.R. Margulis. Tokyo: Igaku-Shoin.
52. Dreizen, P. 2004. The Nobel Prize for MRI: a wonderful discovery and a sad controversy. *The Lancet*, 363, 78.

12 The Future

Making predictions for the future of current imaging technology, or for the development of new technology, is always going to be difficult. If we consider CT scanning as having been introduced in 1972, by the 1980s many were saying that CT was now dead, this being the result of the impressive developments in MRI technology.[1] This would have been a reasonable prediction since there had been no significant development of CT technology during the 1980s. However, in the 1990s, and perhaps unpredictably, CT scanning went through a renaissance with a period of rapid innovation. CT transitioned from the slow process of scanning individual slices into a rapid scanning of a complete volume as developed by Willi Kalender and others. Not only did multislice or spiral CT introduce technical advances, but it also brought in new clinical applications, particularly related to vascular imaging. Conditions that had previously been difficult to diagnose, such as aortic dissection, could be shown readily using the new scanners.

Tom Hills, who was the director of radiology at Guy's Hospital in London, presented a thoughtful paper on the future developments in radiology in 1971, the year before CT scanning was announced.[2] Hills noted that to introduce something really new, any attempt to improve on the current systems and methods must be abandoned. We will learn nothing from simply observing a modern department and studying the current methods. For example, if we look at a "flight of the imagination" made by Sebastian Gilbert Scott showing the London Hospital X-Ray Treatment Department in AD 2000, as rendered by the English ecclesiastical architect Adrian Gilbert Scott (1882–1963), we see a design very much of its time and bearing little resemblance to any development of 100 years later (Figure 12.1). Hills observed that there was a general belief among those most involved in any subject that they know all that there is to know, and that they cannot understand how further progress could be made. This is not unlike the attitude of many towards the end of the 19th century regarding both science and medicine. Hills noted that new techniques usually came as a response to clinical needs; however, he most perceptively said that it was essential to keep a close watch on any development in completely remote fields of engineering in the hope that they may have an application of use to us. Unknown to Hills, Godfrey Hounsfield was working on the new scanner at that time, and indeed scanned his first patient on 1 October 1971. Radiology was transformed by the EMI scanner; however, it was not transformed by radiologists who were seeking new developments, but rather by Godfrey Hounsfield who, having made a development, was looking for a medical application. It is salutary to observe that when the National Hospital for Nervous Diseases, as it then was, in London's Queen Square was approached by Hounsfield and EMI, the proposal for the new scanner was not taken seriously and the scanner went instead to Atkinson Morley's Hospital where James Ambrose and his colleagues were more receptive.

DOI: 10.1201/9780429325748-12

FIGURE 12.1 A flight of the imagination: the X-ray department of the London Hospital in the year 2000, as imagined by Sebastian Gilbert Scott and designed by Adrian Gilbert Scott. Author's collection. Gift of Michael Gilbert Scott.

The attitude at the end of the 19th century just prior to Röntgen's discovery is shown in the writings of the future radiology pioneer Charles Thurstan Holland (1863–1941) (Figure 12.2). Holland was a general practitioner in Liverpool and on 4 October 1895 gave his presidential address to the Liverpool Medico-Literary Society.[3] It is interesting to read Holland's words, when he says that

> In the case of operative surgery we have, I take it, almost reached the acme of the art. It is difficult to see in what way it can make any further great advances. Every part of the human body, both inside and out, can be subjected to operation with a minimum of risk and a maximum of benefit. And although no doubt improvements will be made in the manner of operating, and in many technical details, it is difficult to see where new operations and great advances are coming from, from the merely operative point of view. That advances in the practice of surgery will come, and are coming every day, and which will be far reaching in their effects, is obvious; but they are not advances, so to speak, in the way of new, unthought of operations.

Who today would have his self-assurance? Holland made major contributions to radiology in the next four decades, and he must have looked back on his words of 1895 with more than a little embarrassment.

And so, when an elderly and distinguished scientist states that something is possible, they are almost certainly right. When they state that something is impossible, they are almost certainly wrong. This is the first of Arthur C. Clarke's (1917–2008) three laws which he published in 1973.[4] As an example, Michael Marshall has written

FIGURE 12.2   Charles Thurstan Holland (1863–1941). Author's collection.

on Lord Kelvin's 1895 statement that heavier-than-air flying machines are impossible, and it was to be only a few years later that the Wright Brothers made their first manned flight.[5] However, in 1848, John Stringfellow (1799–1883) made the first unmanned flight using a monoplane having a lightweight wooden frame, a close fitting fabric cover, and powered by a steam-driven engine, and this machine successfully made short flights in an empty room in Oram's lace mill in Chard.[6] Stringfellow of Chard was therefore first to achieve recorded, powered flight (Figure 12.3). Lord Kelvin and Marshall were both unaware of Stringfellow's achievement, which illustrates the difficulties with communication before the Internet. Whilst Stringfellow's work is remarkable, in 1848 there was little need for an unmanned and unguided monoplane powered by a steam engine; however, he did show that powered flight was possible. Predictions therefore need to be made with caution, and what we predict may have already occurred.

Clarke's second law is that the only way of discovering the limits of the possible is to venture a little way past them into the impossible. Perhaps this law means that we should not have too many preconceptions about what is impossible, and so when we are approached by a Godfrey Hounsfield or a Torsten Almén, we should listen to them. A degree of wisdom is needed, and doctors should always err on the side of caution. The story of N-rays is salutary. Prosper-René Blondlot (1849–1930) was a senior and well-respected French physicist at the University of Nancy, and had worked on electromagnetic radiation. In 1903, he announced a novel form of radiation, the N-rays, and there was considerable scientific interest with many working on these new rays.[7] Neither William Crookes nor Lord Kelvin was able to repeat

FIGURE 12.3 John Stringfellow (1799–1883) holding his monoplane, outside Chard Museum, Chard. Author's photograph.

Blondlot's results and the N-rays were discredited. Blondlot wanted his new rays to be true and was probably biased in his experimental methods, observing what he wanted to see. Blondlot's advice on how to view the N-rays is interesting. He says that when viewing the screen or luminous object to observe the N-rays, no attempt at eye accommodation should be made. Indeed, the observer should accustom themselves to look at the screen just as a painter, and in particular how an impressionist painter would look at a landscape. Blondlot said that to attain this required some practice, and that it was not an easy task with some people never succeeding. When his images are observed (Figure 12.4), it is difficult to see anything other than smudges. Blondlot was obviously sincere, but sincerity is not enough.

Clarke's third law is that any sufficiently advanced technology is indistinguishable from magic. This is most certainly true. If Tom Hills from 1971 could see modern medical imaging, then he would obviously be fascinated but he would not view it as magical. However, someone from 100 years earlier in 1871 would find modern technology entirely inexplicable, and would not have any theoretical basis for explaining their observations.

Regarding the future, Clarke presented two hazards of forecasting, which he described as a failure of nerve and a failure of imagination. In reality, we have plenty of nerve and plenty of imagination. So often our imaginations for the future have proven as unfounded in fact as Blondlot's N-rays. There are many predictions that have not come to pass, and as an example in the 1960s, it was stated that by the

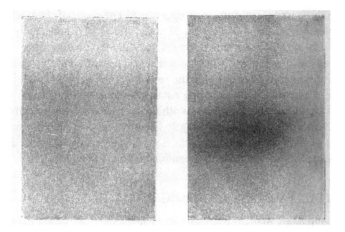

FIGURE 12.4 "N" rays: reproduction of photographs by Blondlot showing that "N" rays issuing from a Crookes tube are polarised. From Blondlot, 1905, see Note 7. Public Domain.

present day energy would be so cheap that it would not be necessary to meter it. Most developments take place in the application and development of known technology. CT and MRI were unexpected, as were the discoveries of X-rays and radioactivity. As Clarke has emphasised, the collapse of "classical" physics began with the discovery of X-rays in 1895, and led "into realms where no human mind had ever ventured before".

It is likely that the easy gains have been made, and that new discoveries will require ever-increasing resources. In 1895, Röntgen made his discovery working by himself, using a simple evacuated glass bulb and home-made apparatus. The discovery of the Higgs boson which was announced on 4 July 2012 needed the resources of the Large Hadron Collider at CERN and teams of researchers. CERN, the European Organisation for Nuclear Research, is the world's largest centre for particle research, and was founded by 12 European countries.[8]

The major challenges for radiology in the 2020s and beyond are related to the environmental impact of modern imaging, to equitable access to medical imaging, to the relationship of image and reality, to the nature of modern medicine, and to the role of the radiologist.

THE ENVIRONMENTAL IMPACT OF MODERN IMAGING

There is an increasing concern about the environmental impact of human activity, and radiology makes its own contributions. The traditional concerns over the release of silver into the environment from photographic processing are now in the past. In the UK, it is estimated that 62% of carbon emission sources in the NHS are the result of medicines, medical equipment, and other supply chain medicines. Many companies are taking seriously their environmental impact and publishing recommendations.[9] An innovative attitude towards X-ray equipment may result in better recycling or the possibility of reuse or upgrading. The reuse of medical equipment can be problematic because of the risk of cross-infection. Traditionally, much was re-sterilised and reused, including syringes and angiographic needles – these have been replaced by single-use disposable units because of concerns over infection. Having appropriate radiology closer to the patient's residence will result in fewer and shorter patient journeys to the imaging centre. A cultural change within radiology departments should be easy to achieve with the avoidance of waste, and the encouragement and rewarding of environmentally friendly practices, and the summer 2021 magazine of the American Roentgen Ray Society has an article on the contribution of radiology to climate change with recommendations for actions.[10]

THE EQUITABLE ACCESS TO MEDICAL IMAGING

Very-high-technology medicine in centres of excellence may benefit the few, but it is only when the technology is widely available that there will be an impact on public health. How the technology is used is as important as the nature of the technology, as was shown with the introduction of the EMI CT scanner. The CT scanner had its maximum impact when the scanner was available in the district hospital and not just in the neurosciences centres. Modern technology is expensive and providing equitable access is not easy. There is a difference between urban and rural communities and persistent rural–urban disparities continue to raise concerns regarding access to treatment.[11] As Jan Eberth from South Carolina has documented, living in a rural area is associated with cancer diagnosis at a later stage, and also with inappropriate and/or underuse of treatments, and an overall poorer survival.[12] This has been a concern for a long time, and in 1937 R. Maitland Beath from Belfast was expressing his concerns about how unreasonable it is that someone might be denied the advantages of radiology simply because they live in a remote country district.[13] The use of medical imaging should not be restricted to the wealthy and privileged. As FitzPatrick and his colleagues have stated, physicians have a duty to tackle the widening health inequalities in society, and should resemble the population that it serves, and this needs to be addressed in both undergraduate and postgraduate curricula.[14] Radiology needs to be a part of this process.

THE RELATIONSHIP OF IMAGE AND REALITY

The medical image needs to be viewed with caution and no small degree of wisdom, and as radiological technology has advanced, the problem has perhaps intensified. What is the difference between the patient and the image of the patient, and what is the relationship between reality and virtual reality? The painting *La Trahison Des Images* (The Treachery of Images) of 1929 by Belgian surrealist artist René Magritte (1898–1967) is well known and appeared in multiple versions indicating its significance for the artist. In his work, Magritte explored the relationship between reality and representation. The 1929 version of *La Trahison Des Images* is a painting of a pipe with the words "Ceci n'est pas une pipe" (This is not a pipe). Michel Foucault (1926–1984) has written about the first version of 1926, which was a drawing of a pipe. Foucault sees each element as holding an apparently negative discourse, since it denies, along with resemblance, the assertion of reality that resemblance conveys. The image is basically affirmative, and affirms the simulacrum, and is an affirmation of the element within the network of the similar.[15] Such images are designed to upset our complacent self-assurance in the logic of a stable and predictable world. Figure 12.5 is a homage to Magritte showing a head MRI with the words "Ceci n'est pas un cerveau" (This is not a brain). The MRI scan is not a brain, but then neither are functional images. Magritte was interested in words and their relationships to images and then to the object itself, for example in his *Les Mots et Les Images* (Words and Images) of 1937. For the radiologist, there is a complex relationship between the patient, the image of the patient, and the words about the image, that is, the radiological report. René Magritte can be seen as the key artist for radiologists.

The introduction of photography in the 19th century changed the nature of human vision as Corey Keller, the curator of photography at the San Francisco Museum of Modern Art, affirms, and contributed to a radical re-evaluation on how we look at reality.[16] Her book is the catalogue to the exhibition *Brought to Light, Photography and the Invisible 1840–1900 that* was presented at the Albertina gallery in Vienna in 2009, coinciding with the European Congress of Radiology of that year. The X-ray images in the Albertina were viewed actively by visitors, and their meanings were sought and were not necessarily obvious on first viewing. The modern radiological images presented in the conference centre were viewed far more passively, and their significance was assumed both by the speakers and the radiological audiences. Keller comments that the certain faith that we used to have in the eye has been replaced by near-total dependence on a technologically inflected vision. This results in a disconnection between seeing and knowing, and in a strange way the image may appear more real than the reality. In the medical environment, is the patient being treated or only the image of the patient? What is the relationship between physical examination and radiological examination, and is one more true than another? Is physical examination even necessary before medical imaging, as occurred during

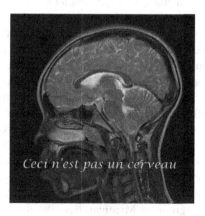

FIGURE 12.5   Ceci n'est pas un cerveau (This is not a brain). A homage to René Magritte. Author's artwork.

the time of COVID-19 pandemic where there was commonly a virtual clinical consultation followed by imaging? In many clinical scenarios, MRI has been used to replace clinical assessment and physical examination, for example when a surgeon requests an MRI or a CT scan before the clinical consultation.

That medical imaging could show an abnormality in an otherwise well patient with no physical signs was a difficult concept in the early days of radiology. The chest radiograph might reveal the presence of tuberculosis in a patient with no physical signs and no symptoms, and this was difficult for both the physician and patient to accept, with the image showing an unexpected reality.

It is therefore always useful to consider the relationship of image to reality, and to consider if it is indeed the case that medical imaging merely depicts the inside of the body in a photographic manner, thereby rendering the body transparent.[17] Medical imaging from its earliest days has done more than simply reveal anatomy and pathology, as it has also profoundly changed how we view ourselves, how we perceive disease, and how doctors and patients interact. The contemporary patient now expects an MRI scan for musculoskeletal pain, and indeed may not feel that they have been adequately examined or treated without an obligatory scan. The image replaces reality. As medicine becomes progressively digitised, the ability to interact with and interpret patient-derived data will become increasingly more important. However, this is only the intensification of a problem that has existed since the introduction of medical imaging. In 1912, Alfred Barclay (1876–1949) with fellow Manchester radiologist William Bythell (1872–1950) commented on the need for doctors to be educated in radiology in order to know when radiology would be of help.[18] Of particular interest is their comment that the rays were a new sense added to the other five; and did not relieve the doctor from the necessity of training in the use of the old-established methods of observation that may be used equally well at the cottage bedside or in the wards. Bythell and Barclay state that information derived from medical imaging becomes more useful when combined with the information available from every source. An intelligent combination of clinical findings with the radiological results will lead to optimum clinical care, no matter what techniques may develop in the future. Perhaps one of the greatest dangers of medical technologies is when they become isolated from the patient and the actual problem that the patient has. So, the MRI scan does not necessarily reveal the truth about the body as might be initially believed. Whilst the contemporary radiological image might be considered as showing anatomy in more realistic manner than the earliest radiographs, to some extent the reverse is the case. The earliest images were directly obtained using a simple photographic plate or film in direct contact with the patient, and were of the same size as the patient. The size of the contemporary image bears no necessary relationship to the size of the patient. The fluoroscopic image on the monitor is also more removed from directly viewing the patient on the early fluorescent screens.

With the advent of virtual reality (VR) it should be questioned how long viewing a radiological image presented on a flat screen will continue, even if that image is presented in 3-D. The group in Michigan have developed a VR tool to educate patients about MRI and to simulate the experience of actually being scanned. The tool is totally immersive and incorporates both visual and auditory sensations that the patient might encounter during a scan.[19] Entering a virtual world in an immersive fashion has been prefigured in films such as *Lawnmower Man* of 1992 directed by Brett Leonard, and the more recent *Ready Player One* of 2018 directed by Steven Spielberg. Considerable effort is currently being made to build the next computing platform, which is "a new phase of interconnected virtual experiences using technologies like virtual and augmented reality".[20] At its heart of the concept of the metaverse is that by creating a greater sense of virtual presence, that interacting online will become much closer to interacting in person. A future scenario could, and probably will, exist where an MRI or CT dataset is loaded into the metaverse, and the radiologist puts on a VR headset and VR suit instead of scrubs and examines the patient's virtual presence in a similar fashion to how the traditional physician performed a physical examination in the real world. And as time progresses the virtual world and physical world will become increasingly blurred as predicted by Ray Kurzweil. In *The Age of Spiritual Machines* the relationship between human and computers is considered by Kurzweil.[21] A more direct linkage of the human and computer is predicted with the use of brain chips, of perhaps even a direct physical link into the nervous system as depicted in film *eXistenZ* of 1999 and written by David Cronenberg. What is then more real, actual reality or virtual reality, and how can we tell the difference? We can easily have a scenario when virtual reality is so much more interesting and vibrant than the real world, as shown in *Ready Player One*. It is noteworthy that radiographs were initially called skiagrams or shadowgrams, and we must always remember that shadows, no matter how technologically advanced, are not reality. In Plato's allegory of the cave we are prisoners and see the world as shadows cast on a wall rather than directly appreciating reality.[22] Plato shows the prisoners as leaving the cave and passing to the upper world and then seeing directly for themselves. When we become immersed in VR it is as if we re-enter a cave world of shadows. This will not necessarily be good, particularly if we forget that they are shadows, and at some future point humanity will need to come out of the virtual neoplatonic cave and enter the real world again.

THE NATURE OF MODERN MEDICINE

The history of medicine and radiology since 1945 has been one of the major developments and improvements previously not imaginable. The changes in radiology since 1970 have been dramatic and have transformed investigative medicine. It might be thought that the history of medicine in general, and the history of radiology in particular, might be presented as one of continued development and improvement resulting in increasing optimism for the future. That this is not the case is a paradox and needs some explanation. One of the first to raise a concern was Ivan Illich (1926–2002) who was a critic of both institutional education and institutional medicine. His book *Medical Nemesis* was published in 1975 and was profoundly influential.[23] Illich saw the medical establishment as a major threat to health, and was concerned about the disabling impact that professional control was having on health. The traditional doctor and teacher were artisans interacting with a known person. Illich saw the doctor as becoming a technician applying scientific rules to classes of patients.

Just as teachers are now instructed what to teach by a defined national curriculum, so doctors are now informed how to treat their patients by defined protocols. One model is not necessarily better or worse than another, but we must be aware that the paradigm has changed. Other voices include Vernon Coleman who has written many books critical of modern medical practices.[24] Coleman describes an increasing lack of trust of doctors, laying the blame on outside influences such as drug companies, politicians, and lawmakers, resulting in a situation where trusting your doctor may be hazardous to health. Coleman also states that one-fifth of radiology examinations are unhelpful. Whilst Illich and Coleman may overstate their cases, they represent a trend. The influential book by James Le Fanu with the message in its title, *The Rise and Fall of Modern Medicine*, is more difficult to ignore (Figure 12.6).[25] In his chapter on technology's failings, Le Fanu discusses the 1980s when there were major developments in radiology. Nonetheless, and against the background of the remarkable innovations, Le Fanu presents a general and probably correct perception of a medical technology that is out of control. He then discusses over investigation (the overuse of diagnostic technology), various false premises and promises, and the role of intensive care in needlessly prolonging the process of dying. Le Fanu presents this over-investigation with the performance of large numbers of tests in patients whose medical problems are straightforward, as far from being a trivial matter. The tests and investigations are costly, and adds an alien element into the medical consultation, replacing wisdom and experience with a spurious objectivity. Illich's use of the expression medical nemesis is worth exploring. In the religion of the ancient Greeks,

FIGURE 12.6 James Le Fanu, as the 2010 Orator of the Osler Club of London. Author's photograph.

we find Nemesis as the goddess who administers retribution against those who suffer from hubris, that is, an arrogance before the gods. The name of Nemesis also has a meaning of "distribute, allot, apportion one's due" and "just indignation, righteous anger".[26] Medical nemesis then becomes a direct result of medical hubris. Hubris can be seen as a condition of excessive pride or one of a dangerous overconfidence, and primarily one of acting against the natural order of the world. Medical pride carries with it the seeds of its own failure, and at least a failure to live up to the expectations that the public has been promised.

In the UK, the British Medical Association has led campaigns to avoid overdiagnosis,[27] with a particular focus on radiological investigations.[28] The "Too Much Medicine" campaign is designed to highlight medical conditions in which overdiagnosis exposes patients to unnecessary and potentially harmful care.[29] As many more radiological examinations are performed with a lower threshold for referral, the interpretation of the results becomes ever more difficult. The campaign particularly highlights incidental radiological findings which would have not been detected on traditional studies. Le Fanu may also be overstating his case, but to what extent is public health improved by ever-increasing volumes of medical imaging? The question remains, is medicine still good for us? This is the title of the thoughtful 2019 book by Julian Streather.[30] Streather whilst noting that modern does the most remarkable things, holds that for all its wonders that there is a gathering sense that medicine is heading in the wrong direction. And so, change is both appropriate and necessary, however will also be difficult to implement. There are many vested interests in keeping the medical model as it stands.

In part, the problem with the growth of high-technology medicine is that it can lead to a neglect or distrust of the older and simpler skills. In his "Letter from the Editor" of 1981 in his influential journal *Seminars in Roentgenology*, Benjamin Felson describes how the clinician in order to save time or money will commonly go straight to special procedures whilst ignoring conventional studies.[31] Felson commented somewhat dryly that this omission may prove detrimental to the patient, and discussed appropriate pathways. Felson described patients who had been subjected to special procedures which would have been superfluous had the initial conventional radiographs been carefully assessed. A similar point was made by Charles Essex in 2005.[32] Essex was a consultant neurodevelopmental paediatrician, and said that the hardest things to do in medicine is to say nothing (or "I don't know") and to do nothing. This brings to mind the words of Socrates as recounted by Plato who said "I know that I know nothing". This is the Socratic paradox, and true wisdom is to recognise our own ignorance. The temptation for the doctor is to perform yet another test. And Essex describes the patient as being caught in a web of further investigations, referrals, and sometimes treatment before finally being recognised as healthy, which is what they were in the first place. With a colleague, Essex reviewed the children on the waiting list for a head MRI scan and found that a half of them had not even had their head circumference measured, which is a basic indicator of brain growth. The expensive MRI scans had been requested, not a minor event for a child, and Essex noted that few of these scans would alter clinical management, and some would lead to further referrals and investigations because of results that would turn out to be normal variations. Parental anxiety would certainly be increased. Performing a small number of investigations for strong indications with a high likelihood of a relevant abnormality is different from preforming a large number of investigations for weaker indications. If a large number of tests are performed, the significance of abnormal findings is more difficult to interpret. In the normal distribution curve, the upper and lower 2.5% are usually described as abnormal, and as a result 5% of the results are called abnormal even though the patients could be healthy. If the doctor requests an unnecessary examination, what is the response if the results are slightly outside the normal range? The temptation is to do another test to be on the safe side. This results in further tests and outpatient appointments. Essex sees such needless procedures as a side effect of unnecessary investigations and an uncritical clinical practice. In the past, the CT scan used to be performed only if a request form was signed by a consultant as a means of limiting access to a valuable resource, and now the CT scan is now part of a series of tests to assess a confused patient. Andrew ElHabr and colleagues in Atlanta reviewed the increasing use of emergency department neuroimaging from 2007 to 2017, and found that age-adjusted rates increased by 72% overall (with a compound annual growth rate of 5%).[33] They concluded that the growth of head and neck CT angiography far outpaced other imaging modalities, and that the unenhanced head CT scan remained the dominant emergency department neuroimaging examination. The authors state that the appropriateness of this growth should be monitored as the indications for CT angiography expand. The use of MRI also increased and the study concluded that conscious or unconscious bias regarding financial incentives might have influenced imaging decisions. The study's findings suggested that an older and sicker population was being investigated more intensively. There was no comment regarding any impact that this large increase in medical imaging was having on patient outcomes, or indeed if overall outcomes for the patients during the course of the study changed.

In 1976, Thomas McKeown examined the death rates from the 1840s to the present time.[34] It had previously been thought that the development of modern medicine had been the cause of the dramatic fall in mortality during that period. McKeown's thesis was that the fall in mortality and increase in life expectancy was the result of social changes, and he stated that external influences and personal behaviour were the predominant determinants of health. If we consider pulmonary tuberculosis as an example, then there was an incidence of 4,000 per million of population in 1838, falling to 350 per million by 1945 when drug treatment was introduced, and falling still further by 1960. Mass miniature radiography as population screening was introduced at a time when tuberculosis rates were already falling. So, it needs to be asked to what extent therefore did radiological screening significantly influence the rates of tuberculosis? Le Fanu notes that 92% of the fall in tuberculosis rates was secondary to social factors, and only 8% to the use of antibiotics. This is supported by the views of the Canadian cardiothoracic surgeon Norman Bethune (1890–1939) who was an early supporter of socialised medicine. Bethune maintained that whilst tuberculosis might be caused by *Mycobacterium tuberculosis*, it is equally caused by poverty and poor living conditions.[35] It was Denis Burkitt (1911–1993), whilst working in rural Africa, who noted that certain diseases were characteristic of modern

Western civilisation, but were unknown in communities that had deviated little from their traditional way of life,[36] and that a rise in their frequency followed the adoption of Western customs. Further support was given to Burkitt's thesis by the fact that such diseases were rare or uncommon in the West a century earlier. Therefore, to approach the treatment of disease with simply a medical model will not be successful, although medical intervention for the individual patient is obviously necessary.

At some point, the increasing radiological activity will become unsustainable, either for financial reasons or for a lack of staff even when augmented with artificial intelligence (AI). Whilst AI is being presented as an aid to managing an increasing workload, the reverse is likely to be the case, since in any given system if the efficiency increases, then activity will increase. This is the paradox stated by the economist William Stanley Jevons (1835–1882), after whom it is named. In the 1860s, there had been contemporary concerns about the availability of coal, and measures were undertaken to increase the efficiency of its use. Jevons observed that technological improvements to increase efficiency paradoxically led to the increased consumption of coal. The Jevons' paradox would predict that as the availability of medical imaging has expanded with an increase in efficiency, there would be a concomitant increase in activity, and this is exactly what has been observed. It should be remembered that from the earliest times, radiologists have been complaining about inadequate resources. In 1897, the Electrical Pavilion opened at the Glasgow Royal Infirmary only shortly after the X-rays had been discovered. The X-ray rooms were fitted with state-of-the-art apparatus, and electricity was supplied throughout the hospital. The radiological services included plain film radiography, radiotherapy, foreign body localisation, and stereoscopic radiography. By 1903, the departmental head John Macintyre (1857–1928) was already exclaiming that the new buildings of 1897 were being fully utilised and that the staffing numbers had needed to be increased to cater for the increasing numbers of patients.[37] The Electrical Pavilion had been built as large as possible in 1897, and so it's interesting that by 1903, the demands upon it were far in excess of what had been anticipated, or that could then be accommodated. Like many radiologists since, Macintyre requested that "still greater facilities will be placed at the disposal of the staff". In June 1970, a three-day conference on "The Future of Diagnostic Radiology" was held in London.[38] The participants were 12 British and 14 American radiologists. After discussions on staffing and funding, it was concluded that in order to achieve excellence in patient care as well as in academic pursuits, higher radiology staffing levels were necessary than were then available in most British and many American hospitals. This was at a time before the huge changes that took place in radiology in the 1970s and 1980s, and subsequently. It is remarkable that the meeting said that consideration should be given to having technicians and "super-technicians" handle some of the simpler tasks presently performed by radiologists. This extended role has been successfully implemented in the UK with, for example, the introduction of clinical reporting by radiographers starting with the reporting of plain films.[39] The delegation of reporting by radiologists to radiographers had been proposed by the Royal College of Radiologists in 1994.[40] With the development of modern imaging, the pressure on reporting has continued to increase. In 2021, the Royal College of Radiologists in the UK published its workforce census[41] and noted that by 2025 the clinical radiology workforce shortfall will be 44%, nearly half of hospital trusts will not have the staff or transfer networks needed to provide safe interventional radiology care, and demand for complex medical imaging will be growing faster than the clinical radiology workforce. The clinical radiology workforce is said to be operating at two-thirds of adequate capacity and needs at least another 1,939 consultants just to keep up with pre-coronavirus levels of demand for scans and surgery. It is the case that at least as far back as 1903, the demands on radiology have been in excess of what could be accomplished. It may be the case that now is the time to review the current model for providing radiology services and it may be useful to reconsider earlier models. Having a separate radiology department with radiologists who are trained in radiology but not in a clinical specialty may not be the best model of care. The early radiologists were firstly clinicians who then became interested in radiology. Florence Stoney was first of all a clinical doctor and only later became a radiologist.[42] In the 1920s at Guy's Hospital in London, there were separate surgical and medical X-ray departments, and the senior surgical radiologist was a qualified surgeon.[43] Traditionally, many of the invasive procedures were performed by a physician or surgeon and were then reported by a radiologist. If the imaging services were to be devolved to the clinical departments with separate or shared facilities, this might encourage a better match between requests and resources. One junior surgeon could decide to specialise in minimally invasive interventional radiology, another in MRI, and a third in operative surgery. The professional lines are already blurred between surgeons and radiologists involved in intervention, and radiology rooms are commonly shared with non-radiologists who use imaging for diagnosis or treatments.

Part of the concerns about modern medicine is many believe that it is based on a reductionist paradigm and on scientism, believing that only science can provide true answers. Essentially, doctors are not mechanics and the hospital is not a factory, and the factory model results in alienation for both staff and patients. The hospital easily becomes an alienated space, and this view is expressed in the Danish television series *The Kingdom* (Riget) created by Lars von Trier in 1994. The Riget Hospital is superficially a normal and rational hospital, and yet not far below the surface is irrational, supernatural, and deeply unscientific. The patients seem coincidental to the events that unfold, and the tension between the scientific and non-scientific, the rational and irrational can be profound. One of the most interesting and disturbing depictions of modern medicine is the film *The Doctor* of 1991 directed by Randa Haines. William Hurt plays an insensitive heart surgeon who undergoes a transformation when he becomes ill himself, and enters the hospital system. The heart surgeon finds his heart and starts to see the patients as fellow human beings, and not simply as grist to the hospital mill. Aidan Halligan has reviewed the importance of values in healthcare.[44] Halligan reflected on the attitudes of kindness, caring, good communication, honesty, reliability, and trust: the interpersonal parts of doctoring, which are critical to patient perception. Halligan notes that these attitudes have been eroded through a "pernicious dilution by our preoccupation with the rise of scientific medicine", and noted a large survey from 2000 that reported that only 60% of doctors believed that those personal attributes were important to the point that if they were not done properly, something should be done about it. Forty percent of doctors did not believe that those attributes were important. Halligan then notes that those values struggle to survive in a culture which is overmanaged and under-led and the mania for setting targets has, over time, exerted a profoundly corrosive effect. It's not that targets are unimportant, but that they need to be used with wisdom. This has been further elaborated by Della Fish and others.[45] Fish describes two models of professional practice and associated values: the technical and rational view and the professional artistry view. Radiology has become increasingly wedded to the technical and rational view and this may have a harmful effect since it makes the professional purely about having efficient skills, within bounds that are set by other people. Perhaps this is part of the cause of the malaise that is seen in some radiologists.

The designs seen in our modern world have become impersonal, although there are some signs of change. In many ways, the Victorians were better at design than we are. The New Hospital for Women (NHW) in London, which became the Elizabeth Garrett Anderson Hospital, was built in 1889 of red and white brick, and was described as having a somewhat unprofessional air (Figure 12.7).[46] One of the patients described the NHW as more like a gentleman's house than a hospital. The internal decoration was the design and work of Agnes Garrett, the sister of the pioneer woman doctor Elizabeth Garrett Anderson (1836–1917), with the aim of providing a beautiful and restful environment. In 1874, Agnes Garrett with Millicent Fawcett had started the first all-woman interior decorating business A&R Garrett House Decorators. The four wards were decorated with Italian bas-reliefs, including

FIGURE 12.7 The New Hospital for Women in London, now part of the headquarters of the trade union UNISON. The ground floor contains a museum. Author's photograph.

casts from work by Luca della Robia and Donatello. For the poor women who attended the hospital, it must have seemed like a palace. In contrast, many modern hospitals and radiology rooms are impersonal. Horace Sweet in his *Handy Book of Cottage Hospitals* of 1870 describes

> a hospital near enough to (the) home for ease of visiting, and in familiar surroundings, but where the room is not overcrowded, but open to air and sunshine, with well cooked food, suitable medicines, peace and quiet and regular care where the chance of safe and speedy recovery is enhanced.[47]

To what extent are these values found in contemporary hospitals?

WELLNESS AND WELL-BEING

The future of radiology will be as much about the culture in radiology departments as about the nature of the technology. The contemporary hospital environment seems almost designed to produce staff stress. In the July 2021 RSNA News, Kevin Rees reviewed the current situation.[48] Tom Vaughan is a radiologist and the Chief Wellness Officer at Bayhealth Healthcare in Delaware. Vaughan notes that though the typical working shift for a radiologist was significantly longer even a decade ago, the overall stress and burnout levels are much higher today. Lim and Pinto have evaluated work stress, job satisfaction, and burnout in New Zealand radiologists in both public hospital and private practices.[49] In their study, they found that radiologists in the public hospital environment experience more work stress, a lower level of job satisfaction, and higher rates of burnout compared to private practice. There was a trend towards a higher rate of psychiatric morbidity among radiologists who practised in public hospitals. Lim and Pinto identified various aspects of work stress that were important to radiologists, so that they could be addressed to improve their mental health. The study raised implications for workforce planning, recruitment, and retention of radiologists in the public health system.

Vaughan attributed radiologist stress to work overload, a lack of breaks between cases, and also the risk of malpractice that could occur many years after a scan was reported. Part of the solution may lie in appointing wellness officers in radiology departments, or at least in discussing the problems openly.

Vaughan states that many radiologists struggle with isolation due to limited interaction with colleagues and patients, and that some radiologists feel their work lacks meaning. The modern radiologist is faced with an increasing volume of cross-sectional imaging, that is, CT and MRI scans, and the radiologist has seldom seen the patient, and was not usually present when the examination was performed,. The radiologist then reports using voice recognition while sitting by themselves into the radiology information system in a darkened reporting room, and often for whole sessions or days. In the 1970s, the pattern of radiologist working was different. Reporting was done by direct dictation to a secretary with a typewriter who sat next to the radiologist. There were very many more procedures with direct patient contact which have now been replaced. For example, intravenous urography involved meeting the patient, injecting the contrast medium, and finally reporting the study. There were many more staff in radiology departments in the 1970s and the digital environment made many jobs unnecessary. In the 1970s and 1980s the lunch hour was still commonplace, and hospital and community medical staff would often meet to eat and talk together. It was the character Gordon Gekko in the 1987 film *Wall Street* who uttered the phrase "Lunch is for Wimps". Lunch is now commonly eaten sitting in front of a computer screen, all adding to atomisation and isolation.

It is to be expected that as patient contact has reduced and reporting has become a solitary activity, there is a diminished sense of meaning. Vaughan noted that traditionally when surveys were undertaken on physician happiness, radiologists were near the top. Today radiologists consistently rank near the bottom. In the UK, the NHS has identified five steps to mental well-being.[50] These are to connect with other people, to be physically active, to learn new skills, to give to others, and to pay attention to the present moment (that is, mindfulness). It is the first two and the last one that are particularly problematic. The present moment has now become one of continuous reporting radiological studies but in an impersonal way, and sitting for long periods with limited human contact. Whilst reporting can be performed in a relatively isolated reporting room, the reporting may also be performed remotely at home, and this has been encouraged in the UK during the COVID-19 pandemic. This has been reviewed by the Informatics Adviser for the Royal College of Radiologists, who noted that there have been enormous cultural changes within the NHS regarding the use of technology for remote working.[51] The review presented these changes as having only long-lasting positive effects on NHS radiologists, and in the future as leading to more flexible remote working. The only hazard suggested was the potential loss of training opportunities. In the discussion following a session on remote working at the Annual Radiology Congress in the UK in 2021, it was stated that many radiologists now preferred to work from home. There is little discussion about the effect of remote reporting on mental health, and the consequences of increased isolation of radiologists from other staff, and the consequences of even less patient contact. These issues are important since changes in working practices will affect both recruitment and retention of radiologists. It is known that workloads in radiology departments have constantly increased in previous decades. Increasing stress and fatigue not only has an adverse effect on the individual radiologist, but will also potentially harm the patient since Jan Vosshenrich has shown that the resulting radiologist fatigue is considered a rising problem and one that affects diagnostic accuracy.[52] In a poll performed in the UK in April 2021, it was found that 41% of radiologists were moderately or severely demoralised in their jobs post-pandemic, and 12% considering leaving the UK's National Health Service.[53] This enforces the view that the future of radiology will be as much about how the service is provided than about new technical advances.

THE FUTURE OF INTERVENTION

The future of image guided interventions is difficult to predict. The 1966 film *Fantastic Voyage*, directed by Richard Fleischer, has previously been discussed. The submarine, the Proteus, which with its crew is shrunk to "about the size of a microbe" and injected into a sick scientist for the purpose of thrombolysis.[54] The dedication of the film to "doctors, technicians and research scientists, whose knowledge and insight helped guide this production", suggesting that the concepts might be fact one day. The film was shown to medical students in the 1970s. Today this film suggests the use of injectable nanotechnology for therapeutic aims.[55] Syringe-injectable biomaterials are currently receiving a great deal of attention as minimally invasive implants that can be used for diagnosis, therapy, and for regenerative medicine. The basic principles for access to the body have been established by Sven Seldinger and others, and the major challenge for the future will be whether an intervention should be performed, rather than if such an intervention is technically achievable. Early radiology was a *mememto mori* and showed us our mortality, and the Röntgen rays change a happy beach scene with carefree bathers into a grotesque scene of dancing skeletons, the dance of death. Neither modern medicine nor modern radiology can give us immortality, and indeed immortality in this world is traditionally seen as a curse and not as a blessing.

THE FUTURE OF IMAGING

It is perhaps too easy to present the history of radiology as one of continual improvement from simple beginnings to our modern excellence, and extending without a perceptible end into the future. Reality is more nuanced, and the concept of unceasing progress is relatively recent and originated in the Enlightenment. The two most prominent proponents of progress are Karl Marx (1818–1883) and Charles Darwin (1809–1882). In both classical Marxism and Darwinism, the passage of time is presented as producing a necessary improvement; however, the events of the 20th century have placed a large question mark over such 19th century optimism. The industrial revolution with its associated scientific and technological advances which led to the discovery of X-rays had profound environmental consequences, the significance of which we are only now realising. Whilst industrial development brought wealth and prosperity, at least to some, in South Wales it came at a price. The copper industry resulted in mountains of slag and toxic smoke.[56] Whilst there might have been jobs created for humans, the farmers complained about damage to crops and the death of livestock. Sidney Pollard in his study of progress sees the idea of progress as having a central position in our lives and as the source of our sense of meaning.[57] Pollard stated that our scientific history convinces us that our purpose, and our freedom of action, can be served only by discovering the laws of scientific development and following them. Pollard wrote these words in 1968, but who would agree today? Linear models of development have been dominant since the Enlightenment, and yet other models have existed. In writing a story or a history, a linear approach is not always the most appropriate. Part of the problem with a simple linear chronology is that it implies a development that is inevitable, and like a row of dominoes. And so one thing leads to another and to another and so on in a simple series. Paul Petter Waldenström (1838–1917) in telling his allegorical story of Squire Adamsson of 1862 has his hero taking a distinctly non-linear path. Mark Safstrom in his introduction to the book discusses the linear nature of the pursuit of truth in the scientific method of the Enlightenment and in the Judeo-Christian historical worldview.[58] Truth is seen as discoverable and absolute, and this view was emphasised following the Enlightenment with a reliance on the scientific method. Safstrom comments that this can be seen as creating a blindness, with philosophers and others believing that there is a vertical movement towards absolute truth, when in reality there is simply a movement from one paradigm to another. When one paradigm is replaced by another, is it helpful to say that one is more true than another? They are simply different viewpoints and in time the new viewpoint will also be replaced. Richard Rorty in his 1994 essay "Truth without correspondence to reality" discusses the pragmatic approach to truth and the confusion of truth and justification.[59] Rorty noted that it is actually those who state that pragmatists confuse truth and justification who are the ones who are confused. They present truth as being something towards which we are moving, and as something to which as we get closer, the more justification we have. The pragmatist thinks that there is much that can be said about justification to any given audience, but nothing can be said about justification in general. This is why for Rorty, nothing general can be said about the nature or limits of human knowledge, nor is there anything to be said about a connection between justification and truth. This is not because truth is atemporal and justification temporal, but because "the *only* point in contrasting the true with the merely justified is to contrast a possible future with the actual present" (Rorty's emphasis).

In looking at the history of the radiological sciences, various models and paradigms have been used over the decades. There is an objective reality that we may observe, with the limiting factor being our perception. Reality is more complex than simple models. The two fictional characters, Dr. Xavier and David Winn are both driven mad by their X-ray vision, and both illustrate that at some point science may show us a depth of reality that is no longer comprehensible. As radiology advances further into technological solutions, will artificial intelligence or radiomics cause more problems than they solve? Radiomics was introduced as a term in 2012, and translates medical images into quantitative data to yield biological information for diagnosis, therapy, decision support, and monitoring.[60] At some point, it's possible, and even likely, that the technology will show a radiological reality that is no longer comprehensible, and how will Dr. Xavier the radiologist then respond?

Our visions of the future are increasingly dystopian, and the utopian visions of the past seem increasingly unattainable. In 1895, when Röntgen discovered his new rays, his invisible light, H.G. Wells (1866–1946) published his vision of a future, *The Time Machine*. In the epilogue, his time traveller

> thought but cheerlessly of the Advancement of Mankind, and saw in the growing pile of civilisation only a foolish heaping that must inevitably fall back upon and destroy its makers in the end. If that is so then it remains for us to live as though it were not so.

This need not be so, and the story of radiology shows human ingenuity making discoveries and finding novel solutions. The two models of professional practice, the technical rationale and the professional artistry, are not mutually incompatible.

A solution is not so much either/or but rather both/and. Using a different model, as humans we need whole brain actions and not simply right brain or left brain. The answer to a technology out of control is not to abolish technology, were this possible, but to promote an appropriate technology as promoted by E.F. "Fritz" Schumacher (1911–1977).[61] Schumacher's book *Small Is Beautiful* of 1973 remains as relevant today as when it was written, and pointed a way forward. The decisions that we make now will determine the future.

As technology has progressively been imported into radiology departments, many of the traditional professions and roles have disappeared. Martin Ford has discussed this in his book *The Rise of the Robots* and comes to the possibly surprising conclusion that increasing technological progress with loss of human jobs carries within the seeds of its own failure.[62] The staff that remain in the 2020s seem significantly more unhappy than those in the 1970s and 1980s. Ford discusses the history of mechanisation in the West with displaced farm labourers entering the manufacturing sector, and more recently into new service jobs. With the rise of technology, Ford shows how new jobs were created and the dispossessed workers were able to find new opportunities. These new jobs needed upgraded skills and also paid better wages. Ford stated that at no time was this more true than in the 25 years after the Second World War, and this was a golden age for the West with a seemingly perfect symbiosis between rapid technological progress and the welfare of the workforce. The machines improved, production increased, and salaries and wages rose. This coincided with the 1970s as a golden decade for radiology, continuing into the 1980s, with rapid technological advances and a transformation of medical care. And so, Ford describes a forthcoming era that is defined by a fundamental shift in the relationship between workers and machines. Instead of machines as tools to increase the productivity of workers., the machines themselves become the workers. The changes that took place in X-ray technology during the 20th century made the equipment easier to use and productivity increased. The examinations were guided and controlled by a radiographer and radiologist. At some point in the future, the radiograph will be obtained using an AI protocol with the patient guided, if at all, by an assistant with simple training, and the radiograph will be reported using another AI protocol. This has happened with the electrocardiogram, so why should a similar process not apply to radiography? The public is now used to shopping and paying for goods without any human contact, so why should healthcare be any different? These are vital and important concerns and questions, and the decisions that we make now will affect the future.

PHRONESIS AND CONCLUSION

At a time of increasing concerns about scientific medicine, we need wisdom as well as knowledge. Whilst knowledge is relatively easy to define, and facts can be learnt and examinations passed, the acquisition and indeed definition of wisdom requires more thought. Aristotle, the founder of the Peripatetic School of philosophy, who was a student of Plato, writing in the sixth book of his *Nicomachean Ethics* discusses the intellectual virtues. He distinguishes between *sophia* and *phronesis*. *Sophia* is a theoretical wisdom, and is knowledge that is logical and can be taught, and resonates with the discipline of science. *Phronesis* can be seen as the supreme intellectual virtue, and is practical wisdom, which is reasoned, and is capable of action with regard to the things that are good or bad for us. *Phronesis* is about living well, that is living the good life, and in harmony with our environment. It is a skilled deliberation, and has an ethical significance. With a combination of theoretical and practical wisdom, radiology can be assured of a confident future.

NOTES

1. Kalender, W.A. 2000. *Computed Tomography, Fundamentals, System Technology, Image Quality, Applications.* Munich: Publicis MCD Werbeagentur GmbH.
2. Hills, T.H. 1971. A radiologists thoughts on future developments in radiography. *The Journal of Photographic Science*, 19, 140–142.
3. Holland, C.T. 1895. *The Healing Art.* Unpublished manuscript.
4. Clarke, A.C. 2013. *Profiles of the Future: An Inquiry into the Limits of the Possible.* Gateway.
5. https://www.newscientist.com/article/dn13556-10-impossibilities-conquered-by-science/ (accessed 19 June 2021).
6. https://www.chardmuseum.co.uk/john-stringfellow (accessed 19 June 2021).
7. Blondlot, R. 1905. *"N" Rays.* Trans. J. Garcin. London: Longmans, Green & Co.
8. Editorial Team: Alvarez-Gaume, L., et al. 2004. *Infinitely CERN, Memories of Fifty Years of Research, 1854–2004.* Editions Suzanne Hurter.
9. *A Green Guide for UK Diagnostic Imaging: How to Start Making Progressive Steps towards the NHS Carbon Zero Targets.* https://uk.medical.canon/wp-content/uploads/sites/8/2021/06/Canon-Medical-Systems-Green-Guide.pdf (Accessed 1 August 2021).
10. Schoen, J.H., Theil, C.L., Gross, J.S. 2021. Climate change and radiology: a primer. *ARRS in Practice*, 15, 4–5.
11. Klenske, N. 2021. Rural areas face imaging obstacles on the road to health care equity. *RSNA News*, 31, 12–13.
12. Hung, P., Deng, S., Zahnd, W.E., Adams, S.A., Olatosi, B., Crouch, E.L. Eberth, J.M. 2020. Geographic disparities in residential proximity to colorectal and cervical cancer care providers. *Cancer*, 126, 1068–1076.
13. Beath, R.M. 1937. Radiology: its background and its future. *Ulster Medical Journal*, April 1937.
14. FitzPatrick, M.E.B., Badu-Boateng, C., Huntley, C., Morgan, C. 2021. "Attorneys to the poor": training physicians to tackle health inequalities. *Future Healthcare Journal*, 8, 12–18.
15. Foucault, M. 1983. *This Is Not a Pipe.* Trans. J. Harkness. Berkley: University of California Press.
16. Keller, C. 2009. Sight unseen: picturing the invisible. In: *Brought to Light, Photography and the Invisible 1840–1900.* Ed. C. Keller. New Haven: Yale University Press.
17. van Dijck, J. 2005. *The Transparent Body: A Cultural Analysis of Medical Imaging.* Seattle: University of Washington Press.
18. Bythell, W.J.S., Barclay, A.E. 1912. *X-Ray Diagnosis and Treatment: A Handbook for General Practitioners and Students.* London: Henry Frowde (Oxford Medical Publications).
19. Brown, R.K.J., Petty, S., O'Malley, S., Stojanovska, J., Davenport, M.S., Kazerooni, E.A., Fessahazion, D. 2018. Virtual Reality Tool Simulates MRI Experience. *Tomography*, 4, 95-98

20. Investing in European Talent to Help Build the Metaverse https://about.fb.com/news /2021/10/creating-jobs-europe-metaverse/ (accessed 25 November 2021).
21. Kurzweil, R. 2000. *The Age of Spiritual Machines: When Computers Exceed Human Intelligence.* Harmondsworth: Penguin Books.
22. Cornford, F.M. 1941. *The Republic of Plato* (Translated with introduction and notes by F.M.Cornford). Oxford: The Clarendon Press.
23. Illich, I. 1977. *Limits to Medicine. Medical Nemesis: The Expropriation of Health.* Harmondsworth: Penguin Books.
24. Coleman, V. 2006. *Coleman's Laws: The Twelve Medical Truths You Cannot Live Without.* Barnstaple: European Medical Journal.
25. Le Fanu, J. 1999. *The Rise and Fall of Modern Medicine.* London: Little, Brown and Company (UK).
26. Online Etymological Dictionary. https://www.etymonline.com/word/nemesis (accessed 20 September 2021).
27. Godlee, F. 2012. Preventing overdiagnosis. *British Medical Journal,* 344, e3783.
28. Godlee, F. 2013. Suspect a PE? Less may be more. *British Medical Journal,* 347, f4314.
29. https://www.bmj.com/too-much-medicine (accessed 2 August 2021).
30. Streather, J. 2019. *Is Medicine Still Good for Us? A Primer for the 21st Century.* London: Thames & Hudson.
31. Felson, B. 1981. Letter from the editor. *Seminars in Roentgenology,* 16, 1–2.
32. Essex, C. 2005. Ulysses syndrome. *British Medical Journal,* 1, 1269.
33. ElHabr, A., Merdan, S., Ayer, T., Prater, A., Hanna, T., Horný, M., Duszak, R., Hughes, D.R. 2021. Increasing utilization of emergency department neuroimaging from 2007 to 2017. *American Journal of Roentgenology,* 10.2214/AJR.21.25864.
34. McKeown, T. 1979. *The Role of Medicine.* Oxford: Blackwell.
35. Gordon, S., Allan, T. 1954. *The Scalpel, the Sword: The Story of Dr Norman Bethune.* London: Robert Hale Limited.
36. Burkitt, D.P. 1973. Some diseases characteristic of modern western civilization. *British Medical Journal,* 1, 274–278.
37. Macintyre, J. 1903. The electrical pavilion, Glasgow Royal Infirmary. *Archives of the Roentgen Ray,* 7, 101–102 (with additional plates).
38. Margulis, A.R., Steiner, R.E. 1970. Editorial: British-American symposium: a conference on the future of diagnostic radiology. *British Journal of Radiology,* 43, 833–834.
39. Paterson, A.M., Price, R.C., Thomas, A., Nuttall, L. 2004. Reporting by radiographers: a policy and practice guide. *Radiography,* 10, 205–212.
40. Faculty of Clinical Radiology of the Royal College of Radiologists. 1994. *Röntgen's Progress: A Discussion Paper on the Future of Clinical Radiology in the UK.* London: The Royal College of Radiologists.
41. The Royal College of Radiologists. 2021. *Clinical Radiology UK Workforce Census 2020 Report.* London: The Royal College of Radiologists.
42. Thomas, A., Duck, F. 2019. *Edith and Florence Stoney, Sisters in Radiology (Springer Biographies).* Switzerland: Springer Nature.
43. Redding, J.M. 1926. *X-Ray Diagnosis.* London: Cassell and Company, LTD.
44. Halligan, A. 2008. The importance of values in healthcare. *Journal of the Royal Society of Medicine,* 101, 480–481.
45. Fish, D., de Cossart, L. 2007. *Developing the Wise Doctor: A Resource for Trainers and Trainees in MMC.* London: Royal Society of Medicine Press, Ltd.
46. ACS. 1892. The new hospital for women. *The Illustrated London News,* 10 February.
47. Burdett, H.C. 1896. *Cottage Hospitals, General, Fever, and Convalescent.* 3rd ed. London: The Scientific Press, Limited.

48. Beese, K. 2021. Chief wellness officers take on a critical new role in radiology, health care. *RSNA News*, 31, Issue 7, 6.
49. Lim, R.C.H., Pinto, C. 2009. Work stress, satisfaction and burnout in New Zealand radiologists: comparison of public hospital and private practice in New Zealand. *Journal of Medical Imaging and Radiation Oncology*, 53, 194–199.
50. https://www.nhs.uk/mental-health/self-help/guides-tools-and-activities/five-steps-to-mental-wellbeing/ (accessed 5 August 2021).
51. Dugar, N. 2020. Radiology remote working. *RCR Newsletter*, Autumn 2020, Issue 135, 10.
52. Vosshenrich, J., Brantner, P., Cyriac, J., Boll, D.T., Merkle, E.M., Heye, T. 2021. Quantifying radiology resident fatigue: analysis of preliminary reports. *Radiology*, 298, 632–639.
53. Anon. 2021. RCR census reveals potential consultant exodus looming. *RAD Magazine*, 47(552), 1–2.
54. Asimov, A. 1966. *Fantastic Voyage: A Novel*. London: Dennis Dobson.
55. https://drug-dev.com/injectable-nanomedicines-new-developments-in-long-acting-injectable-nanoformulations/ (accessed 7 August 2021).
56. Rees, R. 2000. *King Copper. South Wales and the Copper Trade*. Cardiff: University of Wales Press.
57. Pollard, S. 1971. *The Idea of Progress*. Harmondsworth: Pelican Books.
58. Safstrom, M. 2013/2014. Introduction. In: *Squire Adamsson, Or, Where Do You Live?: An Allegorical Tale from the Swedish Awakening*. Ed. Waldenström, Paul Peter. Seattle: Pietisten.
59. Rorty, R. 1999. *Philosophy and Social Hope*. London: Penguin Books.
60. Rogers, W., Thulasi Seetha, S., Refaee, T.A.G., Lieverse, R.I.Y., Granzier, R.W.Y., Ibrahim, A., et al. 2020. Radiomics: from qualitative to quantitative imaging. *British Journal of Radiology*, 93, 20190948.
61. Schumacher, E.F. 1973. *Small Is Beautiful: A Study of Economics as if People Matter*. London: Blond & Briggs.
62. Ford, M. 2015. *The Rise of the Robots*. London: Oneworld Publications.

Index

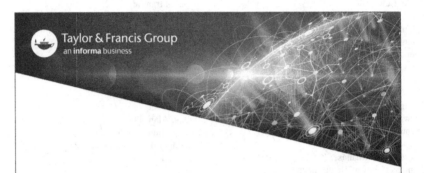